Advances in

APPLIED MICROBIOLOGY

VOLUME 66

Advances in APPLIED MICROBIOLOGY

VOLUME 66

Edited by

ALLEN I. LASKIN
Somerset, New Jersey, USA

SIMA SARIASLANI
Wilmington, Delaware, USA

GEOFFREY M. GADD
Dundee, Scotland, UK

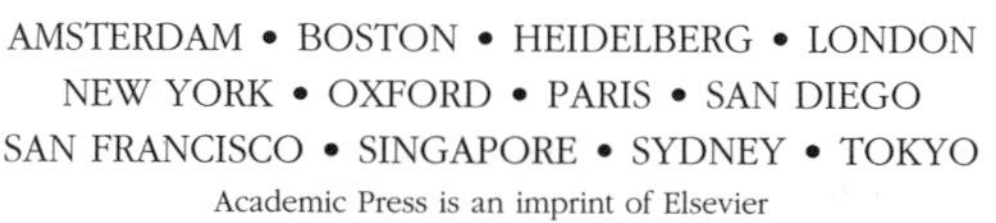
AMSTERDAM • BOSTON • HEIDELBERG • LONDON
NEW YORK • OXFORD • PARIS • SAN DIEGO
SAN FRANCISCO • SINGAPORE • SYDNEY • TOKYO
Academic Press is an imprint of Elsevier

Academic Press is an imprint of Elsevier
525 B Street, Suite 1900, San Diego, CA 92101-4495, USA
30 Corporate Drive, Suite 400, Burlington, MA 01803, USA
32, Jamestown Road, London NW1 7BY, UK

First edition 2009

ISBN: 978-0-12-374788-4
ISSN: 0065-2164

For information on all Academic Press publications
visit our website at elsevierdirect.com

Printed and bound in USA
09 10 11 12 10 9 8 7 6 5 4 3 2 1

CONTENTS

CONTRIBUTORS

Gladys Alexandre
Department of Biochemistry, Cellular and Molecular Biology, and Department of Microbiology, The University of Tennessee, Knoxville, Tennessee 37996.

Kyla Driscoll Carroll
Department of Antibody Technology ImClone Systems, a wholly-owned subsidiary of Eli Lilly & Co. New York, NY 10014.

Mostafa S. Elshahed
Department of Microbiology and Molecular Genetics, Oklahoma State University, 1110 S Innovation Way, Stillwater, Oklahoma 74074.

Sam Foggett
School of Agriculture, Food and Rural Development, and School of Biomedical Sciences, Faculty of Medical Sciences, Newcastle University, Newcastle upon Tyne, NE1 7RU, United Kingdom.

Christine Gaylarde
Departamento de Microbiología Ambiental y Biotecnología, Universidad Autónoma de Campeche, Campeche, Campeche, México.

Gabriela Alves Macedo
Food Science Department, Faculty of Food Engineering, Campinas State University (UNICAMP), 13083970 Campinas, SP, Brazil.

Paulo Cesar Maciag
Department of Antibody Technology ImClone Systems, a wholly-owned subsidiary of Eli Lilly & Co. New York, NY 10014.

Michael J. McInerney
Department of Microbiology and Molecular Genetics, Oklahoma State University, 1110 S Innovation Way, Stillwater, Oklahoma 74074.

Lance D. Miller
Department of Biochemistry, Cellular and Molecular Biology, The University of Tennessee, Knoxville, Tennessee 37996.

Otto Ortega-Morales
Departamento de Microbiología Ambiental y Biotecnología, Universidad Autónoma de Campeche, Campeche, Campeche, México.

Yvonne Paterson
Department of Microbiology, University of Pennsylvania, Philadelphia, Pennsylvania 19104.

Tatiana Fontes Pio
Food Science Department, Faculty of Food Engineering, Campinas State University (UNICAMP), 13083970 Campinas, SP, Brazil.

Sandra Rivera
Department of Antibody Technology ImClone Systems, a wholly-owned subsidiary of Eli Lilly & Co. New York, NY 10014.

Matthew H. Russell
Department of Biochemistry, Cellular and Molecular Biology, The University of Tennessee, Knoxville, Tennessee 37996.

Stefanie Scheerer
Cardiff School of Biosciences, Cardiff University, Cardiff CF10 3TL, United Kingdom.

Vafa Shahabi
Department of Antibody Technology ImClone Systems, a wholly-owned subsidiary of Eli Lilly & Co. New York, NY 10014.

Olivier Sparagano
School of Agriculture, Food and Rural Development, Newcastle University, Newcastle upon Tyne, NE1 7RU, United Kingdom.

Anu Wallecha
Department of Antibody Technology ImClone Systems, a wholly-owned subsidiary of Eli Lilly & Co. New York, NY 10014.

Noha Youssef
Department of Microbiology and Molecular Genetics, Oklahoma State University, 1110 S Innovation Way, Stillwater, Oklahoma 74074.

CHAPTER 1

Multiple Effector Mechanisms Induced by Recombinant *Listeria monocytogenes* Anticancer Immunotherapeutics

Anu Wallecha,* Kyla Driscoll Carroll,* Paulo Cesar Maciag,* Sandra Rivera,* Vafa Shahabi,* and **Yvonne Paterson†**

Contents

* Department of Antibody Technology ImClone Systems, a wholly-owned subsidiary of Eli Lilly & Co. New York, NY 10014
† Department of Microbiology, University of Pennsylvania, Philadelphia, Pennsylvania 19104

Advances in Applied Microbiology, Volume 66
ISSN 0065-2164, DOI: 10.1016/S0065-2164(08)00801-0

Abstract *Listeria monocytogenes* is a facultative intracellular gram-positive bacterium that naturally infects professional antigen presenting cells (APC) to target antigens to both class I and class II antigen processing pathways. This infection process results in the stimulation of strong innate and adaptive immune responses, which make it an ideal candidate for a vaccine vector to deliver heterologous antigens. This ability of *L. monocytogenes* has been exploited by several researchers over the past decade to specifically deliver tumor-associated antigens that are poorly immunogenic such as self-antigens. This review describes the preclinical studies that have elucidated the multiple immune responses elicited by this bacterium that direct its ability to influence tumor growth.

I. INTRODUCTION

Listeria monocytogenes is a gram-positive facultative intracellular bacterium responsible for causing listeriosis in humans and animals (Lecuit, 2007; Lorber, 1997; Vazquez-Boland *et al.*, 2001). *L. monocytogenes* is able to infect both phagocytic and nonphagocytic cells (Camilli *et al.*, 1993; Gaillard *et al.*, 1987; Tilney and Portnoy, 1989). Due to its intracellular growth behavior, *L. monocytogenes* triggers potent innate and adaptive immune responses in an infected host that results in the clearance of the organism (Paterson and Maciag, 2005). This unique ability to induce efficient immune responses using multiple simultaneous and integrated mechanisms of action has encouraged efforts to develop this bacterium as an antigen delivery vector to induce protective cellular immunity against cancer or infection. This review describes the multiple effector responses induced by this multifaceted organism, *L. monocytogenes*.

II. MOLECULAR DETERMINANTS OF *L. monocytogenes* VIRULENCE

To survive within the host and cause the severe pathologies associated with infection such as crossing the intestinal, blood-brain, and fetoplacental barriers, *L. monocytogenes* activates a set of virulence genes. The virulence genes of *L. monocytogenes* have been identified mainly through biochemical and molecular genetic approaches. The majority of the genes that are responsible for the internalization and intracellular growth of *L. monocytogenes* such as *actA, hly, inlA, inlB, inlC, mpl, plcA,* and *plcB* are regulated by a pluripotential transcriptional activator, PrfA (Chakraborty *et al.*, 1992; Freitag *et al.*, 1993; Renzoni *et al.*, 1999; Scortti *et al.*, 2007). Thus, *prfA* defective *L. monocytogenes* are completely avirulent as they lack the ability to survive within the infected host's phagocytic cells such as dendritic cells (DC), macrophages, and neutrophils (Leimeister-Wachter *et al.*, 1990; Szalay *et al.*, 1994).

A. Virulence factors associated with *L. monocytogenes* invasion

A set of *L. monocytogenes* surface proteins known as invasins interact with the receptors present on host cell plasma membranes to subvert signaling cascades leading to bacterial internalization. The internalins (InlA and InlB) were the first surface proteins that were identified to promote host cell invasion (Braun *et al.*, 1998; Cossart and Lecuit, 1998; Lecuit *et al.*, 1997). InternalinA is a key invasion factor that interacts with the epithelial cadherin (E-cadherin), which is expressed on the surface of epithelial cells and thus promotes epithelial cell invasion and crossing of the gastrointestinal barrier. The efficiency of the interaction between InlA with its receptor E-cadherin is variable in different mammalian hosts. For example, mice are resistant to intestinal infection with *L. monocytogenes* because of a single amino acid difference between mouse and human E-cadherin (Lecuit *et al.*, 1999). InlA is also suggested to be important for crossing the maternofetal barrier since E-cadherin is expressed by the basal and apical plasma membranes of synciotrophoblasts and villous cytotrophoblasts of the placenta (Lecuit *et al.*, 1997, 2001). However, the precise role of InlA in crossing the fetoplacental barrier remains to be demonstrated since, fetoplacental transmission occurs in mice that lack the *inlA* receptor and also occurs in guinea pigs that are infected with an *inlA* deletion mutant *L. monocytogenes* (Lecuit *et al.*, 2001, 2004).

InternalinB promotes *L. monocytogenes* entry into a variety of mammalian cell types including epithelial cells, endothelial cells, hepatocytes, and fibroblasts. The hepatocyte growth factor receptor (Met/HGF-R) has been identified as the major ligand for InlB and is responsible for causing the entry of *L. monocytogenes* into nonphagocytic cells (Bierne and Cossart, 2002). Met belongs to the family of receptor tryosine kinases, one of the most important families of transmembrane signaling receptors expressed by a variety of cells. The activation of Met by InlB is also species specific; indeed InlB fails to activate rabbit and guinea pig Met, but activates human and murine Met (Khelef *et al.*, 2006). *In vivo* virulence studies in mice have shown that InlB plays an important role in mediating the colonization of *L. monocytogenes* in the spleen and liver (Gaillard *et al.*, 1996). InlB is also considered important for crossing the fetoplacental barrier due to the observation that in the absence of InlB, InlA expressing *L. monocytogenes* invaded placental tissue inefficiently (Lecuit *et al.*, 2004). It has also been suggested that InlB is involved in crossing the blood-brain barrier as InlB is necessary for *in vitro* infection of human brain microvascular endothelial cells (Greiffenberg *et al.*, 1998).

Twenty four additional internalins are present in the *L. monocytogenes* genome and could potentially contribute to host cell invasion (Dramsi *et al.*, 1997). It is plausible that these internalins might cooperate with each other in order to facilitate entry into host cells, for example, InlA mediated entry is enhanced in the presence of InlB and InlC. However, additional studies are required to understand the contributions of each internalin and how these proteins participate in the bacterial entry to establish the successful infection of various cell types.

In addition to the internalins, several other proteins such as Ami, Auto, and Vip are also implicated in the ability of *L. monocytogenes* to enter host cells. In the absence of InlA and InlB, it has been shown that Ami digests the *L. monocytogenes* cell wall and mediates the adherence of a Δ*inlAB* bacterial strain to mammalian cells (Milohanic *et al.*, 2001). Auto is another autolysin that regulates the bacterial surface architecture required for adherence (Cabanes *et al.*, 2004). Vip is a cell wall anchored protein that is involved in the invasion of various cell lines. The endoplasmic reticulum resident chaperone gp96 has been identified as a cellular ligand for this protein (Cabanes *et al.*, 2005). Thus, these *L. monocytogenes* cell surface proteins contribute to the ability of *L. monocytogenes* to infect multiple cell types.

B. *L. monocytogenes* survival in the macrophage

Upon infection of host cells such as macrophages and DC, a majority of the bacteria are killed in the phagolysosome of the host cell with less than 10% of the *L. monocytogenes* escaping into the host cell cytosol. This escape from the phagolysosome is mediated by the expression of Listeriolysin O (LLO), a pore forming hemolysin, which is the product of the *hly* gene and phospholipases (PlcA and PlcB) (Fig. 1.1). LLO is the first identified major virulence factor of *L. monocytogenes* and is a member of the cholesterol-dependent cytolysin family (CDC) (Portnoy *et al.*, 1992a,b; Tweten, 2005).

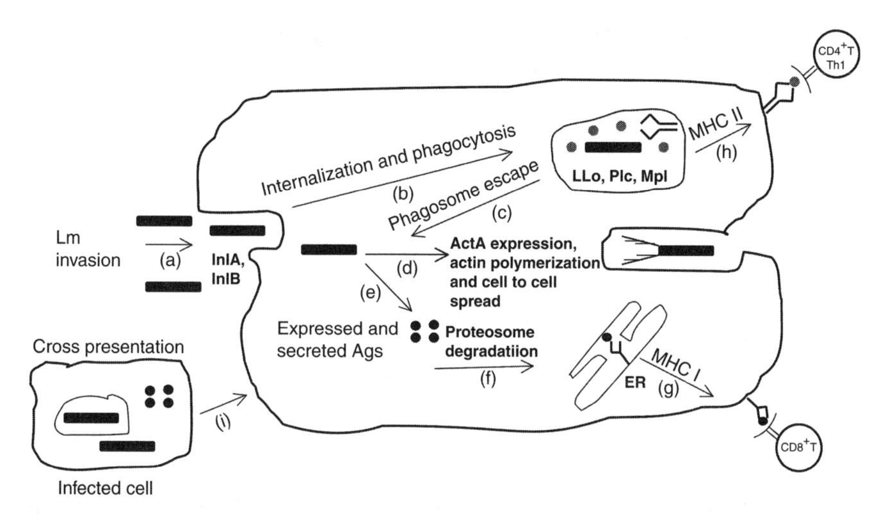

FIGURE 1.1 Intracellular growth of *L. monocytogenes* in an antigen-presenting cell and antigen presentation. Internalization of *L. monocytogenes* on the host cell is mediated by phagocytosis in macrophages but in other host cells such as epithelial and endothelial cells it requires invasins such as InlA and InlB (a). After cellular entry *L. monocytogenes* escape the phagolysosome by secreting Listeriolysin O (LLO), phospholipase (Plc), and metalloprotease (Mpl) resulting in the lysis of the vacuolar membrane, releasing the bacteria in the host cytosol (b and c). Cytosolic bacteria express protein ActA that polymerizes actin filaments and mediates cell to cell spread of *L. monocytogenes* (d). Cytosolic antigens produced after *L. monocytogenes* escape from phagosome are degraded by the proteosome to antigenic epitopes and presented by MHC class I molecules (e, f, and g). Bacterial antigens inside the phagosome are processed as exogenous antigens and epitopes are presented on the membrane surface in the context of MHC class II molecules (h). An alternate route for antigen presentation involves cross presentation with the antigens derived from an *L. monocytogenes* infected cell (i).

LLO binds to the host cell membrane initially as a monomer but then forms oligomers composed of up to 50 subunits, which are inserted into the membrane to form pores of diameter ranging 200–300Å (Walz, 2005). The function of LLO is very crucial for the cellular invasion of *L. monocytogenes* in both phagocytic and nonphagocytic cells.

After entry into the cytosol, another *L. monocytogenes* secreted protein called ActA enables bacterial propulsion in the cytosol leading to the invasion of neighboring uninfected cells by a process called cell to cell spreading (Alvarez-Dominguez *et al.*, 1997; Suarez *et al.*, 2001). In the cytoplasm, *L. monocytogenes* replicates and uses ActA to polymerize host cell actin to become motile enabling spread from cell to cell (Dussurget *et al.*, 2004; Fig. 1.1). As a result, the deletion of *actA* from *L. monocytogenes* results in a highly attenuated bacterium and thus establishes that ActA is a major virulence factor.

III. IMMUNE RESPONSE TO *L. monocytogenes* INFECTION

A. Innate immunity

Innate immunity plays an essential role in the clearance of *L. monocytogenes* and control of the infection at early stages. Mice deficient in T and B cell responses, such as *SCID* and *nude* mice, have normal early resistance to sublethal *L. monocytogenes* infection. However, SCID and nude mice eventually succumb to infection because complete clearance of *L. monocytogenes* requires T-cell mediated immunity (Pamer, 2004). Upon systemic inoculation of *L. monocytogenes*, circulating bacteria are removed from the blood stream primarily by splenic and hepatic macrophages (Aichele *et al.*, 2003). In the spleen, the bacteria localize within macrophages and DC of the marginal zone, between the white and red pulp (Conlan, 1996). Within the first day of infection, these cells containing live bacteria migrate to the T-cell zones in the white pulp, establishing a secondary focus of infection and attracting neutrophils. Interestingly, this process has been associated with lymphocytopenia in this compartment (Conlan, 1996), as T cells undergo apoptosis induced by the *L. monocytogenes* infection in an antigen-independent manner (Carrero and Unanue, 2007).

Both macrophages and neutrophils have essential roles in controlling *L. monocytogenes* infection at early time points. Recruitment of monocytes to the site of infection is an important characteristic of *L. monocytogenes* infection. In the liver, the Kupffer cells clear most of the circulating bacteria. As early as 3 h after systemic injection, *L. monocytogenes* can be found inside the Kupffer cells, followed by granulocyte and mononuclear cell infiltration and formation of foci of infection (Mandel and Cheers, 1980). Neutrophils are rapidly recruited to the site of infection by the cytokine IL-6 and other chemo-attractants, which secrete IL-8 (Arnold and Konig, 1998), CSF-1 and MCP-1. These chemokines are important in the inflammatory response and for attracting macrophages to the infection foci. In the following few days, granulocytes are gradually replaced by large mononuclear cells and within 2 weeks the lesions are completely resolved (Mandel and Cheers, 1980). Further studies have shown that mice depleted of granulocytes are unable to control *L. monocytogenes* infection (Conlan and North, 1994; Conlan *et al.*, 1993; Czuprynski *et al.*, 1994; Rogers and Unanue, 1993). In murine listeriosis, *L. monocytogenes* replicates inside hepatocytes, which are lysed by the granulocytes recruited to the infection foci, releasing the intracellular bacteria to be phagocytosed and killed by neutrophils (Conlan *et al.*, 1993). Although neutrophils are very important in fighting *L. monocytogenes* infection in the liver, depletion of neutrophils does not significantly change the infection course in the spleen (Conlan and North, 1994). Interestingly, mice depleted of mast cells have significantly higher titers of *L. monocytogenes* in the spleen and liver and are considerably impaired in neutrophil mobilization (Gekara *et al.*, 2008). Although not directly infected by *L. monocytogenes*, mast cells can be activated by the bacteria and rapidly secrete TNF-α and induce neutrophil recruitment (Gekara *et al.*, 2008).

At the cell surface, toll like receptors (TLRs) play a role in the recognition of *L. monocytogenes*. TLRs are important components of innate immunity, recognizing conserved molecular structures on pathogens, and signaling through adaptor molecules, such as MyD88, to induce NF- κB activation and transcription of several proinflammatory genes. NF-κB is a heterodimeric transcription factor composed of p50 and p65 subunits and activates several genes involved in innate immune responses. Mice lacking the p50 subunit of NF-κB are highly susceptible to *L. monocytogenes* infections (Sha *et al.*, 1995).

In particular, TLR2 seems to play a role during *L. monocytogenes* infection because mice deficient in TLR2 are slightly more susceptible to listeriosis (Torres *et al.*, 2004). TLR2 recognizes bacterial peptidoglycan, lipoteichoic acid, and lipoproteins present in the cell wall of gram-positive bacteria, including *L. monocytogenes*. TLR5, which binds bacterial flagellin, however, is unlikely to be involved in *L. monocytogenes* recognition since flagellin expression is downregulated at 37 °C for most *L. monocytogenes* isolates. In addition, TLR5 is not required for innate immune activation against this bacterial infection (Way and Wilson, 2004).

The presence of unmethylated CpG dinucleotides in the bacterial DNA also has stimulatory effects on mammalian immune cells. CpG motifs present in bacterial DNA act as pathogen associated molecular patterns (PAMPs) (Hemmi *et al.*, 2000; Tsujimura *et al.*, 2004) interacting with TLR-9 to trigger an innate immune response in which lymphocytes, DC, and macrophages are stimulated to produce immunoprotective cytokines and chemokines (Ballas *et al.*, 1996; Haddad *et al.*, 1997; Hemmi *et al.*, 2000; Ishii *et al.*, 2002; Tsujimura *et al.*, 2004).

Although TLRs are important in bacterial recognition, a single TLR has not been shown to be essential in innate immune responses to *L. monocytogenes*. On the other hand, the adaptor molecule MyD88, which is used by signal transduction pathways of all TLRs, except TLR-3, is critical to host defense against *L. monocytogenes* and infection with *L. monocytogenes* is lethal in MyD88-deficient mice. Additionally, MyD88$^{-/-}$ mice are unable or severely impaired in the production of IL-12, IFN-γ, TNF-α, and nitric oxide (NO) following *L. monocytogenes* infection. MyD88 is not required for MCP-1 production and monocyte recruitment following *L. monocytogenes* infection but is essential for IL-12 and TNF-α production and monocyte activation (Serbina *et al.*, 2003). The NOD-LRR receptor interacting protein 2 (RIP2) kinase, identified as immediately downstream of NOD-1, is also required for full signaling through TLR2, 3, and 4. Mice deficient in RIP2 are impaired in their ability to defend against *L. monocytogenes* infection and have decreased IFN-γ production by NK and T cells, which is partially attributed to a defective interleukin-12 signaling (Chin *et al.*, 2002). In addition, Portnoy and associates have recently shown that cytosolic Listerial peptidoglycans generated in the phagosome induce IFN-β in macrophages by a TLR-independent, NOD-1 dependent pathway (Leber *et al.*, 2008).

Overall, several components of the innate immune response participate in early defenses against infection with *L. monocytogenes*. Although there is a critical role of innate immunity in listeriosis, complete eradication of wild type *L. monocytogenes* requires antigen-specific T cell responses against this pathogen.

B. Cellular immune responses

Earlier studies using the mouse as a model of *L. monocytogenes* infection clearly demonstrated the cell mediated nature of the immune responses to the bacterium (Mackaness, 1962). Subsequently, it has been shown that *L. monocytogenes* elicits both class I and class II MHC responses that are essential for controlling infection and inducing long term protective immunity (Ladel *et al.*, 1994).

1. MHC class Ia and Ib restricted T cell responses to *L. monocytogenes*

L. monocytogenes specific $CD8^+$ T cell responses fall into two groups: One recognizes peptides generated by cytosolic degradation of secreted bacterial proteins (class Ia MHC); the other recognizes short hydrophobic peptides that contain *N*-formyl methionine at the amino terminus (class Ib MHC).

MHC-class Ia restricted peptide antigens derived from *L. monocytogenes* are generated from the degradation of secreted proteins (Finelli *et al.*, 1999). *In vitro* labeling studies have shown that *L. monocytogenes* secretes a limited number of proteins into the cytosol of the host cell (Villanueva *et al.*, 1994). Bacterially secreted proteins in the cytosol of macrophages are rapidly degraded by proteosomes. Some secreted proteins such as p60 and LLO are rapidly degraded because their amino termini contain destabilizing residues as defined by the N-end rule (Schnupf *et al.*, 2007; Sijts *et al.*, 1997). LLO is also degraded in a proteosome-dependent fashion as it contains a PEST-like sequence (Decatur and Portnoy, 2000). LLO and p60 are the most antigenic of the secreted proteins in terms of induction of a $CD8^+$ T cell response. On the other hand, ActA has enhanced stability in the cytosol as it contains a stabilizing amino acid at the amino terminus (Moors *et al.*, 1999). The rapid proteosome mediated degradation of a potentially toxic protein such as LLO enhances host cell survival and generates peptide fragments that enter the MHC class I antigen processing pathway.

MHC class Ia restricted T cell responses to *L. monocytogenes* reach peak frequencies approximately 8 days after intravenous inoculation (Busch *et al.*, 1998). The magnitude of T cell responses that are generated for specific antigenic peptides is independent of the quantity or the duration of *in vivo* antigen presentation. This finding is supported by experiments in which mice were treated with antibiotics to curtail the duration of the infection (Badovinac *et al.*, 2002; Mercado *et al.*, 2000). Despite significant differences in the number of viable bacteria and inflammatory responses, the expansion and contraction of $CD8^+$ T cells is similar in mice treated with antibiotics 24 h after infection and in mice that are untreated, indicating that T cells are programmed during the first few days of infection (Wong and Pamer, 2001). This is consistent with *in vitro* studies of *L. monocytogenes* specific $CD8^+$ T cell proliferation, which showed that transient antigen presentation is followed by prolonged proliferation and do not require further exposure to antigen (Wong and Pamer, 2001). This suggests that innate immune responses that occur after the first 24 h of infection have a very small impact on the kinetics and magnitude of $CD8^+$ T cell responses. The reason for antigen independent proliferation of $CD8^+$ T cells remain unclear, although one hypothesis is that antigen independent T cell proliferation is driven by cytokines such as IL-2. However, studies by Wong *et al.* (Wong and Pamer, 2001) showed that endogenous IL-2 production by $CD8^+$ T cells is required for Ag-independent expansion following TCR stimulation *in vitro*, but not *in vivo*. Thus, there are other factors in addition to IL-2 that regulate antigen-independent proliferation of $CD8^+$ T cells *in vivo*.

The magnitude of *in vivo* $CD8^+$ T cell responses following *L. monocytogenes* infection is also influenced by the cytokines IFN-γ and perforin. *L. monocytogenes* infection of mice deficient in both IFN-γ and perforin results in an increased magnitude of *L. monocytogenes* specific $CD8^+$ T cell responses, and shifting of the immunodominance hierarchy (Badovinac and Harty, 2000). This suggests that neither perforin nor IFN-γ is absolutely necessary for the development of anti-*L. monocytogenes* immune responses.

L. monocytogenes infection of mice lacking MHC class Ia molecules induces $CD8^+$ T cell immunity equivalent to that seen in normal mice. These $CD8^+$ T cells are restricted by MHC class Ib. H2-M3 MHC class Ib molecules selectively bind peptides with *N*-formyl methionine at the N-terminus. H2-M3 restricted T cells are cytolytic and produce IFN-γ and TNF-α and can mediate protective immunity (Finelli *et al.*, 1999).

Transfer of H2-M3 restricted CTL into TAP (transporter for antigen presentation) deficient mice confers partial protection, indicating that TAP dependent and TAP independent antigen processing pathways are operative. Processing and presentation of *L. monocytogenes* *N*-formyl-methionine peptides by infected cells are poorly defined. In uninfected cells, most H2-M3 molecules remain in the ER because endogenous *N*-formyl-peptides are scarce. Some *L. monocytogenes* derived *N*-formyl-peptides are bound by gp96 prior to association with H2-M3. The number of *L. monocytogenes* specific H2-M3 T cells peak 5–6 days post infection (Finelli *et al.*, 1999). Contraction of H2-M3 restricted T cells results in the generation of a pool of memory cells, but they only have some of the characteristics of traditional memory cells. When rechallenged with a second *L. monocytogenes* infection, these cells upregulate surface expression of activation markers, but do not proliferate. This suppression of proliferation is mediated by the expansion of the MHC class Ia response, which limits available DC for antigen presentation. However, these cells do play a role in the control of primary infection since, H2-M3 knock out mice have a defect in bacterial clearance suggesting that early expansion and IFN-γ production by these cells cannot be compensated by other T cell subsets. Recently, it was demonstrated that MHC class Ib-restricted T cells also help in the enhancement of Ag-specific $CD4^+$ T cell responses (Chow *et al.*, 2006).

Infection of mice intraperitoneally with *L. monocytogenes* has been shown to cause a site-specific induction of γ/δ T cells in the peritoneal cavity (Skeen and Ziegler, 1993). However, no changes are observed in the splenic or lymph node T cell populations after these injections. Moreover, when peritoneal T cells from *L. monocytogenes*-immunized mice are restimulated *in vitro*, the induced γ/δ T cells exhibited a greater expansion potential than the α/β T cells. Significant increase in peritoneal $CD3^+$ cells expressing the γ/δ T cell receptor is observed for 8 days after *L. monocytogenes* injection and the population remains elevated for 6–7 weeks. Both, the induced γ/δ T cells or γ/δ T cells from the normal mice were not found to express $CD4^+$ or $CD8^+$ on the cell surface. The modifications that abrogate the virulence of *L. monocytogenes* such as heat killed *L. monocytogenes* or *hly* negative mutants, also results in elimination of the inductive effect for γ/δ T cells. The *in vivo* depletion of either α/β or γ/δ T cells using a monoclonal antibody in mice results in an impairment in resistance to primary infection with *L. monocytogenes*. However, the memory response is virtually unaffected by the depletion of γ/δ T cells, supporting the hypothesis that this T cell subset forms an important line of defense in innate, rather than adaptive immunity to *L. monocytogenes* (Skeen and Ziegler, 1993).

2. Class II MHC restricted T cells responses

In addition to $CD8^+$ T cell responses, infection with *L. monocytogenes* results in the generation of robust $CD4^+$ T cell responses. Expansion of $CD4^+$ T cells has been shown to be synchronous with the expansion of $CD8^+$ T cells (Skoberne *et al.*, 2002). During the course of infection, $CD4^+$ T cells produce large amounts of Th1 cytokines that are thought to contribute to clearance of *L. monocytogenes*. Immunization with *L. monocytogenes* results in the activation of $CD4^+$ T cells that coexpress dual cytokines such as IFN-γ and TNF-α on day 6 post infection and triple positive cells, TNF-α^+ IFN-γ^+ IL-2^+ on day 10–27 (Freeman and Ziegler, 2005), indicating the generation of memory $CD4^+$ T cell responses. Adoptive transfer studies using *L. monocytogenes* specific $CD4^+$ and $CD8^+$ T cells have shown that $CD4^+$ T cell-mediated protective immunity requires T-cell production of IFN-γ, whereas $CD8^+$ T cells mediate protection independently of IFN-γ (Harty and Bevan, 1995; Harty *et al.*, 1992). It is probable that production of IFN-γ from $CD4^+$ T cells activates macrophages to become more bactericidal, which is supported by *in vitro* studies showing that treatment of macrophages with IFN-γ prevents bacterial escape from the phagosome (Portnoy *et al.*, 1989).

3. Cell-mediated immune responses to heat-killed and irradiated *L. monocytogenes*

T cells primed with live *L. monocytogenes* undergo prolonged division, become cytolytic and produce IFN-γ. By contrast, infection with heat-killed *L. monocytogenes* does not induce a protective immune response. For years, one hypothesis to explain this finding was that killed bacteria do not enter the cytosol of macrophages following phagocytosis, thereby resulting in insufficient antigen presentation. Surprisingly, the immunization of mice with heat-killed *L. monocytogenes* results in the proliferation of antigen specific $CD8^+$ T cells, but does not induce full differentiation of the primed T cells into effector cells (Lauvau *et al.*, 2001). Therefore, T cells that are primed with heat-killed *L. monocytogenes* undergo attenuated division and do not acquire effector functions. In contrast, infection with live bacteria provides a stimulus that remains highly localized and induces T-cell differentiation. On the other hand, irradiated *L. monocytogenes* efficiently activates DC and induces protective T cell responses when used for vaccination (Datta *et al.*, 2006). Therefore, irradiated bacteria could serve as a better vaccine platform for recombinant antigens derived from other pathogens, allergens, and tumors when compared to heat-killed *L. monocytogenes*. However, infection with live *L. monocytogenes* provides the most potent stimulus that remains highly localized and induces T cell differentiation.

IV. RECOMBINANT L. monocytogenes AS A VACCINE VECTOR

L. monocytogenes has been used as a vaccine vector to generate cell mediated immunity against a wide range of viral or tumor antigens such as influenza nucleoprotein, LCMV nucleoprotein, HPV16 E7, HIV gag, SIV gag and env, tyrosinase-related protein (Trp2), high molecular weight melanoma associated antigen (HMW-MAA), ovalbumin, prostrate specific antigen (PSA), and HER-2/neu (Gunn *et al.*, 2001; Ikonomidis *et al.*, 1994; Shahabi *et al.*, 2008; Singh *et al.*, 2005).

A. Construction of recombinant *L. monocytogenes* strains

A variety of viral and tumor antigens such as HPV16E7, HER-2/neu, HMW-MAA, NP, and PSA that are expressed by *L. monocytogenes* as a fusion protein with LLO have been shown to generate antigen specific $CD4^+$ and $CD8^+$ T cell responses in mice. These antigens can be expressed in *L. monocytogenes* by an episomal or chromosomal system. Plasmid based strategies have the advantage of multicopy expression but rely on complementation for the maintenance of the plasmid *in vivo* (Gunn *et al.*, 2001). Chromosomal integration techniques involve either allelic exchange into a known chromosomal locus (Mata *et al.*, 2001) or a phage-based system, which utilizes a site-specific integrase to stably integrate plasmid into the genome (Lauer *et al.*, 2002). Most of the episomal expression systems are based on fusion of the antigen of interest to a nonhemolytic fragment of *hly* (truncated LLO) (Gunn *et al.*, 2001). The retention of plasmid by *L. monocytogenes in vivo* is achieved by the complementation of the *prfA* gene from the plasmid in a *prfA* mutant *L. monocytogenes* background (Gunn *et al.*, 2001). A *prfA* mutant *L. monocytogenes* (XFL7) cannot escape the phagolysosome and is destroyed by host cell macrophages and neutrophils. Thus, due to the lack of intracellular growth, a *prfA* mutant *L. monocytogenes* cannot deliver and present antigenic peptides to the immune cells. Including a copy of *prfA* in the plasmid ensures the *in vivo* retention of the plasmid in *L. monocytogenes* strain XFL7 (Pan *et al.*, 1995a, b). An alternate approach described by Verch *et al.* is based on the retention of a plasmid (pTV3) by complementation of D-alanine racemase in both *Escherichia coli* and *L. monocytogenes* strains that are deficient in D-alanine racemase and D-alanine amino transferase *in vitro* and *in vivo* (Verch *et al.*, 2004). The plasmid pTV3 is devoid of antibiotic resistance and therefore, this recombinant *L. monocytogenes* strain expressing a foreign antigen is more suitable for use in the clinic (Verch *et al.*, 2004).

B. LLO and ACTA as adjuvants in *L. monocytogenes* based immunotherapy

The genetic fusion of antigens to a nonhemolytic truncated form of LLO results in enhanced immunogenicity and *in vivo* efficacy (Gunn *et al.*, 2001; Singh *et al.*, 2005). The immunogenic nature of LLO has been attributed to the presence of PEST sequences close to the N-terminus of the protein that targets LLO for ubiquitin proteosome mediated degradation (Sewell *et al.*, 2004a). Removal of the PEST sequence from LLO used in the fusion constructs partially abrogates the ability of vaccine to induce full tumor regression in mice (Sewell *et al.*, 2004a). Recently, Schnupf *et al.* (2007) have shown that LLO is a substrate of the ubiquitin-dependent N-end rule pathway, which recognizes LLO through its N-terminal Lys residue. The N-end rule pathway is an ubiquitin-dependent proteolytic pathway that is present in all eukaryotes. Thus, the fusion of antigens to LLO may facilitate the secretion of an antigen (Gunn *et al.*, 2001; Ikonomidis *et al.*, 1994), increase antigen presentation (Sewell *et al.*, 2004a), and help to stimulate the maturation of DC (Peng *et al.*, 2004).

Fusion of LLO to tumor antigens in other immunotherapeutic approaches such as viral vectors (Lamikanra *et al.*, 2001) and DNA vaccines (Peng *et al.*, 2007) also enhances vaccine efficacy. Studies using DNA based vaccines have demonstrated that genetic fusion of antigens to LLO is essential for this adjuvant effect as there is a difference in the therapeutic efficacy of chimera or bicistronic vaccines (Peng *et al.*, 2007). However, high levels of specific $CD4^{+}$ T cell immune responses for the passenger antigen are obtained using bicistronic expression of LLO and antigen (Peng *et al.*, 2007). Recently, Neeson *et al.* (2008) have shown that LLO has adjuvant properties when used in the form of a recombinant protein. In this study, the chemical conjugation of LLO to lymphoma immunoglobulin idiotype induces a potent humoral and cell-mediated immune response and promoted epitope spreading after lymphoma challenge. Thus, LLO is a global enhancer of immune responses in various vaccination studies.

The reasons why LLO potentiates immune responses are only partially understood. LLO is a potent inducer of inflammatory cytokines such as IL-6, IL-8, IL-12, IL-18, and IFN-γ (D'Orazio *et al.*, 2006; Nomura *et al.*, 2002; Yamamoto *et al.*, 2006) that are important for innate and adaptive immune responses. Since, a related pore-forming toxin, anthrolysin, is reported to be a ligand of Toll-like receptor 4 (TLR4) (Park *et al.*, 2004), the proinflammatory cytokine-inducing property of LLO may be a consequence of the activation of the TLR4 signaling pathway (Park *et al.*, 2004). In addition to $CD8^{+}$ T cell responses, LLO also modulates $CD4^{+}$ T cell responses. LLO is capable of inhibiting a Th2 immune response by shifting the differentiation of antigen-specific T cells to Th1 cells (Yamamoto *et al.*, 2005, 2006). Due to the high Th1 cytokine-inducing activity of LLO, protective immunity to *L. monocytogenes* is induced when mice are immunized with killed or avirulent *L. monocytogenes* together with LLO, whereas protection is not generated in mice immunized with killed or avirulent *L. monocytogenes* alone (Tanabe *et al.*, 1999). These results demonstrate that LLO potentiates a strong Th1 response, leading to highly effective cell mediated immunity.

In addition to LLO, the proline-rich listerial virulence factor ActA also contains PEST-like sequences. To test whether ActA could also act as an adjuvant, an *L. monocytogenes* strain was constructed that secreted a fusion protein of the first 390 residues of ActA, which contains four PEST sequences, fused to HPV-16 E7 (Sewell *et al.*, 2004b). This strain enhanced immunogenicity and *in vivo* efficacy, similar to LLO, and was effective at eliminating established E7 expressing tumors in wild type mice (Sewell *et al.*, 2004b) and mice transgenic for E7 (Souders *et al.*, 2007).

V. THE PLEIOTROPIC EFFECTS OF *L. monocytogenes* ON THE TUMOR MICROENVIRONMENT

A. Protective and therapeutic tumor immunity

A number of tumor antigens associated with various types of cancer have shown promise as a target for immunotherapy using *L. monocytogenes* based vaccine strategies. For example, preclinical studies using a recombinant *L. monocytogenes* strain expressing HPV16 E7 has demonstrated both prophylactic and therapeutic efficacy against E7 expressing tumors (Gunn *et al.*, 2001). In addition, *L. monocytogenes* vaccine strains expressing fragments of HER-2/neu are able to induce anti-Her2/neu CTL responses in mice with prolonged stasis in tumor growth (Singh *et al.*, 2005). Very recently, Advaxis has described a recombinant *L. monocytogenes* expressing PSA, *L. monocytogenes*-LLO-PSA that induced the regression of more than 80% of tumors expressing PSA (Shahabi *et al.*, 2008). HMW-MAA, also known as melanoma chondrotin sulfate proteoglycan, is overexpressed on over 90% of the surgically removed benign nevi and melanoma lesions, basal cell carcinoma tumors of neural crest origin and some forms of childhood leukemia and lobular breast carcinoma lesions (Chang *et al.*, 2004). In addition, HMW-MAA is expressed at high levels on both activated pericytes and pericytes involved in tumor angiogenic vasculature (Campoli *et al.*, 2004; Chang *et al.*, 2004). Maciag *et al.* (2008) have shown that recombinant *L. monocytogenes* expressing LLO-HMW-MAA used to target pericytes present within the tumor vasculature has potent antiangiogenic effects in the tumors that express HMW-MAA. The recombinant *L. monocytogenes* expressing HMW-MAA not only destroyed the cells that support tumor formation such as pericytes but also impacted on the frequency of tumor-infiltrating lymphocytes. *L. monocytogenes* based vaccines have also been studied in melanoma models using TRP-2 as the target antigen (Bruhn *et al.*, 2005). Tumor protection induced by *L. monocytogenes*-TRP2 was long lasting and therapeutic, conferring tumor protection against both tumor subcutaneous tumors and metastatic tumor nodules in the lungs (Bruhn *et al.*, 2005).

The detailed analyses of the T cell responses generated by recombinant *L. monocytogenes* suggest that both $CD4^+$ and $CD8^+$ T cells are important for the regression of established tumors and protection against subsequent challenge in some models (Fig. 1.2). In addition to generating CTLs against the tumor specific antigens, immunization with recombinant *L. monocytogenes* can also impact the growth of tumors that do not contain vaccine epitopes, presumably by means of epitope spreading (Liau *et al.*, 2002). Epitope spreading refers to the development of an immune response to epitopes distinct and non cross-reactive with the disease-causing epitope (Fig. 1.2). This phenomenon is thought to occur following the release of antigens from the tumor cells killed by vaccine induced T cells. These antigens are then phagocytosed by APCs and presented to naïve T cells of different specificities. Epitope spreading correlates with tumor regression in patients undergoing immunotherapy and could therefore, potentially be harnessed to broaden the immune responses to unidentified tumor antigens in the context of therapeutic vaccines (Liau *et al.*, 2002).

B. *L. monocytogenes* promotes a favorable intratumoral milieu

For immunotherapies to be effective vaccination must result in robust generation of a high number of cytolytic T cells followed by their significant infiltration into the tumor microenvironment. Thus, the major challenge in developing a cancer vaccine is not only to generate the right T cells but also to create conditions for them to migrate, infiltrate, and eliminate tumor cells.

Studies from the Paterson lab have suggested that *L. monocytogenes* vaccines are effective agents for tumor immunotherapy because they result in the accumulation of activated $CD8^+$ T cells within tumors (Hussain and Paterson, 2005). While the reasons for this accumulation of $CD8^+$ cells in tumors is not known, Hussain *et al.* have speculated that it may be due to the ability of the vaccine to induce a specific chemokine profile in the $CD8^+$ cells (Hussain and Paterson, 2005). Specifically, studies have shown that the PEST region of LLO is required for the high numbers of $CD8^+$, antigen-specific TILs, which are in turn critical for vaccine efficacy (Sewell *et al.*, 2004a).

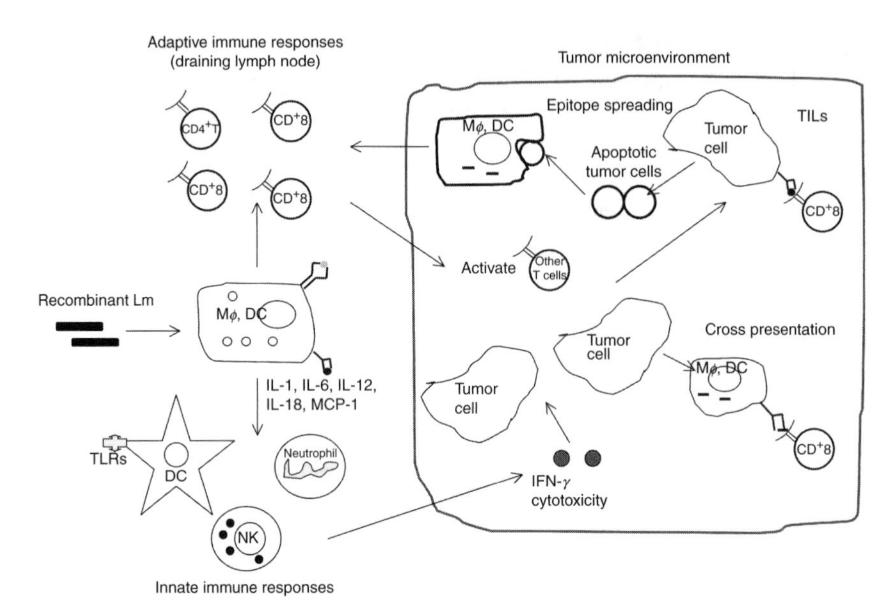

FIGURE 1.2 Multiple effects of *L. monocytogenes* based immunotherapy on the tumor microenvironment. Recombinant *L. monocytogenes*, which is expressing and secreting a target antigen, will be taken up by an antigen presenting cell (APC) such as a macrophage or dendritic cell. This will result in the activation of innate immune responses resulting in the production of various cytokines such as IL-1, IL-6, IL-12, IL-18, and chemokines such as MCP-1 that will attract other immune cells such as dendritic cells, neutrophils, and NK cells to the site of infection. The amplification of immune responses and secretion of these inflammatory cytokines will influence the tumor microenvironment directly or indirectly resulting in the lysis of tumor cells. Additionally, proteins secreted by recombinant *L. monocytogenes* will gain entry into both class I and class II MHC pathways for $CD8^+$ and $CD4^+$ T cell responses. The $CD8^+$ T cells specific for the tumor antigen will lyse the tumor cells presenting the antigen due to their cytotoxic activity. Additionally, recombinant *L. monocytogenes* has the ability to elicit an immune response to epitopes distinct and non cross-reactive with, the disease-causing epitope, referred to as epitope spreading. In this process, the antigens released from the dying tumor cell are taken up by an APC. The mature APC will present those tumor cell antigens to naive $CD8^+$ T cells in the draining lymph node with the activation and expansion of T cells to tumor antigens not shared by the *L. monocytogenes* vaccine. These $CD8^+$ T cells may infiltrate into the tumors and this cycle may continue. Also, there will be cross presentation of the antigens derived from a dying tumor cell to other $CD8^+$ T cells.

Due to its unique life cycle, *L. monocytogenes* also triggers a potent $CD4^+$ T cell response in addition to the cell mediated $CD8^+$ T cell response. Accordingly, tumor specific $CD4^+$ helper cells are produced and migrate to the tumor, similar to CTLs (Beck-Engeser *et al.*, 2001; Pan *et al.*, 1995a) following *L. monocytogenes* vaccination (Fig. 1.2). The fact that

$CD4^+$ cells can lyse antigen/MHC-II expressing tumor cells (Echchakir *et al.*, 2000; Neeson *et al.*, 2008; Ozaki *et al.*, 1987; Yoshimura *et al.*, 1993) is of little consequence since most tumors only express MHC class I molecules. Therefore, the ability of $CD4^+$ T helper cells to promote rejection of MHC-II negative tumors likely occurs via the production of paracrine factors or cytokines (Beck-Engeser *et al.*, 2001; Greenberg, 1991). In fact, the $CD4^+$ T cell response to *L. monocytogenes* infection has been shown to be primarily of the Th1 type with production of the antitumoral cytokines IFN-γ, TNFα, and IL-2.

In addition to targeting exogenous antigens, *L. monocytogenes* vaccines have also been shown to break tolerance in a transgenic mouse model for E6/E7 (Sewell *et al.*, in press; Souders *et al.*, 2007) and HER-2/neu (Singh and Paterson, 2007b). *L. monocytogenes*-based constructs expressing E7 such as *L. monocytogenes*-LLO-E7 and *L. monocytogenes*-ActA-E7 are able to impact the growth of autochthonous tumors that arise in E6/E7 transgenic mice (Sewell *et al.*, in press; Souders *et al.*, 2007). However, the tumor-regression and CTL responses observed following vaccination in transgenic mice was weaker than that observed in the wild type mice. Similarly, in HER-2/neu transgenic mice, all of the *L. monocytogenes* vaccines are capable of slowing or halting the tumor growth despite the fact that $CD8^+$ T cells from the transgenic HER-2/neu mice are of lower avidity than those that arise from the wild-type mice. *L. monocytogenes*-based HER-2/neu constructs also delayed the appearance of spontaneous tumors in the transgenic HER-2/neu mice (Singh and Paterson, 2007b). Interestingly, the tumors that emerged had developed mutations within the CTL epitopes of the HER-2/neu protein. These mutations resided in the exact regions that were targeted by the *L. monocytogenes*-based vaccines suggesting that the rate of generation of escape mutants is a significant factor in the efficacy of these vaccines (Singh and Paterson, 2007a). Based on these findings, it appears that *L. monocytogenes* can overcome tolerance to self antigens and expand autoreactive T cells by activating cells that are usually too low in number and avidity, leading to antitumor responses.

As well as adaptive T cell immunity, multiple cytokines released during innate immune phases play a role in the ability of *L. monocytogenes* to function as an effective immunotherapeutic agent. IFN-γ, for example, plays an especially important role in effective *L. monocytogenes* antitumor responses. Although the majority of IFN-γ is produced by NK cells, $CD4^+$ T-helper cells may also contribute to the IFN-γ levels (Beatty and Paterson, 2001). Using a tumor that is insensitive to IFN-γ (TC1mugR), Dominiecki *et al.* (2005) have shown that *L. monocytogenes* vaccines require IFN-γ for effective tumor regression. Interestingly, the authors demonstrate that IFN-γ is specifically required for tumor infiltration of lymphocytes but not for trafficking to the tumor

(Dominiecki *et al.*, 2005). Additionally, IFN-γ can inhibit angiogenesis at the tumor site in the early effector phase following vaccination (Beatty and Paterson, 2001).

C. Effect of *L. monocytogenes* vaccination on regulatory T cells in the tumors

The accumulation of T regulatory cells (Tregs) represents a formidable challenge to traditional cancer immuno-therapeutics. Frequently, the tumors have evolved to exploit the suppressive properties of these regulatory cells in order to promote their growth and persistence within the host. Furthermore, vaccine strategies may be hampered by their inability to prevent Treg accumulation within tumors. *L. monocytogenes* based vaccines; however, seem to function by decreasing the population of Tregs in the tumors.

L. monocytogenes based vaccines, which express antigen-LLO fusion proteins, have been shown to uniquely prevent large infiltrates of Tregs within tumors. For example, immunization with recombinant *L. monocytogenes*-LLO-E7 fusion protein resulted in fewer Tregs ($CD4^+$, $CD25^+$ cells) in the tumors when compared to recombinant *L. monocytogenes*-E7 that secretes the antigen, but not the LLO-antigen fusion protein (Hussain and Paterson, 2004). Interestingly, immunization with a nonspecific recombinant *L. monocytogenes* expressing LLO-irrelevant antigen vaccine also results in the reduction of Tregs within tumors (Shahabi *et al.*, 2008). The reduction in Tregs, however, was further enhanced when the vaccine was antigen specific suggesting that the mechanism is at least partially antigen-dependent for a maximal effect (Nitcheu-Tefit *et al.*, 2007; Shahabi *et al.*, 2008). Interesting, there is no effect on the population of Tregs in spleens, implying that *L. monocytogenes* selectively reduces Tregs within the tumors. This is an important observation since other therapies (including antibody-mediated depletion of Tregs) that targets Tregs are associated with extensive side effects in humans. *L. monocytogenes*-LLO based vaccines thus may seem superior to other vaccine strategies due, at least in part, to their ability to inhibit Tregs accumulation only within the tumors. Coexpression of LLO with other antigens in different bacterial vectors also enhances the efficacy of the vaccines through the inhibition of Tregs (Nitcheu-Tefit *et al.*, 2007). The combination of *L. monocytogenes* ability to induce MHC I, MHC II pathways coupled with the fact that LLO is a very potent immunogenic molecule likely may have important implications for antitumor vaccination strategies in humans.

D. Implication of the immune response to *L. monocytogenes* infection: *L. monocytogenes* within the tumor

Adaptive immune cells clearly play an important role in modulating the microenvironment of tumors following *L. monocytogenes* vaccination. However, in addition to $CD4^+$ and $CD8^+$ T cells, a number of other regulatory factors can be found within tumors of *L. monocytogenes* vaccinated mice. Most strikingly, *L. monocytogenes* itself can be found within the tumor for up to 7 days while being cleared from the spleen and the liver after just 3 days (Huang *et al.*, 2007; Paterson *et al.*, unpublished data). The persistence of *L. monocytogenes* within tumors suggests that immune responses to the infection itself at the site of the tumor, independent of antigen-specific effects, may play a role in the potent antitumoral effect of these vaccines. For example, macrophages activated by *L. monocytogenes* may home to the tumor and secrete a variety of tumoricidal cytokines including IL-6, IL-12, IL-1, and TNFα. In addition, infected macrophages would serve as a source of LLO which in turn induces a Th1-type cytokine profile with secretion of the proinflammatory cytokines IL-12, IL-18, IFNγ as well as IL-1, IL-6, and TNFα. Interestingly, in *L. monocytogenes* based vaccines, partial depletion of macrophages has no effect on the tumor recall response after vaccination (Weiskirch *et al.*, 2001). In addition to macrophages, mast cells are activated by *L. monocytogenes* and are required to clear the bacteria from the spleen and the liver. Once activated, mast cells secrete TNF-α and induce neutrophil recruitment. Neutrophils are known to play an essential role in controlling *L. monocytogenes* infection at early time points. Once activated, neutrophils secrete IL-8, CSF-1, and MCP-1. These cytokines in turn attract and activate additional macrophages and propagate the antitumoral effects of these cells. It is conceivable that all of these cells could be recruited to tumors and aid in *L. monocytogenes* vaccine efficacy.

VI. CONCLUSIONS AND FUTURE PROSPECTS

Several aspects of *L. monocytogenes* make it a uniquely attractive vaccine vector candidate as compared to other live vectors such as vaccinia virus, *Salmonella*, *Shigella*, *Legionella*, *Lactococcus*, and *Mycobacterium* (BCG), since *L. monocytogenes* can be grown under standard BSL2 laboratory conditions and genetic manipulation of this organism is well-established allowing construction of recombinant vaccine strains. In addition, a single recombinant *L. monocytogenes* strain can be manipulated to express multiple gene products using plasmid or chromosomal systems. There is extensive knowledge about the life cycle, genetics, and immunological characteristics of *L. monocytogenes*. This provides a rationale for the design of potent, specific and safe vaccine platforms. Results with the various attenuated strains have been very promising and therefore, safety issues are being well addressed.

The ability of *L. monocytogenes* to generate strong innate and adaptive immune responses in the periphery and tumor microenvironment has been exploited for the design of suitable vaccines. Combination of *L. monocytogenes* with other vaccine strategies such as protein, DNA, or peptide coated DC in a heterologous prime-boost strategy could also significantly improve the immune responses for therapeutic studies. Much work remains to be done to identify the combination regimens necessary to obtain optimal responses. In addition, vaccination strategies exploiting epitope spreading may enhance the efficacy of antitumor immune responses.

Preclinical studies to evaluate the efficacy of *L. monocytogenes* based vaccines have demonstrated potent and protective immune responses in mouse models. These aspects provide the foundation for testing these vaccines in clinical trials in humans. Recently, Advaxis Inc., a New Jersey based biotechnology company performed a phase I clinical trial using its *L. monocytogenes* based construct expressing the tumor antigen, HPV16-E7 (Lovaxin C) in end stage cervical cancer patients. Lovaxin C was shown to be well tolerated in most patients who received an IV dose and displayed a dose dependent pattern of side effects. A phase II study to evaluate the efficacy of Lovaxin C in the US is currently being discussed with the Food and Drug Administration.

REFERENCES

Aichele, P., Zinke, J., Grode, L., Schwendener, R. A., Kaufmann, S. H., and Seiler, P. (2003). Macrophages of the splenic marginal zone are essential for trapping of blood-borne particulate antigen but dispensable for induction of specific T cell responses. *J. Immunol.* **171,** 1148–1155.

Alvarez-Dominguez, C., Roberts, R., and Stahl, P. D. (1997). Internalized *Listeria monocytogenes* modulates intracellular trafficking and delays maturation of the phagosome. *J. Cell. Sci.* **110**(Pt. 6), 731–743.

Arnold, R., and Konig, W. (1998). Interleukin-8 release from human neutrophils after phagocytosis of *Listeria monocytogenes* and *Yersinia enterocolitica*. *J. Med. Microbiol.* **47,** 55–62.

Badovinac, V. P., and Harty, J. T. (2000). Adaptive immunity and enhanced CD8+ T cell response to *Listeria monocytogenes* in the absence of perforin and IFN-gamma. *J. Immunol.* **164,** 6444–6452.

Badovinac, V. P., Porter, B. B., and Harty, J. T. (2002). Programmed contraction of CD8(+) T cells after infection. *Nat. Immunol.* **3,** 619–626.

Ballas, Z. K., Rasmussen, W. L., and Krieg, A. M. (1996). Induction of NK activity in murine and human cells by CpG motifs in oligodeoxynucleotides and bacterial DNA. *J. Immunol.* **157,** 1840–1845.

Beatty, G. L., and Paterson, Y. (2001). Regulation of tumor growth by IFN-gamma in cancer immunotherapy. *Immunol. Res.* **24,** 201–210.

Beck-Engeser, G. B., Monach, P. A., Mumberg, D., Yang, F., Wanderling, S., Schreiber, K., Espinosa, R., III, Le Beau, M. M., Meredith, S. C., and Schreiber, H. (2001). Point mutation in essential genes with loss or mutation of the second allele: Relevance to the retention of tumor-specific antigens. *J. Exp. Med.* **194,** 285–300.

Bierne, H., and Cossart, P. (2002). InlB, a surface protein of *Listeria monocytogenes* that behaves as an invasin and a growth factor. *J. Cell. Sci.* **115,** 3357–3367.

Braun, L., Ohayon, H., and Cossart, P. (1998). The InlB protein of *Listeria monocytogenes* is sufficient to promote entry into mammalian cells. *Mol. Microbiol.* **27,** 1077–1087.

Bruhn, K. W., Craft, N., Nguyen, B. D., Yip, J., and Miller, J. F. (2005). Characterization of anti-self CD8 T-cell responses stimulated by recombinant *Listeria monocytogenes* expressing the melanoma antigen TRP-2. *Vaccine* **23,** 4263–4272.

Busch, D. H., Pilip, I. M., Vijh, S., and Pamer, E. G. (1998). Coordinate regulation of complex T cell populations responding to bacterial infection. *Immunity* **8,** 353–362.

Cabanes, D., Dussurget, O., Dehoux, P., and Cossart, P. (2004). Auto, a surface associated autolysin of *Listeria monocytogenes* required for entry into eukaryotic cells and virulence. *Mol. Microbiol.* **51,** 1601–1614.

Cabanes, D., Sousa, S., Cebria, A., Lecuit, M., Garcia-del Portillo, F., and Cossart, P. (2005). Gp96 is a receptor for a novel *Listeria monocytogenes* virulence factor, Vip, a surface protein. *EMBO. J.* **24,** 2827–2838.

Camilli, A., Tilney, L. G., and Portnoy, D. A. (1993). Dual roles of plcA in *Listeria monocytogenes* pathogenesis. *Mol. Microbiol.* **8,** 143–157.

Campoli, M. R., Chang, C. C., Kageshita, T., Wang, X., McCarthy, J. B., and Ferrone, S. (2004). Human high molecular weight-melanoma-associated antigen (HMW-MAA): A melanoma cell surface chondroitin sulfate proteoglycan (MSCP) with biological and clinical significance. *Crit. Rev. Immunol.* **24,** 267–296.

Carrero, J. A., and Unanue, E. R. (2007). Impact of lymphocyte apoptosis on the innate immune stages of infection. *Immunol. Res.* **38,** 333–341.

Chakraborty, T., Leimeister-Wachter, M., Domann, E., Hartl, M., Goebel, W., Nichterlein, T., and Notermans, S. (1992). Coordinate regulation of virulence genes in *Listeria monocytogenes* requires the product of the prfA gene. *J. Bacteriol.* **174,** 568–574.

Chang, C. C., Campoli, M., Luo, W., Zhao, W., Zaenker, K. S., and Ferrone, S. (2004). Immunotherapy of melanoma targeting human high molecular weight melanoma-associated antigen: Potential role of nonimmunological mechanisms. *Ann. N. Y. Acad. Sci.* **1028,** 340–350.

Chin, A. I., Dempsey, P. W., Bruhn, K., Miller, J. F., Xu, Y., and Cheng, G. (2002). Involvement of receptor-interacting protein 2 in innate and adaptive immune responses. *Nature* **416,** 190–194.

Chow, M. T., Dhanji, S., Cross, J., Johnson, P., and Teh, H. S. (2006). H2-M3-restricted T cells participate in the priming of antigen-specific CD4+ T cells. *J. Immunol.* **177,** 5098–5104.

Conlan, J. W. (1996). Early pathogenesis of *Listeria monocytogenes* infection in the mouse spleen. *J. Med. Microbiol.* **44,** 295–302.

Conlan, J. W., Dunn, P. L., and North, R. J. (1993). Leukocyte-mediated lysis of infected hepatocytes during listeriosis occurs in mice depleted of NK cells or CD4+ CD8+ Thy1.2+ T cells. *Infect. Immun.* **61,** 2703–2707.

Conlan, J. W., and North, R. J. (1994). Neutrophils are essential for early anti-Listeria defense in the liver, but not in the spleen or peritoneal cavity, as revealed by a granulocyte-depleting monoclonal antibody. *J. Exp. Med.* **179,** 259–268.

Cossart, P., and Lecuit, M. (1998). Interactions of *Listeria monocytogenes* with mammalian cells during entry and actin-based movement: Bacterial factors, cellular ligands, and signaling. *EMBO. J.* **17,** 3797–3806.

Czuprynski, C. J., Brown, J. F., Maroushek, N., Wagner, R. D., and Steinberg, H. (1994). Administration of anti-granulocyte mAb RB6-8C5 impairs the resistance of mice to *Listeria monocytogenes* infection. *J. Immunol.* **152,** 1836–1846.

Datta, S. K., Okamoto, S., Hayashi, T., Shin, S. S., Mihajlov, I., Fermin, A., Guiney, D. G., Fierer, J., and Raz, E. (2006). Vaccination with irradiated *Listeria* induces protective T cell immunity. *Immunity* **25,** 143–152.

Decatur, A. L., and Portnoy, D. A. (2000). A PEST-like sequence in listeriolysin O essential for *Listeria monocytogenes* pathogenicity. *Science* **290,** 992–995.

Dominiecki, M. E., Beatty, G. L., Pan, Z. K., Neeson, P., and Paterson, Y. (2005). Tumor sensitivity to IFN-gamma is required for successful antigen-specific immunotherapy of a transplantable mouse tumor model for HPV-transformed tumors. *Cancer Immunol. Immunother.* **54,** 477–488.

D'Orazio, S. E., Troese, M. J., and Starnbach, M. N. (2006). Cytosolic localization of *Listeria monocytogenes* triggers an early IFN-gamma response by CD8+ T cells that correlates with innate resistance to infection. *J. Immunol.* **177,** 7146–7154.

Dramsi, S., Dehoux, P., Lebrun, M., Goossens, P. L., and Cossart, P. (1997). Identification of four new members of the internalin multigene family of *Listeria monocytogenes* EGD. *Infect. Immun.* **65,** 1615–1625.

Dussurget, O., Pizarro-Cerda, J., and Cossart, P. (2004). Molecular determinants of *Listeria monocytogenes* virulence. *Annu. Rev. Microbiol.* **58,** 587–610.

Echchakir, H., Bagot, M., Dorothee, G., Martinvalet, D., Le Gouvello, S., Boumsell, L., Chouaib, S., Bensussan, A., and Mami-Chouaib, F. (2000). Cutaneous T cell lymphoma reactive CD4+ cytotoxic T lymphocyte clones display a Th1 cytokine profile and use a fas-independent pathway for specific tumor cell lysis. *J. Invest. Dermatol.* **115,** 74–80.

Finelli, A., Kerksiek, K. M., Allen, S. E., Marshall, N., Mercado, R., Pilip, I., Busch, D. H., and Pamer, E. G. (1999). MHC class I restricted T cell responses to *Listeria monocytogenes*, an intracellular bacterial pathogen. *Immunol. Res.* **19,** 211–223.

Freeman, M. M., and Ziegler, H. K. (2005). Simultaneous Th1-type cytokine expression is a signature of peritoneal CD4+ lymphocytes responding to infection with *Listeria monocytogenes. J. Immunol.* **175,** 394–403.

Freitag, N. E., Rong, L., and Portnoy, D. A. (1993). Regulation of the prfA transcriptional activator of *Listeria monocytogenes*: Multiple promoter elements contribute to intracellular growth and cell-to-cell spread. *Infect. Immun.* **61,** 2537–2544.

Gaillard, J. L., Berche, P., Mounier, J., Richard, S., and Sansonetti, P. (1987). *In vitro* model of penetration and intracellular growth of *Listeria monocytogenes* in the human enterocyte-like cell line Caco-2. *Infect. Immun.* **55,** 2822–2829.

Gaillard, J. L., Jaubert, F., and Berche, P. (1996). The inlAB locus mediates the entry of *Listeria monocytogenes* into hepatocytes *in vivo*. *J. Exp. Med.* **183,** 359–369.

Gekara, N. O., Groebe, L., Viegas, N., and Weiss, S. (2008). *Listeria monocytogenes* desensitizes immune cells to subsequent Ca2+ signaling via listeriolysin O-induced depletion of intracellular Ca^{2+} stores. *Infect. Immun.* **76,** 857–862.

Greenberg, P. D. (1991). Adoptive T cell therapy of tumors: Mechanisms operative in the recognition and elimination of tumor cells. *Adv. Immunol.* **49,** 281–355.

Greiffenberg, L., Goebel, W., Kim, K. S., Weiglein, I., Bubert, A., Engelbrecht, F., Stins, M., and Kuhn, M. (1998). Interaction of *Listeria monocytogenes* with human brain microvascular endothelial cells: InlB-dependent invasion, long-term intracellular growth, and spread from macrophages to endothelial cells. *Infect. Immun.* **66,** 5260–5267.

Gunn, G. R., Zubair, A., Peters, C., Pan, Z. K., Wu, T. C., and Paterson, Y. (2001). Two *Listeria monocytogenes* vaccine vectors that express different molecular forms of human papilloma virus-16 (HPV-16) E7 induce qualitatively different T cell immunity that correlates with their ability to induce regression of established tumors immortalized by HPV-16. *J. Immunol.* **167,** 6471–6479.

Haddad, E. K., Duclos, A. J., Antecka, E., Lapp, W. S., and Baines, M. G. (1997). Role of interferon-gamma in the priming of decidual macrophages for nitric oxide production and early pregnancy loss. *Cell Immunol.* **181,** 68–75.

Harty, J. T., and Bevan, M. J. (1995). Specific immunity to *Listeria monocytogenes* in the absence of IFN gamma. *Immunity* **3,** 109–117.

Harty, J. T., Schreiber, R. D., and Bevan, M. J. (1992). CD8 T cells can protect against an intracellular bacterium in an interferon gamma-independent fashion. *Proc. Natl. Acad. Sci. USA* **89,** 11612–11616.

Hemmi, H., Takeuchi, O., Kawai, T., Kaisho, T., Sato, S., Sanjo, H., Matsumoto, M., Hoshino, K., Wagner, H., Takeda, K., and Akira, S. (2000). A toll-like receptor recognizes bacterial DNA. *Nature* **408,** 740–745.

Huang, B., Zhao, J., Shen, S., Li, H., He, K. L., Shen, G. X., Mayer, L., Unkeless, J., Li, D., Yuan, Y., Zhang, G. M., Xiong, H., *et al.* (2007). *Listeria monocytogenes* promotes tumor growth via tumor cell toll-like receptor 2 signaling. *Cancer Res.* **67,** 4346–4352.

Hussain, S. F., and Paterson, Y. (2004). CD4+ CD25+ regulatory T cells that secrete TGFbeta and IL-10 are preferentially induced by a vaccine vector. *J. Immunother.* **27,** 339–346.

Hussain, S. F., and Paterson, Y. (2005). What is needed for effective antitumor immunotherapy? Lessons learned using *Listeria monocytogenes* as a live vector for HPV-associated tumors *Cancer Immunol. Immunother.* **54,** 577–586.

Ikonomidis, G., Paterson, Y., Kos, F. J., and Portnoy, D. A. (1994). Delivery of a viral antigen to the class I processing and presentation pathway by *Listeria monocytogenes. J. Exp. Med.* **180,** 2209–2218.

Ishii, K. J., Takeshita, F., Gursel, I., Gursel, M., Conover, J., Nussenzweig, A., and Klinman, D. M. (2002). Potential role of phosphatidylinositol 3 kinase, rather than DNA-dependent protein kinase, in CpG DNA-induced immune activation. *J. Exp. Med.* **196,** 269–274.

Khelef, N., Lecuit, M., Bierne, H., and Cossart, P. (2006). Species specificity of the *Listeria monocytogenes* InlB protein. *Cell Microbiol.* **8,** 457–470.

Ladel, C. H., Flesch, I. E., Arnoldi, J., and Kaufmann, S. H. (1994). Studies with MHC-deficient knock-out mice reveal impact of both MHC I- and MHC II-dependent T cell responses on *Listeria monocytogenes* infection. *J. Immunol.* **153,** 3116–3122.

Lamikanra, A., Pan, Z. K., Isaacs, S. N., Wu, T. C., and Paterson, Y. (2001). Regression of established human papillomavirus type 16 (HPV-16) immortalized tumors *in vivo* by vaccinia viruses expressing different forms of HPV-16 E7 correlates with enhanced CD8 (+) T-cell responses that home to the tumor site. *J. Virol.* **75,** 9654–9664.

Lauer, P., Chow, M. Y., Loessner, M. J., Portnoy, D. A., and Calendar, R. (2002). Construction, characterization, and use of two *Listeria monocytogenes* site-specific phage integration vectors. *J. Bacteriol.* **184,** 4177–4186.

Lauvau, G., Vijh, S., Kong, P., Horng, T., Kerksiek, K., Serbina, N., Tuma, R. A., and Pamer, E. G. (2001). Priming of memory but not effector CD8 T cells by a killed bacterial vaccine. *Science* **294,** 1735–1739.

Leber, J. H., Crimmins, G. T., Raghavan, S., Meyer-Morse, N. P., Cox, J. S., and Portnoy, D. A. (2008). Distinct TLR- and NLR-mediated transcriptional responses to an intracellular pathogen. *PLoS Pathog.* **4,** e6.

Lecuit, M. (2007). Human listeriosis and animal models. *Microbes Infect.* **9,** 1216–1225.

Lecuit, M., Dramsi, S., Gottardi, C., Fedor-Chaiken, M., Gumbiner, B., and Cossart, P. (1999). A single amino acid in E-cadherin responsible for host specificity towards the human pathogen *Listeria monocytogenes. EMBO. J.* **18,** 3956–3963.

Lecuit, M., Nelson, D. M., Smith, S. D., Khun, H., Huerre, M., Vacher-Lavenu, M. C., Gordon, J. I., and Cossart, P. (2004). Targeting and crossing of the human maternofetal barrier by *Listeria monocytogenes*: Role of internalin interaction with trophoblast E-cadherin. *Proc. Natl. Acad. Sci. USA* **101,** 6152–6157.

Lecuit, M., Ohayon, H., Braun, L., Mengaud, J., and Cossart, P. (1997). Internalin of *Listeria monocytogenes* with an intact leucine-rich repeat region is sufficient to promote internalization. *Infect. Immun.* **65,** 5309–5319.

Lecuit, M., Vandormael-Pournin, S., Lefort, J., Huerre, M., Gounon, P., Dupuy, C., Babinet, C., and Cossart, P. (2001). A transgenic model for listeriosis: Role of internalin in crossing the intestinal barrier. *Science* **292,** 1722–1725.

Leimeister-Wachter, M., Haffner, C., Domann, E., Goebel, W., and Chakraborty, T. (1990). Identification of a gene that positively regulates expression of listeriolysin, the major virulence factor of *Listeria monocytogenes*. *Proc. Natl. Acad. Sci. USA* **87,** 8336–8340.

Liau, L. M., Jensen, E. R., Kremen, T. J., Odesa, S. K., Sykes, S. N., Soung, M. C., Miller, J. F., and Bronstein, J. M. (2002). Tumor immunity within the central nervous system stimulated by recombinant *Listeria monocytogenes* vaccination. *Cancer Res.* **62,** 2287–2293.

Lorber, B. (1997). Listeriosis. *Clin. Infect. Dis.* **24,** (1–9); quiz 10–1.

Mackaness, G. B. (1962). Cellular resistance to infection. *J. Exp. Med.* **116,** 381–406.

Maciag, P. C., Seavey, M. M., Pan, Z. K., Ferrone, S., and Paterson, Y. (2008). Cancer immunotherapy targeting the high molecular weight melanoma-associated antigen protein results in a broad antitumor response and reduction of pericytes in the tumor vasculature. *Cancer Res.* **68,** 8066–8075.

Mandel, T. E., and Cheers, C. (1980). Resistance and susceptibility of mice to bacterial infection: Histopathology of listeriosis in resistant and susceptible strains. *Infect. Immun.* **30,** 851–861.

Mata, M., Yao, Z. J., Zubair, A., Syres, K., and Paterson, Y. (2001). Evaluation of a recombinant *Listeria monocytogenes* expressing an HIV protein that protects mice against viral challenge. *Vaccine* **19,** 1435–1445.

Mercado, R., Vijh, S., Allen, S. E., Kerksiek, K., Pilip, I. M., and Pamer, E. G. (2000). Early programming of T cell populations responding to bacterial infection. *J. Immunol.* **165,** 6833–6839.

Milohanic, E., Jonquieres, R., Cossart, P., Berche, P., and Gaillard, J. L. (2001). The autolysin Ami contributes to the adhesion of *Listeria monocytogenes* to eukaryotic cells via its cell wall anchor. *Mol. Microbiol.* **39,** 1212–1224.

Moors, M. A., Auerbuch, V., and Portnoy, D. A. (1999). Stability of the *Listeria monocytogenes* ActA protein in mammalian cells is regulated by the N-end rule pathway. *Cell Microbiol.* **1,** 249–257.

Neeson, P., Pan, Z. K., and Paterson, Y. (2008). Listeriolysin O is an improved protein carrier for lymphoma immunoglobulin idiotype and provides systemic protection against 38C13 lymphoma. *Cancer Immunol. Immunother.* **57,** 493–505.

Nitcheu-Tefit, J., Dai, M. S., Critchley-Thorne, R. J., Ramirez-Jimenez, F., Xu, M., Conchon, S., Ferry, N., Stauss, H. J., and Vassaux, G. (2007). Listeriolysin O expressed in a bacterial vaccine suppresses CD4+ CD25high regulatory T cell function *in vivo*. *J. Immunol.* **179,** 1532–1541.

Nomura, T., Kawamura, I., Tsuchiya, K., Kohda, C., Baba, H., Ito, Y., Kimoto, T., Watanabe, I., and Mitsuyama, M. (2002). Essential role of interleukin-12 (IL-12) and IL-18 for gamma interferon production induced by listeriolysin O in mouse spleen cells. *Infect. Immun.* **70,** 1049–1055.

Ozaki, S., York-Jolley, J., Kawamura, H., and Berzofsky, J. A. (1987). Cloned protein antigen-specific, Ia-restricted T cells with both helper and cytolytic activities: Mechanisms of activation and killing. *Cell Immunol.* **105,** 301–316.

Pamer, E. G. (2004). Immune responses to *Listeria monocytogenes*. *Nat. Rev. Immunol.* **4,** 812–823.

Pan, Z. K., Ikonomidis, G., Lazenby, A., Pardoll, D., and Paterson, Y. (1995a). A recombinant *Listeria monocytogenes* vaccine expressing a model tumour antigen protects mice against lethal tumour cell challenge and causes regression of established tumours. *Nat. Med.* **1,** 471–477.

Pan, Z. K., Ikonomidis, G., Pardoll, D., and Paterson, Y. (1995b). Regression of established tumors in mice mediated by the oral administration of a recombinant *Listeria monocytogenes* vaccine. *Cancer Res.* **55,** 4776–4779.

Park, J. M., Ng, V. H., Maeda, S., Rest, R. F., and Karin, M. (2004). Anthrolysin *O* and other gram-positive cytolysins are toll-like receptor 4 agonists. *J. Exp. Med.* **200,** 1647–1655.

Paterson, Y., and Maciag, P. C. (2005). Listeria-based vaccines for cancer treatment. *Curr. Opin. Mol. Ther.* **7,** 454–460.

Peng, X., Hussain, S. F., and Paterson, Y. (2004). The ability of two *Listeria monocytogenes* vaccines targeting human papillomavirus-16 E7 to induce an antitumor response correlates with myeloid dendritic cell function. *J. Immunol.* **172,** 6030–6038.

Peng, X., Treml, J., and Paterson, Y. (2007). Adjuvant properties of listeriolysin O protein in a DNA vaccination strategy. *Cancer Immunol. Immunother.* **56,** 797–806.

Portnoy, D. A., Chakraborty, T., Goebel, W., and Cossart, P. (1992a). Molecular determinants of *Listeria monocytogenes* pathogenesis. *Infect. Immun.* **60,** 1263–1267.

Portnoy, D. A., Schreiber, R. D., Connelly, P., and Tilney, L. G. (1989). Gamma interferon limits access of *Listeria monocytogenes* to the macrophage cytoplasm. *J. Exp. Med.* **170,** 2141–2146.

Portnoy, D. A., Tweten, R. K., Kehoe, M., and Bielecki, J. (1992b). Capacity of listeriolysin O, streptolysin *O*, and perfringolysin *O* to mediate growth of *Bacillus subtilis* within mammalian cells. *Infect. Immun.* **60,** 2710–2717.

Renzoni, A., Cossart, P., and Dramsi, S. (1999). PrfA, the transcriptional activator of virulence genes, is upregulated during interaction of *Listeria monocytogenes* with mammalian cells and in eukaryotic cell extracts. *Mol. Microbiol.* **34,** 552–561.

Rogers, H. W., and Unanue, E. R. (1993). Neutrophils are involved in acute, nonspecific resistance to *Listeria monocytogenes* in mice. *Infect. Immun.* **61,** 5090–5096.

Schnupf, P., Zhou, J., Varshavsky, A., and Portnoy, D. A. (2007). Listeriolysin O secreted by *Listeria monocytogenes* into the host cell cytosol is degraded by the N-end rule pathway. *Infect. Immun.* **75,** 5135–5147.

Scortti, M., Monzo, H. J., Lacharme-Lora, L., Lewis, D. A., and Vazquez-Boland, J. A. (2007). The PrfA virulence regulon. *Microbes Infect.* **9,** 1196–1207.

Serbina, N. V., Kuziel, W., Flavell, R., Akira, S., Rollins, B., and Pamer, E. G. (2003). Sequential MyD88-independent and -dependent activation of innate immune responses to intracellular bacterial infection. *Immunity* **19,** 891–901.

Sewell, D. A., Douven, D., Pan, Z. K., Rodriguez, A., and Paterson, Y. (2004b). Regression of HPV-positive tumors treated with a new *Listeria monocytogenes* vaccine. *Arch. Otolaryngol. Head Neck Surg.* **130,** 92–97.

Sewell, D. A., Pan, Z. K., and Paterson, Y. (2008). Listeria-based HPV-16 E7 vaccines limit autochthonous tumor growth in a transgenic mouse model for HPV-16 transformed tumors. *Vaccine* **26,** 5315–5320.

Sewell, D. A., Shahabi, V., Gunn, G. R., III, Pan, Z. K., Dominiecki, M. E., and Paterson, Y. (2004a). Recombinant Listeria vaccines containing PEST sequences are potent immune adjuvants for the tumor-associated antigen human papillomavirus-16 E7. *Cancer Res.* **64,** 8821–8825.

Sha, W. C., Liou, H. C., Tuomanen, E. I., and Baltimore, D. (1995). Targeted disruption of the p50 subunit of NF-kappa B leads to multifocal defects in immune responses. *Cell* **80,** 321–330.

Shahabi, V., Reyes-Reyes, M., Wallecha, A., Rivera, S., Paterson, Y., and Maciag, P. (2008). Development of a *Listeria monocytogenes* based vaccine against prostate cancer. *Cancer Immunol. Immunother.* **57,** 1301–1313.

Sijts, A. J., Pilip, I., and Pamer, E. G. (1997). The *Listeria monocytogenes*-secreted p60 protein is an N-end rule substrate in the cytosol of infected cells. Implications for major histocompatibility complex class I antigen processing of bacterial proteins. *J. Biol. Chem.* **272,** 19261–19268.

Singh, R., Dominiecki, M. E., Jaffee, E. M., and Paterson, Y. (2005). Fusion to Listeriolysin O and delivery by *Listeria monocytogenes* enhances the immunogenicity of HER-2/neu and reveals subdominant epitopes in the FVB/N mouse. *J. Immunol.* **175,** 3663–3673.

Singh, R., and Paterson, Y. (2007a). Immunoediting sculpts tumor epitopes during immunotherapy. *Cancer Res.* **67,** 1887–1892.

Singh, R., and Paterson, Y. (2007b). In the FVB/N HER-2/neu transgenic mouse both peripheral and central tolerance limit the immune response targeting HER-2/neu induced by *Listeria monocytogenes*-based vaccines. *Cancer Immunol. Immunother.* **56,** 927–938.

Skeen, M. J., and Ziegler, H. K. (1993). Induction of murine peritoneal gamma/delta T cells and their role in resistance to bacterial infection. *J. Exp. Med.* **178,** 971–984.

Skoberne, M., Schenk, S., Hof, H., and Geginat, G. (2002). Cross-presentation of *Listeria monocytogenes*-derived CD4 T cell epitopes. *J. Immunol.* **169,** 1410–1418.

Souders, N. C., Sewell, D. A., Pan, Z. K., Hussain, S. F., Rodriguez, A., Wallecha, A., and Paterson, Y. (2007). Listeria-based vaccines can overcome tolerance by expanding low avidity CD8+ T cells capable of eradicating a solid tumor in a transgenic mouse model of cancer. *Cancer Immun.* **7,** 2.

Suarez, M., Gonzalez-Zorn, B., Vega, Y., Chico-Calero, I., and Vazquez-Boland, J. A. (2001). A role for ActA in epithelial cell invasion by *Listeria monocytogenes. Cell Microbiol.* **3,** 853–864.

Szalay, G., Hess, J., and Kaufmann, S. H. (1994). Presentation of *Listeria monocytogenes* antigens by major histocompatibility complex class I molecules to CD8 cytotoxic T lymphocytes independent of listeriolysin secretion and virulence. *Eur. J. Immunol.* **24,** 1471–1477.

Tanabe, Y., Xiong, H., Nomura, T., Arakawa, M., and Mitsuyama, M. (1999). Induction of protective T cells against *Listeria monocytogenes* in mice by immunization with a listeriolysin O-negative avirulent strain of bacteria and liposome-encapsulated listeriolysin O. *Infect. Immun.* **67,** 568–575.

Tilney, L. G., and Portnoy, D. A. (1989). Actin filaments and the growth, movement, and spread of the intracellular bacterial parasite, *Listeria monocytogenes. J. Cell. Biol.* **109,** 1597–1608.

Torres, D., Barrier, M., Bihl, F., Quesniaux, V. J., Maillet, I., Akira, S., Ryffel, B., and Erard, F. (2004). Toll-like receptor 2 is required for optimal control of *Listeria monocytogenes* infection. *Infect. Immun.* **72,** 2131–2139.

Tsujimura, H., Tamura, T., Kong, H. J., Nishiyama, A., Ishii, K. J., Klinman, D. M., and Ozato, K. (2004). Toll-like receptor 9 signaling activates NF-kappaB through IFN regulatory factor-8/IFN consensus sequence binding protein in dendritic cells. *J. Immunol.* **172,** 6820–6827.

Tweten, R. K. (2005). Cholesterol-dependent cytolysins, a family of versatile pore-forming toxins. *Infect. Immun.* **73,** 6199–6209.

Vazquez-Boland, J. A., Kuhn, M., Berche, P., Chakraborty, T., Dominguez-Bernal, G., Goebel, W., Gonzalez-Zorn, B., Wehland, J., and Kreft, J. (2001). Listeria pathogenesis and molecular virulence determinants. *Clin. Microbiol. Rev.* **14,** 584–640.

Verch, T., Pan, Z. K., and Paterson, Y. (2004). *Listeria monocytogenes*-based antibiotic resistance gene-free antigen delivery system applicable to other bacterial vectors and DNA vaccines. *Infect. Immun.* **72,** 6418–6425.

Villanueva, M. S., Fischer, P., Feen, K., and Pamer, E. G. (1994). Efficiency of MHC class I antigen processing: A quantitative analysis. *Immunity* **1,** 479–489.

Walz, T. (2005). How cholesterol-dependent cytolysins bite holes into membranes. *Mol. Cell* **18,** 393–394.

Way, S. S., and Wilson, C. B. (2004). Cutting edge: Immunity and IFN-gamma production during *Listeria monocytogenes* infection in the absence of T-bet. *J. Immunol.* **173,** 5918–5922.

Weiskirch, L. M., Pan, Z. K., and Paterson, Y. (2001). The tumor recall response of antitumor immunity primed by a live, recombinant *Listeria monocytogenes* vaccine comprises multiple effector mechanisms. *Clin. Immunol.* **98,** 346–357.

Wong, P., and Pamer, E. G. (2001). Cutting edge: Antigen-independent CD8 T cell proliferation. *J. Immunol.* **166,** 5864–5868.

Yamamoto, K., Kawamura, I., Tominaga, T., Nomura, T., Ito, J., and Mitsuyama, M. (2006). Listeriolysin O derived from *Listeria monocytogenes* inhibits the effector phase of an experimental allergic rhinitis induced by ovalbumin in mice. *Clin. Exp. Immunol.* **144,** 475–484.

Yamamoto, K., Kawamura, I., Tominaga, T., Nomura, T., Kohda, C., Ito, J., and Mitsuyama, M. (2005). Listeriolysin O, a cytolysin derived from *Listeria monocytogenes,* inhibits generation of ovalbumin-specific Th2 immune response by skewing maturation of antigen-specific T cells into Th1 cells. *Clin. Exp. Immunol.* **142,** 268–274.

Yoshimura, A., Shiku, H., and Nakayama, E. (1993). Rejection of an IA+ variant line of FBL-3 leukemia by cytotoxic T lymphocytes with CD4+ and CD4-CD8- T cell receptor-alpha beta phenotypes generated in CD8-depleted C57BL/6 mice. *J. Immunol.* **150,** 4900–4910.

CHAPTER 2

Diagnosis of Clinically Relevant Fungi in Medicine and Veterinary Sciences

Olivier Sparagano[*,1] and **Sam Foggett**[*,†]

Contents

Abstract

This review focuses on the most economically and epidemiologically important fungi affecting humans and animals. This paper will also summarize the different techniques, either molecular, based on nucleic acid and antibody analysis, or nonmolecular such as microscopy, culture, UV Wood';s lamp, radiology, and

* School of Agriculture, Food and Rural Development, Newcastle University, Newcastle upon Tyne, NE1 7RU, United Kingdom
† School of Biomedical Sciences, Faculty of Medical Sciences, Newcastle University, Newcastle upon Tyne, NE1 7RU, United Kingdom
1 Corresponding author.: School of Agriculture, Food and Rural Development, Newcastle University, Newcastle upon Tyne, NE1 7RU, United Kingdom

Advances in Applied Microbiology, Volume 66
ISSN 0065-2164, DOI: 10.1016/S0065-2164(08)00802-2

spectroscopy used to identify species or group of fungi assisting clinicians to take the best control approach to clear such infections. On the molecular side, the paper will review results on genome sequencing which can help colleagues to identify their own DNA/RNA tests if they are interested in the diagnostic of fungi in medicine and veterinary medicine.

I. INTRODUCTION

A. The general structure of fungi

Fungi are eukaryotic, nonmotile organisms which can either be multicellular (in their filamentous form due to hyphae) or unicellular (in their yeast form).

Filamentous fungi, also known as moulds, are named as such because of the long, branched structures which they produce. These structures (hyphae) are created as a result of sporal germination (Kwon-Chung and Bennett, 1992). The purpose of the hyphae is not only to search for nutrients needed in growth and reproduction, but also to maintain spores for reproduction and their eventual distribution (Walker and White, 2006). The hyphae that search for nutrients grow apically from the large network of hyphae called the mycelium, and can range from 1 to 30 μm in size depending on species and environmental conditions. However, the size of a hypha in an individual fungus is not finite, as theoretically hyphae can grow indefinitely given the correct nutrients and other growth conditions (Walker and White, 2006). Septa divide hyphae into separate cells, but some fungal hyphae contain no septa and the protoplasm is therefore continuous throughout the structure. These septa-lacking hyphae are termed coenocytic (Kwon-Chung and Bennett, 1992). Under the microscope, some fungi may appear to have hyphae, but these are actually pseudohyphae. Pseudohyphae are the result of yeast cells budding and elongating, where the progeny of the parent cell remains attached and the budding process repeats to form long chains. Differentiation between hyphae and pseudohyphae can be made by noting that the latter includes constrictions at the septa and contain smaller cells at the end (Kwon-Chung and Bennett, 1992). The mycelium also contains specialized hyphal cells called conidiophores or sporangium in some cases (names depending on the phylum which the species falls under) which either houses or produces conidia, then releases them. These structures often aid the visual identification of fungal species.

The size of a yeast cell is variable, depending on species, but can be between 2–3 and 20–50 μm in length to 1–10 μm in width (Walker and White, 2006). They are unicellular and reproduce via budding or fission. Morphologically, they are generally oval or elongated.

The different levels of fungal infections are presented in Table 2.1.

TABLE 2.1 The various Tinea infections and the areas which they colonize the human body

Mycoses			
Superficial	Subcutaneous	Endemic systemic	Opportunistic
Dermatophytosis	Mycetoma	Histoplasmosis	Candidiasis
Onychomycosis	Sporotrichosis	Coccidioidomycosis	Aspergillosis
Malassezia infections	Chromoblastomycosis	Blastomycosis	Cryptococcosis
Superficial candidiasis		Paracoccidioidomycosis	Zygomycosis

B. Clinically relevant species of fungi

In terms of prevalence, it is rather difficult to have an idea as studies on animals, humans, or plants are ranging from a few percentages to 100% of the studied populations. However, this paper will only focus on the identification of the more relevant fungal infections in humans and animals.

1. Medical field

Regarding, their ability to initiate infection within a given host, most humans have well developed immune responses to fungi and therefore many fungal diseases are considered to be opportunistic infections, that is, those which can only cause disease in persons with insufficient immune responses. For example, in immunocompromized patients, it is known that *Candida* and *Aspergillus* species are responsible for more than 90% of invasive fungal infections (Baskova *et al.*, 2007). Mycoses can be divided into four main categories, which define how they operate: Superficial (and cutaneous), Subcutaneous, Endemic systemic, and opportunistic (see Table 2.1). As previously mentioned, opportunistic mycoses infect immunocompromized hosts. Superficial mycoses apply to those which appear on the outer surfaces of the body. As they colonize the outer layers of the body, they do not encounter an immune response. However, there are defense mechanisms in place such as sweat, sebum, transferrin, and β-defensins which have antifungal properties (Walker and White, 2006). Deeper infections, termed subcutaneous mycoses, develop when traumatic injuries such as cuts or bites occur. The term Endemic Systemic arises from the group of fungal species which can infect the pulmonary system, circulatory system, and internal organs, and which also occur in specific geographical locations, therefore being endemic. These mycoses are often contracted from inhalation of their conidia.

Dermatophytes are superficial fungal infections which infect keratin containing body parts such as nails, hair wool, horns, and skin. They cause *tinea*, more commonly known as ringworm, which affects humans and a range of animal species. Depending on what their natural reservoir is, dermatophytes are classed as either zoophilic (passed to humans via animals) or anthropophilic (humans to humans). In rural areas more than 50% of human ringworm infections are contracted from zoophilic transmission (Richardson and Elewski, 2000).

Tinea diseases are named as a result of the location of the human body that the dermatophyte has colonized. Clinical symptoms of dermatophyte infections vary, and are determined on where on the body that the infection is. Infections on the skin cause the skin to appear dry or scaly, as in the case of *Tinea pedis* and *Tinea manuum*. *Tinea unguium* (an Onychomycosis) is a cutaneous infection of the nails, mainly the toe nails.

The nail becomes removed from the nail bed and is chipped and discolored in appearance. It is speculated that 10% of individuals contract *Tinea unguium* (Richardson and Elewski, 2000).

Normally found as commensals of the mouth, vaginal mucosa, and gastrointestinal tract, *Candida* species, although better known for being opportunistic pathogens, also can cause superficial candidiasis (also known as thrush). Primarily caused by the yeast, *Candida albicans*, other species also cause infection in humans, such as *C. krusei, C. parapsilosis, C. tropicalis, and C. glabrata* (Midgley *et al.*, 1997). In oral candidiasis, symptoms differ depending on the subtype. A patient with chronic plaque-like candidiasis will have a layer of white plaque on their tongue which cannot be removed, whereas in acute erythematous candidiasis, the mucosal surface of the patient's mouth will look glossy and red. Infants, the elderly, and immunocompromised patients are the main sufferers of oral candidiasis. Vaginal candidiasis is mainly due to *C. albicans* and can cause itching of the vagina, discharge, white plaque to appear, and a red rash.

Mycetoma or "Madura foot," is a chronic subcutaneous infection characterized by large granules which form within abscesses (Midgley *et al.*, 1997). They are caused either by true fungi (Eumycetes) or by filamentous bacteria (Actinomyces) (Lewall *et al.*, 1985). The main fungal etiological agent of eumycetoma in USA is *Pseudoallescheria boydii* (Geyer *et al.*, 2006). Others are *Madurella mycetomatis, M. grisea, Fusarium spp., Leptosphaeria senegalensis*, and *Acremonium spp.*, however, the latter is a very rare cause of eumycetoma (Geyer *et al.*, 2006; Midgley *et al.*, 1997). Color of the granules differs, ranging from black, yellow, and white. It is prevalent in tropical and subtropical areas of the world, frequently occurring in drier parts of the globe such as India where it was first described in 1846 (Kumar *et al.*, 2007). Clinical features involve the legs but most cases predominantly involve infection of the feet, with 70% of mycetoma cases in India having mycetoma of the foot (Pankajalakshmi and Taralakshmi, 1984). Large swellings occur which then develop sinus tracts where the colored granules are discharged.

The dimorphic fungus *Sporothrix schenkii* causes the chronic infection Sporotrichosis and is most common in southern parts of the USA (Greydanusvanderputten *et al.*, 1994). This subcutaneous mycosis affects the skin as cutaneous sporotrichosis when injuries due to trauma occur, and presents itself as a reddish lesion usually on the skin of the face. Other lesions occur nearby the original due to the spread of the pathogen through the lymphatic system. Eventually the lesions discharge pus.

Chromoblastomycosis is a chronic infection prevalent in men above 20 years of age who live in rural areas or who are in agricultural employment (Perez-Blanco *et al.*, 2006). It is a mycotic disease which can be seen throughout the world, from Central and South America, Africa, and the

Far East (Midgley *et al.*, 1997). The two major etiological agents are *Cladosporium carrionii* and *Fonsecaea pedrosoi* (Perez-Blanco *et al.*, 2006; Santos *et al.*, 2007). Clinical symptoms are not apparent until a late stage of infection (Murray *et al.*, 2005), ranging from a time period of 1 month–20 years (Kwon-Chung and Bennett, 1992). Symptoms include small papules on the skin which develop into large verrucous nodules on exposed areas such as the legs and arms (Midgley *et al.*, 1997).

Disseminated and pulmonary histoplasmosis caused by *Histoplasma capsulatum* var *capsulatum* is typically found in Latin America, central and eastern parts of the USA (Midgley *et al.*, 1997; Murray *et al.*, 2005). Symptoms in humans vary depending on the type of histoplasmosis, and even so, cannot be generalized (Kwon-Chung and Bennett, 1992).

It is known to be the most virulent fungal species in humans due to the few numbers of cells required to induce an infection (Murray *et al.*, 2005). Endemic areas are located in the desert areas of the USA, such as California, and parts of Central and South America with common clinical features being skin lesions, fever, and chest pain (Midgley *et al.*, 1997; Murray *et al.*, 2005; Shubitz, 2007). Inhalation of spores leads to respiratory coccidioidomycosis, which can lead to dissemination of the infection in about 5% of cases (Walker and White, 2006).

Infections of blastomycosis mainly occur within the USA (Murray *et al.*, 2005; Walker and White, 2006), and can appear in dogs (Arceneaux *et al.*, 1998). Inhalation of the dimorphic fungus *Blastomyces dermatitidis* is the route of infection. Chronic pulmonary blastomycosis symptoms include chest pains, a cough, and weight loss (Kwon-Chung and Bennett, 1992; Midgley *et al.*, 1997; Murray *et al.*, 2005). Clinical cutaneous features may appear when infection of the lung disseminates, with the appearance of skin lesions with crusted surfaces which emerge on the face, neck, and scalp (Kwon-Chung and Bennett, 1992; Murray *et al.*, 2005).

Endemic to South and Central America, *Paracoccidioides brasiliensis* is the infectious agent of paracoccidioidomycosis (Midgley *et al.*, 1997; Walker and White, 2006). Within this endemic area, it has also been found in wild monkeys (Corte *et al.*, 2007). Pulmonary lesions occur in sufferers of chronic pulmonary paracoccidioidomycosis with additional clinical symptoms being fever, a cough, and chest pain (Murray *et al.*, 2005). Superficial lesions can appear if the infection disseminates from the lungs, due to lack of treatment, which appear on the skin or mucosal regions of the body (Murray *et al.*, 2005).

Within immunocompromized patients, aspergillosis is the most significant opportunistic infection (Barnes, 2007). *Aspergillus fumigatus* is the most common etiological agent of aspergillosis in humans (Chan *et al.*, 2002), but other major species include *A. flavus* and *A. niger* (Midgley *et al.*, 1997). Due to improvements in diagnostic methods, invasive aspergillosis (IA) mortality cases have decreased in recent years (Barnes, 2007; Upton *et al.*, 2007). IA can appear due to bone marrow transplants or blood transfusions (Marr *et al.*, 2004), that is, those receiving immunosuppressive drugs. Lungs are the most common site of IA infection, but it can then be spread to the central nervous system (CNS) (Denning, 1998). Swelling of the face occurs in the infection of the paranasal sinuses (Midgley *et al.*, 1997), with the inclusion of fungus balls in the airways. Invasive pulmonary aspergillosis produces pulmonary infiltrates, fever, and chest pains (Murray *et al.*, 2005).

Candida spp. are commensals which are found in the gastrointestinal tract of most humans, and the mouths of around 50% of healthy individuals (Murray *et al.*, 2005). However, in immunocompromised persons, these species can cause opportunistic mycotic disease. The etiological agents of candidiasis are *C. albicans*, *C. tropicalis*, *C. glabrata*, and *C. krusei*, and the clinical features of each of these species is nondifferential (Midgley *et al.*, 1997). Many of the clinical features of are dependant on the type of candidiasis and the area of infection involved (Midgley *et al.*, 1997). Areas of candidiasis infection include the CNS, eyes, bones, and cardiac area, which are affected due to dissemination of haematogenous candidiasis (Murray *et al.*, 2005).

Although a rare mycotic disease (Midgley *et al.*, 1997), zygomycosis has extremely high mortality rates when contracted, ranging from 70% to 100%. The infectious agents of the disease all come from different genera: *Rhizomucor* spp., *Rhizopus* spp., and *Absidia* spp., with *Rhizopus arrhizus* being the predominant cause of human zygomycosis (Murray *et al.*, 2005). Fungus balls can be seen in radiographs of patients with pulmonary zygomycosis (Murray *et al.*, 2005), and redness of the skin and edema occur on the cheeks of those with rhinocerebral zygomycosis (Midgley *et al.*, 1997).

2. Veterinary medicine field

Within veterinary medicine, *Trichophyton mentagrophytes*, *T. verrucosum*, *T. equinum*, and *Microsporum canis* are the most commonly observed dermatophytes. *M. canis* is accountable for 90% of dermaphytosis infections in cats, and 62% of infections in dogs (Sparkes *et al.*, 1993). *M. equinum* is found in horses, *T. verrucosum* in cattle, and *T. mentagrophytes* in rodents.

Cats and dogs can also contract sporotrichosis (Clinkenbeard, 1991), with lesions on the cat commonly found on the face, legs, and nasal cavity (Welsh, 2003). In 2004, it was reported that there was an outbreak of cat transmitted sporotrichosis between 1998 and 2001. This involved 178 cases from Rio de Janeiro in Brazil (Barros *et al.*, 2004). Systemic sporotrichosis is also possible, mainly due to the inhalation of the pathogen which leads to lung lesions (Kwon-Chung and Bennett, 1992).

Histoplasmosis can affect dogs, cats (Bromel and Sykes, 2005), hedgehogs (Snider *et al.*, 2008), and rats to name a few animal hosts. Two variations of *Histoplasma capsulatum* are the causative agents of Histoplasmosis. They are *H. capsulatum* var *capsulatum* and *H. capsulatum* var *duboisii* which are thermally dimorphic fungi, filamentous in their natural environment, but yeast-like in their host (Bromel and Sykes, 2005). Pulmonary histoplasmosis is contracted by coming into contact with pathogenic cells in large numbers (Midgley *et al.*, 1997) and subsequent inhalation of microconidia (Murray *et al.*, 2005). Such a source can be found in chicken sheds and bat droppings, which are a rich source of nitrogen enabling sporulation to be sped up (Lyon *et al.*, 2004; Wheat and Kauffman, 2003). Cats show signs of fever, weight loss, and lethargy (Davies and Troy, 1996). Sign in dogs are linked to weight loss, but symptoms also include diarrhea (Vansteenhouse and Denovo, 1986).

In veterinary medicine, *Aspergillus fumigatus* is a frequent cause of mycotic abortions in cattle and sheep (Dr. Peter Booth, Veterinary Laboratories Agency, United Kingdom, personal communication). It is also evident within various avian species (Beytut, 2007; Carrasco *et al.*, 1993).

Cryptococcus neoformans, an encapsulated yeast (Midgley *et al.*, 1997), has two variations which both lead to cryptococcosis. They are *C. neoformans* var *neoformans* and *C. neoformans* var *gatii*, with the former being commonly found in pigeon guano (Murray *et al.*, 2005). *C. neoformans* var *gatii* infections are found in tropical and subtropical areas of the world which contain eucalyptus trees (Murray *et al.*, 2005). Infection of the lungs can disseminate to the rest of the body, usually reaching the CNS leading to cryptococcal meningitis (Walker and White, 2006). Superficial manifestations include nodules and ulceration of the skin, and pulmonary features include chest pain, a cough, and fever (Midgley *et al.*, 1997).

Coccidioides spp. are dimorphic fungi which are of hyphal morphology in the environment, but spherical in form within their hosts (Shubitz, 2007). The etiologic agents of coccidioidomycosis are *Coccidioides immitis* or *C. psadassii*, which are capable of infecting humans, chimps, most other mammals, and some reptiles (Hoffman *et al.*, 2007; Shubitz, 2007).

II. NON MOLECULAR METHODS OF FUNGAL DIAGNOSIS

A. Microscopy

Microscopic identification of fungi is mainly based on the distinctive morphological features of the organism which can include hyphal structures, yeasts, conidia, and conidiophores amongst others.

It is generally considered that light microscopy is the most effective method of fungal diagnosis in terms of its cost effectiveness and rapidity

(Murray *et al.*, 2005). Specimens from the area of fungal infection are collected and processed depending on what they are (discharged pus, biopsies, scrapings, etc.). Samples from culture can also be taken via sellotape which reveals the *in situ* arrangement of spores (Midgley *et al.*, 1997). A vast array of preparations and stains can be used to enhance the appearance of fungi on the slide, such as Lactophenol cotton blue, which when added, dyes the fungal cells therefore enhancing the contrast of fungal cells against background material (Midgley *et al.*, 1997; Murray *et al.*, 2005). Another contrast technique is done by applying India Ink, which instead of dying the fungal cells, dyes the background material. This is the primary method for identifying the encapsulated yeast of *Cryptococcus neoformans* from cerebrospinal fluid, and about 50% of CSF cryptococcosis cases are diagnosed this way (Grossgebauer, 1980). In order to eradicate excess nonfungal cells and allow the presence of hyphae to be identified, 10% potassium hydroxide (KOH) is added. Fungal cells are unaffected by this addition due to their cell walls being alkali resistant. Calcofluor white binds to chitin, a cell wall component of fungi, and causes it to brightly fluoresce (a fluorochrome, as described later). The results of a study by Hageage and Harrington (1984) revealed that with a combination of KOH preparation and calcofluor, 13 out of 17 culture positive samples could be identified, compared to the Gram stain which could only identify 7 out of the 17 specimens (Hageage and Harrington, 1984). However, Abdelrahman *et al.* (2006) found that using calcofluor as opposed to KOH preparation had increased specificity and sensitivity for diagnosing dermatophytes. Using calcofluor is entirely dependant on the use of a fluorescence microscope (as described later in this chapter), which is one of its shortcomings. As yeasts are Gram positive, the Gram stain is used to detect *Candida* spp. but filamentous fungi such as *Aspergillus* can also be detected this way. Gomori methenamine silver (GMS) acts on hyphae and yeasts, staining them black on a green background, and is the greatest stain for detecting all fungi (Murray *et al.*, 2005). Giemsa stain is used primarily to detect intracellular *Histoplasma capsulatum* in bone marrow and peripheral blood smears (Kwon-Chung and Bennett, 1992) and typically stains blue (Murray *et al.*, 2005). Within histopathology, the Periodic acid-Schiff stain is the most widely used, and along with GMS are the most fungi specific stains (Kwon-Chung and Bennett, 1992; Murray *et al.*, 2005). Although being one of the primary methods of dermatophyte identification, microscopy has been found to provide false negative results up to 15% of the time (Weitzman and Summerbell, 1995).

Fluorescent microscopy utilizes compounds which absorb ultraviolet light and subsequently emit light at a higher wavelength which is more visible (Murray *et al.*, 2005). The chosen pathogen is stained with a fluorescent compound (a fluorochrome) and viewed under a fluorescence microscope. This piece of equipment includes filters which eradicate infrared light and heat produced by the lamp. The correct wavelength for exciting the fluorochrome is also chosen. Depending on the fluorochrome selected, the fluorescent color of the pathogen will vary when viewed, but will be clearly visible against a black background.

B. Culture

Macroscopic identification of fungi is accomplished by inoculating the pathogen onto a specific growth medium. Features of colonies which aid in diagnosis include the color of the apical surface, color of the colony on reverse side of the plate, and the texture of the colony. The method is used when appropriate numbers of pathogenic cells in a given tissue sample are lacking. Sabourand dextrose agar is the most routinely used medium for the growth of fungi, and is slightly acidic at a pH of 6.5 which inhibits the growth of many bacteria. Fungi which are dimorphic in nature require a richer medium like Brain heart infusion agar. Two plates also needed to be incubated at 26 and 37 °C to show the pathogen in yeast and hyphal form. Chromogenic agars such as CHROMagar™ Candida (Becton–Dickinson Microbiology Systems, Sparks, Mary Land, USA) can be used to identify *Candida* spp. by the specific color colonies that a species will produce on the agar (Toubas *et al.*, 2007; Walker and White, 2006). The Veterinary Laboratories Agency uses blood agar to grow cultures of *Candida albicans* taken from cows suffering from bovine mastitis (Dr. Peter Booth, Veterinary Laboratories Agency, United Kingdom, personal communication). The development time of cultures varies depending on type. Dermatophytes take between 7 and 14 days to produce colonies, whereas subcutaneous and systemic pathogens take between 4 and 12 weeks to develop colonies (Midgley *et al.*, 1997). Antibiotics such as chloramphenicol and cycloheximide are added to the agar to prevent contamination of the culture by unwanted microbes. As cycloheximide inhibits the growth of *Candida* spp. and *Aspergillus* spp., it is important to make two cultures, one containing cycloheximide, and one without (Murray *et al.*, 2005).

A standard operating procedure manual used in NHS microbiology laboratories suggests that cultures intended to detect dermatophytes, *Candida* spp. from nail samples, *Fusarium* spp., and *Acremonium* spp. are incubated in an aerobic atmosphere, and are kept at a temperature of 30 °C to ensure optimal growth (the incubation period is about 28 days). Specimens from dermatophytosis are collected from the area of infection. Pathogenic specimens from *Tinea capitis* are taken by brush samples that are collected from the scalp and can be directly transferred to plates for culture growth. Hairs can also be plucked from the follicle and grown onto plates. Nail samples must contain a sufficient amount of crumbly material, but the distal edge of the nail should not be taken as a sample (Richardson and Elewski, 2000). For skin samples, a 1 cm × 1 cm scraping is appropriate for culture.

C. UV wood's Lamp

An ultraviolet Wood's lamp is a diagnostic tool used for the veterinary and medical diagnosis of dermatophytes by using hair samples (Kwon-Chung and Bennett, 1992 and Dr. Peter Booth, Veterinary Laboratories Agency, United Kingdom, personal communication). Hairs are placed under the lamp and tested for fluorescence. The production of certain metabolites in *Microsporum audouinii, M. canis,* and *M. ferrugineum* (Kwon-Chung and Bennett, 1992) causes them to fluoresce bright green on the hair under a UV Wood's lamp. If no fluorescence is detected, then this does not necessarily mean the absence of a dermatophyte infection, due to the fact some do not fluoresce. This technique can also suffer from false-positives if certain topical treatments have been applied prior to UV Wood's lamp examination (Dr. Peter Booth, Veterinary Laboratories Agency, United Kingdom, personal communication).

D. Radiology

Radiological diagnosis of invasive fungal infections includes the use of computed tomography (CT) and magnetic resonance imaging (MRI).

CT, also known as computed axial tomography (CAT), is the creation of a 2D cross-sectional image of the body produced when X-rays of the body are taken at different angles (NHS, 2007a). The patient may need to ingest or be injected with a contrast dye for optimal image production, before entering the CT machine on a motorized bed (CT-Chest 2007).

MRI utilizes the magnetic field of hydrogen atoms (protons) in the body (MRI, 2007). The magnet of the MRI machine aligns the protons in the direction of its magnetic field. The machine then emits radio waves into the body, causing the change in direction of the magnetic field of the protons (MRI, 2007). The image produced by the MRI is a result of the emission of radio waves from the protons once the scanner ceases emitting its own radio waves (MRI, 2007). Depending on the body part being scanned, the intensity of the image can vary due to different body parts having different water contents (NHS, 2007b).

MRI and CT can both be used in the diagnosis of IA in the CNS where the clinical sign is a hypodense area of brain edema (an increase in interstitial fluid; Dods *et al.*, 2007). Macronodules and the halo sign are detected via chest CT, and are early signs of an invasive fungal infection (Barnes, 2007). Reported cases of increased survival rates have been associated with CT diagnosis at an early stage of pulmonary aspergillosis in combination with antifungal drugs (Greene *et al.*, 2007).

As described earlier, mycetoma can be characterized by the different grains which the etiological agents produce. Kumar *et al.* (2007) argue that diagnosis via the appearance of sinuses and grains is too late (Kumar *et al.*, 2007). Diagnosis of the causative agents of mycetoma via culture can be difficult as it is rare for them to produce structures which lead to their identification (Borman *et al.*, 2006); therefore, MRI is a useful tool for the early detection of soft tissues masses characteristic of mycetoma. The dot-in-circle feature of an MRI scan proposed by Sarris *et al.* (2003), shows high-intensity circular lesions containing a miniscule hypointense focus (Sarris *et al.*, 2003). The circular lesions were determined to be the area of a granulomatous inflammatory response, whereas the small hypointense focus was representative of the mycetomal grains (Sarris *et al.*, 2003).

Abnormalities of radiograph images are used to diagnose histoplasmosis in cats and dogs (Bromel and Sykes, 2005). Davies and Troy (1996) found that out of 31 cats with histoplasmosis, 87% had abnormalities in the radiograph images taken of their thorax.

CT and MRI are also used to detect lesions within cryptococcosis-associated diseases. CT can detect cryptococcomas, but MRI is a more sensitive method for doing this (Kwon-Chung and Bennett, 1992).

However, radiograph imaging has been proven to misdiagnose. Despite pathogenic organisms being present in the lungs of one cat with histoplasmosis, no radiographic abnormalities of the lungs were found (Gabbert *et al.*, 1984).

E. Spectroscopy

Fourier-transform infrared (FTIR) microscopy measures the infrared light that is absorbed and transmitted by the pathogen, which is subsequently shown in graph form. The spectral readout is representative of the unique molecular fingerprint which the pathogen has (Spires, 2001).

Erukhimovitch *et al.* (2005) discussed the possibilities of using FTIR as a potential diagnostic tool for the detection of bacteria and fungi, due to it being a sensitive and inexpensive method with rapid turnover of results (about 1 h to obtain results; Erukhimovitch *et al.*, 2005). It is also a straightforward technique and only requires small amounts of sample material. They found that despite the spectral readouts looking similar for the three species of fungi used (*Penicillium* spp., *Memnoniella* spp., and *Fusarium* spp.); there were marked differences at certain parts on close inspection. At a wavenumber (the inverse of a wavelength) of 1377 cm^{-1}, the fungal species each had unique peaks, allowing specific discrimination between the three. Erukhimovitch *et al.*, however, do admit that in order for FTIR to be considered as a standard diagnostic tool, more studies and results need to be obtained.

A study by Toubas *et al.* (2007) was conducted concerning the use of FTIR for differentiation between three *Candida* spp. (*C. albicans*, *C. krusei*, and *C. glabrata*). Their results were similar to the findings of Erukhimovitch *et al.* (2005) whereby each species seemed to produce similar spectra, but with distinct discriminations at certain peaks. The area of discrimination in this study was at a wavenumber region between 900 and 1200 cm^{-1}, the absorption area of polysaccharides in the samples (Toubas *et al.*, 2007). The spectral peak of discrimination at 1377 cm^{-1} in the study by Erukhimovitch *et al.* (2005) was found to be indicative of the "C–H bending mode of CH_2" (Brandenburg and Seydel, 2001; Erukhimovitch *et al.*, 2005). They concluded that the absorption region of 900–1200 cm^{-1} was an appropriate marker region because of the variability it provided between the three *Candida* spp., and that highly conserved polysaccharides in each species was the reason for this spectral peak (Toubas *et al.*, 2007). A wavenumber of 400–4000 cm^{-1} was used to create the spectra, and at a spectral resolution of 6 cm^{-1}. Erukhimovitch *et al.* produced their spectra with a wavenumber range of 600–4000 cm^{-1} at a spectral resolution of 4 cm^{-1} (Erukhimovitch *et al.*, 2005). These parameters need to be standardized if FTIR is to become an established diagnostic procedure, otherwise results would be deemed invalid due to marked differences in procedure.

Toubas *et al.* propose that not only can FTIR be used for diagnosis, but also for species typing to determine the causative species or strain in an epidemic. They state that there are also possibilities for the technique to become a tool for studying virulence factors.

III. MOLECULAR TECHNIQUES FOR FUNGAL DIAGNOSIS

Several fungus species have now been completely sequenced. Readers can refer to a summary in Table 2.2.

A. Polymerase chain reaction (PCR)

PCR is a clinical diagnostic method which is used to identify the presence of a particular DNA sequence found within the genomes of pathogens. Because species-specific primers exist, detection of a variety of fungal species via PCR is achievable. The amplification of these specific DNA sequences is the basis of PCR pathogen detection. Primers, small segments of DNA complementary to the target DNA in the genome, hybridize to the denatured sample of DNA once it is cooled. The sample is then heated to allow *Taq* polymerase (a heat stable enzyme) to begin copying

TABLE 2.2 Completed fungal genome sequencing projects

Species	Genome size (Mb)	Chromosomes (*n*=)
Ascomycota		
Pezizomycotina		
Aspergillus clavatus	27.9	8[a]
	35.0	8[b]
Aspergillus flavus Af293	36.8	8[a]
Aspergillus fumigatus	30.0	8[b,c]
	29.4	8[a]
Aspergillus nidulans	30.1	8[a]
Aspergillus niger	37.2	8[a]
	37.0	8[b]
Aspergillus oryzae	37.1	8[a]
Aspergillus terreus	35.0/29.3	8[b]/8[a]
Gibberella zeae PH-1	40.0	4[b]
Magnaporthe grisea 70–15	40.0	7[b]
Neurospora crassa OR74A	43.0	7[b]
Saccharomycotina		
Ashbya gossypii ATCC 10895	9.2	7[b]
Candida albicans SC5314	16.0	8[b,d]
Candida albicans WO1	14.0	12[a]
Candida glabrata CBS 138	12.3	13[b]
Candida guilliermondii	10.6	12[a]
Candida lusitaniae	12.1	15[a]
Candida parapsilosis	13.0	7[c]
Candida tropicalis	14.6	17[a]
Debaryomyces hansenii CBS767	12.2.	7[b]
Kluyveromyces lactis NRRL Y-1140	10.6	6[b]
Pichia stipitis CBS 6054	15.4	8[b]
Saccharomyces cerevisiae S288c	12.0	16[b]
YJM789	12.0	16[d]
Yarrowia lipolytica CLIB122	20.5	6[b]
Eremothecium gossypii ATCC10895	9.2	7[b]
Schizosaccharomycetes		
Schizosaccharomyces pombe 972 h	14.0	3[b]
	14.1	3[c]
Basidiomycota		
Cryptococcus neoformans var. neoformans B-3501A	20.0	14[b]

(*continued*)

TABLE 2.2 (*continued*)

Species	Genome size (Mb)	Chromosomes (n=)
Cryptococcus neoformans var. neoformans JEC21	20.0	14[b]
Cryptococcus neoformans B-3501A	18.5	14[d]
Ustilago maydis 521	20.5	23[b]
Microporidia		
Encephalitozoon cuniculi GB-M1	2.9	11[b]

[a] http://www.broad.mit.edu/annotation/fungi
[b] http://www.ncbi.nlm.nih.gov
[c] http://www.sanger.ac.uk/Projects/
[d] http://www-sequence.stanford.edu/group/C.neoformans/overview.html

the required sequence of DNA. The sample is then heated again to separate the newly synthesized strands of DNA, and the process is repeated again with copies of the required sequences increasing exponentially. The method requires the use of a thermal cycler which at given times, increases then decreases the temperature of the sample.

DNA used in the PCR process is extracted from a colony of fungal culture. The colony is homogenized in lyticase lysis buffer, incubated for an hour at 37 °C, and then sample is vortexed for 2 min with the inclusion of acid washed glass beads, and finally, the supernatant is processed by a MagNa Pure LC machine (from Roche Diagnostics) with the function of automated extraction of fungal DNA (Baskova *et al.*, 2007). A much more efficient simple method of DNA extraction is by using Whatman FTA filter paper. It contains agents which have lytic effects on microbes that come into contact with it (Borman *et al.*, 2007). Large fragments of DNA, released from lysed cells, are trapped within the fibrous matrices of the FTA paper (Borman *et al.*, 2007; Bowman *et al.*, 2001). About 96.5% of the fungi tested with this method of DNA extraction released DNA which was suitable for PCR amplification, and the procedure only takes around 15 min (Borman *et al.*, 2007), therefore increasing the rapidity of the already fast diagnosis of PCR.

Real-time PCR involves a dual fluorochrome oligonucleotide probe which is labeled with a reporter dye at the 5′ end and a quencher dye at the 3′ end (Baskova *et al.*, 2007). Frequently used reporters and quenchers are FAM and TAMRA, respectively (Atkins *et al.*, 2005). Due to the attachment of FAM and TAMRA, the probe remains nonfluorescent. The probe, used in addition to the regular primers specific to the target DNA sequence, hybridizes to the accumulating PCR product, and only becomes fluorescent when *Taq* polymerase activity releases the reporter-containing nucleotide from the 5′ end of the probe during amplification (Bowman *et al.*, 2001). When baseline fluorescence (C_T: cycle threshold) is exceeded at an early stage of cycling, this is indicative of large increasing amounts of target DNA in the sample. No fluorescence will be visualized when the target DNA sequence required is not present in the sample. Bowman *et al.* (2001) developed a real-time PCR to detect *A. fumigatus* in a murine model with disseminated aspergillosis (Bowman *et al.*, 2001). The 18S rRNA gene of *A. fumigatus* was used as their target sequence, and test controls found that no fluorescence was observed when real-time PCR was carried out on genomic DNA extracted from *C. albicans*, genomic DNA from the kidneys of uninfected mice, and "a commercial source of mouse genomic DNA."

The prevalence of aspergillosis and candidiasis caused by species other than *A. fumigatus* and *C. albicans* respectively, is increasing (Bille *et al.*, 2005; Coleman *et al.*, 1998; Torres *et al.*, 2003), prompting a need for the rapid diagnosis of an array of pathogenic fungal agents. Baskova *et al.* (2007) suggest that there are no current techniques which are sufficiently sensitive and economically viable for the detection of clinically relevant fungi, therefore they propose a cost effective and standardized method of diagnosing invasive fungal infections caused by *Aspergillus* spp. and *Candida* spp. The Pan-AC assay (pan-*Aspergillus* and pan-*Candida*) enables the quantitative detection of a variety *of Aspergillus* spp. (*A. niger*, *A. flavus*, *A. fumigatus*, *A. versicolor*, *A. nidulans*, etc.) and *Candida* spp. (*C. glabrata*, *C. tropicalis*, *C. albicans*, *C. krusei*, etc.) by real-time PCR (Baskova *et al.*, 2007). The target sequence used for the Pan-AC assay was a highly conserved region of the 28S rDNA, existing in ten *Candida* spp. and six *Aspergillus* spp. This target sequence was successfully amplified in all of the subject species. The Pan-AC assay encountered no cross-reactivity with nonfungal pathogens, but there were complications regarding slight cross-reactivity when test controls were carried out on DNA from healthy human blood. This was due to the homology of certain human genes with the fungal target sequence. Using a 1% concentration of formamide in the assay proved to successfully eradicate these problems. The Pan-AC assay is not species-specific in its diagnosis, as it tests for a broad range of *Aspergillus* spp. and *Candida* spp. However, it does enable rapid detection and quantification of a variety of fungal species from just one sample, and can allow patients being tested positive to be given broad spectrum antifungal dugs as an initial treatment. It could be suggested that supplementary diagnosis is carried out via further PCR with the target DNA sequence being the internal transcribed spacer (ITS) region.

The ITS region is a frequently used target DNA sequence due to its variability, allowing for species-specific detection of fungi via PCR. A nested PCR was used to identify histoplasmosis in dogs by detecting the

ITS region of the ribosomal RNA gene in *Histoplasma capsulatum var. farciminosum* (Ueda *et al.*, 2003). It was also used in the real-time PCR detection of *Paecilomyces lilacinus*, which can be an opportunistic fungus in man (Groll and Walsh, 2001), and also cause disease in cats (Elliott *et al.*, 1984) and turtles (Posthaus *et al.*, 1997).

PCR is a highly sensitive technique and has been known to detect genomic material from as few 10 *Candida* cells per milliliter of blood (Walker and White, 2006), less than 10 *H. capsulatum* cells (Bracca *et al.*, 2003), and less than 10 cells per PCR run in the Pan-AC assay (Baskova *et al.*, 2007). Atkins *et al.* (2005) stated that the quantitative data which PCR provides, allows the quick and efficient identification of pathogens in a sample, which can lead to accurate and rapid treatment. Although highly sensitive and rapid, PCR techniques lack standardization (Barnes, 2007) due to the personal specifications included by the user. However, a UK PCR consensus group has correlated the molecular features of PCR and developed a real-time PCR which is standardized and fit for use in diagnostic laboratories (White *et al.*, 2006).

B. Serological methods

The presence and detection of antigens, and antibodies against fungal pathogens via assays, forms the basis for the serological diagnosis of fungi. Several immunologic techniques are already commonplace within fungal diagnosis. Tests include enzyme-linked immunosorbant assays (ELISAs), complement fixation test, latex agglutination, immunodiffusion, and immunofluorescence. In regards to serological diagnosis, sensitivity is the ability of a test to detect the lowest possible amount of antigen. High sensitivity denotes the positive detection of a very small concentration of antigen in a sample.

Immunodiffusion is the term applied to precipitation reactions carried out on agar gels (Madigan and Martinko, 2006). It works on the principle that when antigen–antibody structures are created from cross-linkage, they precipitate as a consequence of becoming too large, at a mark called their equivalence point (Murray *et al.*, 2005). The amount of precipitation is dependant on the ratio between antibody and antigen, whereby if their equivalence point is reached, then complexes will form (Janeway *et al.*, 2005). Ouchterlony double immunodiffusion is used to test for fungal pathogens in coccidioidomycosis (Kwon-Chung and Bennett, 1992), histoplasmosis (Bromel and Sykes, 2005), and blastomycosis (Murray *et al.*, 2005). Agar gel immunodiffusion (AGID) is the common mode of diagnosis of coccidioidomycosis, but is also the preferred method of serological testing in animals and man (Shubitz, 2007). Wells are cut into agar gel with antigen and antibody placed into them. Where the antigen and antibodies have diffused towards each other, precipitin lines appear, indicative that they have reached their equivalence point (Murray *et al.*, 2005). M and H bands of precipitin to *H. capsulatum* are detected when AGID is carried out on patients with histoplasmosis (Bromel and Sykes, 2005; Kwon-Chung and Bennett, 1992). Immunodiffusion has proven to be less sensitive than complement fixation tests in relation to the diagnosis of histoplasmosis (Wheat, 2003).

Two types of immunofluorescence exist, direct and indirect (Murray *et al.*, 2005). In the former, a fluorochrome is covalently attached to an antibody which binds to fungal cell antigens and the latter uses antibodies to bind to fungal cell antigens, which are subsequently bound by anti-immunoglobulins linked to fluorochromes (Janeway *et al.*, 2005). Direct immunofluorescence is used to detect *Pneumocystis jirovecii*, the etiological agent of *Pneumocystis pneumonia* in humans (Redhead *et al.*, 2006), within NHS microbiology laboratories (personal communication via email with Dr. Richard Ellis, South Tyneside District General Hospital, United Kingdom) and diagnostic laboratories (Huggett *et al.*, 2008). This technique requires the use of a fluorescence microscope which includes filters to ensure that light from only the fluorochrome is visualized.

In latex agglutination, small latex beads are coated with specific antifungal antibody which causes the particles to coagulate. If the detection of antifungal antibodies is required, then the opposite of the above is carried out, where fungal antigen is coated onto latex beads. *Cryptococcus neoformans* can be diagnosed this way (Madigan and Martinko, 2006). The antigen being detected from *C. neoformans* is derived from polysaccharide capsule of the pathogen (Kwon-Chung and Bennett, 1992).

The detection of antibodies or antigens in serum can be performed with ELISA. Antigen from the required pathogen is coated onto a surface and exposed to its complimentary antibody which binds to it. The preparation is then washed to remove any unbound antibodies. A second antibody, attached to an enzyme, is then added, which is specific to the antifungal antibody. When a specific reagent is added to the preparation, the enzyme will react with it produce a color change. This color change can be quantified by using a spectrophotometer (Murray *et al.*, 2005). A variation of the ELISA is the sandwich ELISA. Antibodies are coated onto a surface and exposed to the patient's serum. Any antigen complementary to the affixed antibody will bind. Excess unbound antigen is washed away. Antibody specific to the antigen, and attached to an enzyme, is then added. The enzyme will produce a color change once a reagent is added. Sandwich ELISAs for the detection of *Aspergillus* spp. are achieved by using galactomannan antigen as the diagnostic marker (Aydil *et al.*, 2007; Marr *et al.*, 2004). Galactomannan is a component of the cell walls of *Aspergillus* spp. which is released during hyphal growth (Marr *et al.*, 2004). The monoclonal antibody for galactomannan, β 1–5 galactofuranose-specific EBA2 monoclonal antibody, acts as the "sandwich" elements in the Sandwich ELISA (Stynen *et al.*, 1992). For the detection of *Candida* spp. via sandwich ELISA, mannan is the diagnostic marker (Barnes, 2007). It should be noted however, that galactomannan ELISAs have variable sensitivity, with some cases reporting sensitivity from 17% to 100% (Buchheidt *et al.*, 2004).

Complement fixation occurs when a sample of the patient's serum reacts with provided antigen from the laboratory. Complement is activated and used up (fixed) by antigen–antibody complexes which bind to it (Murray *et al.*, 2005). It is and established serological technique for the detection of antibodies to *H. capsulatum* in human histoplasmosis (Wheat, 2003).

The radioimmunoassay is similar to ELISA, but isotopes as opposed to enzymes are used as the conjugative probe, and radioactivity is measured instead of color change (Madigan and Martinko, 2006). Due to antigens and antibodies having the ability to be iodinated, and not have their binding specificity affected as a result, ^{125}I (a radioactive iodine isotope) is used (Janeway *et al.*, 2005). Firstly, known concentrations of antigen are bound to wells of a microtiter plate, then radio labeled antibodies are added to the wells (Madigan and Martinko, 2006). Washing removes unbound antibodies, and the radioactivity of each well is measured to set up a standard curve of antigen concentration against radioactivity (Madigan and Martinko, 2006). The patient's serum is then added to a separate well, and the same process is carried out. Radioactivity is then measured, and the standard curve is used to determine antigen concentration in the serum sample. Wheat *et al.* (1989) used a radioimmunoassay to detect *Histoplasma* polysaccharide antigen in AIDS patients, finding that 37 of 47 serum samples and 59 of 61 urine samples returned positive results (Wheat *et al.*, 1989).

IV. CONCLUSION

Microscopy and culture are the most heavily relied upon diagnostic techniques of today (Aydil *et al.*, 2007). This is because they are simple to perform, and are standardized so that each diagnostic laboratory will adhere to the same procedural steps. Although culturing of fungi can be a time-consuming process, taking weeks to produce morphologically distinct features at times, it is necessary so that samples can be provided for other techniques such as PCR and FTIR. However, these traditional diagnostic tests based on morphological features (i.e., culture and microscopy) have numerous disadvantages. They lack accuracy, take too much time, and require costly laboratory reagents. They also necessitate the use of experienced staff (Toubas *et al.*, 2007).

Molecular methods based on DNA such as PCR, although rapid, are reported to have false positive or false negative results, and are fairly expensive (Vaneechoutte and VanEldere, 1997). The main limitations of PCR include its uneconomical costs and that it requires the creation of specific primers to detect the target DNA sequence of the fungi in question (Toubas *et al.*, 2007). The living state of the fungi in their host is also not determined in PCR techniques (Atkins *et al.*, 2005).

In their comparison of PCR and the galactomannan ELISA of invasive pulmonary aspergillosis, Becker *et al.* (2000) concluded that in their rat model, PCR was a less sensitive diagnostic technique than ELISA (Becker *et al.*, 2000). There are counter-suggestions to this however, as recent statements have said that microscopy and PCR are the most sensitive techniques for fungi diagnosis (Aydil *et al.*, 2007).

FTIR is a very rapid process (Erukhimovitch *et al.* (2005): 1 h, Toubas *et al.* (2007): 30 minutes) and very precise, allowing for diagnosis at the level of species and even subspecies (Toubas *et al.*, 2007). It is also able to distinguish between different types of microbes (fungi and bacteria) (Erukhimovitch *et al.*, 2005). Due to this level of accuracy, it may be possible that FTIR may become a standardized diagnostic tool in the future, with hospital diagnostic laboratories incorporating its uses.

This paper was focusing on pathogenic fungi in medicine and veterinary medicine only. However, readers should also consider that some fungi are used as pest control to reduce other pathogen impacts.

REFERENCES

Abdelrahman, T., Bru, V. L., Waller, J., Noacco, G., and Candolfi, C. (2006). Dermatomycosis: Comparison of the performance of calcofluor and potassium hydroxide 30% for the direct examination of skin scrapings and nails. *J. Mycol. Med.* **16,** 87–91.

Arceneaux, K. A., Taboada, J., and Hosgood, G. (1998). Blastomycosis in dogs: 115 cases (1980–1995). *J. Am. Vet. Med. Assoc.* **213,** 658–664.

Atkins, S. D., Clark, I. M., Pande, S., Hirsch, P. R., and Kerry, B. R. (2005). The use of real-time PCR and species-specific primers for the identification and monitoring of *Paecilomyces lilacinus*. *FEMS Microbiol. Ecol.* **51,** 257–264.

Aydil, U., Kalkancl, A. E., Ceylan, A., Berk, E., Kustimur, S., and Uslu, S. (2007). Investigation of fungi in massive nasal polyps: Microscopy, culture, polymerase-chain reaction, and serology. *Am. J. Rhinol.* **21,** 417–422.

Barnes, R. A. (2007). Early diagnosis of fungal infection in immunocompromised patients. *J. Antimicrob. Chemother.* **61**(Suppl. 1), I3–I6.

Barros, M. B. D., Schubach, A. D., do Valle, A. C. F., Galhardo, M. C. G., Conceicao-Silva, F., Schubach, T. M. P., Reis, R. S., Wanke, B., Marzochi, K. B. F., and Conceicao, M. J. (2004). Cat-transmitted sporotrichosis epidemic in Rio de Janeiro, Brazil: Description of a series of cases. *Clin. Infect. Dis.* **38,** 529–535.

Baskova, L., Landlinger, C., Preuner, S., and Lion, T. (2007). The Pan-AC assay: A single-reaction real-time PCR test for quantitative detection of a broad range of *Aspergillus* and *Candida* species. *J. Med. Microbiol.* **56,** 1167–1173.

Becker, M. J., de Marie, S., Willemse, D., Verbrugh, H. A., and Bakker-Woudenberg, I. A. J. M. (2000). Quantitative galactomannan detection is superior to PCR in diagnosing and monitoring invasive pulmonary aspergillosis in an experimental rat model. *J. Clin. Microbiol.* **38,** 1434–1438.

Beytut, E. (2007). Immunohistochemical diagnosis of aspergillosis in adult turkeys. *Turk. J. Vet. Anim. Sci.* **31,** 99–104.

Bille, J., Marchetti, O., and Calandra, T. (2005). Changing face of health-care associated fungal infections. *Curr. Opin. Infect. Dis.* **18,** 314–319.

Borman, A. M., Linton, C. J., Miles, S. J., Campbell, C. K., and Johnson, E. M. (2006). Ultra-rapid preparation of total genomic DNA from isolates of yeast and mould using Whatman FTA filter paper technology—a reusable DNA archiving system. *Med. Mycol.* **44,** 389–398.

Borman, A. M., Linton, C. J., Miles, C. K., and Johnson, E. M. (2007). Molecular identification of pathogenic fungi. *J. Antimicrob. Chemother.* **61**(Suppl. 1), I7–I12.

Bowman, J. C., Abruzzo, G. K., Anderson, J. W., Flattery, A. M., Gill, C. J., Pikounis, V. B., Schmatz, D. M., Liberator, P. A., and Douglas, C. M. (2001). Quantitative PCR assay to measure *Aspergillus fumigatus* burden in a murine model of disseminated aspergillosis: Demonstration of efficacy of caspofung in acetate. *Antimicrob. Agents Chemother.* **45,** 3474–3481.

Bracca, A., Tosello, M. E., Girardini, J. E., Amigot, S. L., Gomez, L., and Serra, E. (2003). Molecular detection of *Histoplasma capsulatum var. capsulatum* in human clinical samples. *J. Clin. Microbiol.* **41,** 1753–1755.

Brandenburg, K., and Seydel, U. (2001). *In* "Handbook of Vibrational Spectroscopy." pp. 3481–3507. Wiley, Chichester.

Bromel, C., and Sykes, J. E. (2005). Histoplasmosis in dogs and cats. *Clin. Tech. Small Anim. Pract.* **20,** 227–232.

Buchheidt, D., Hummel, M., Schleiermacher, D., Spiess, B., Schwerdtfeger, R., Cornley, O. A., Wilhelm, S., Reuter, S., Kern, W., Sudhoff, T., Morz, H., and Hehlmann, R. (2004). Prospective clinical evaluation of a LightCycler (TM)-mediated polymerase chain reaction assay, a nested-PCR assay and a galactomannan enzyme-linked immunosorbent assay for detection of invasive aspergillosis in neutropenic cancer patients and haematological stem cell transplant recipients. *Br. J. Haematol.* **125,** 196–202.

Carrasco, L., Bautista, M. J., Delasmulas, J. M., and Jensen, H. E. (1993). Application of enzyme-immunohistochemistry for the diagnosis of Aspergillosis, Candidiasis, and Zygomycosis in 3 Lovebirds. *Avian Dis.* **37,** 923–927.

Chan, C. M., Woo, P. C. Y., Leung, A. S. P., Lau, S. K. P., Che, X. Y., Cao, L., and Yuen, K. Y. (2002). Detection of antibodies specific to an antigenic cell wall galactomannoprotein for serodiagnosis of *Aspergillus fumigatus* aspergillosis. *J. Clin. Microbiol.* **40,** 2041–2045.

Clinkenbeard, K. D. (1991). Diagnostic cytology—sporotrichosis. *Compend. Contin. Educ. Pract. Vet.* **13,** 207–211.

Coleman, D. C., Rinald, M. G., Haynes, K. A., Rex, J. H., Summerbell, R. C., Anaissie, E. J., Li, A., and Sullivan, D. J. (1998). Importance of Candida species other than *Candida albicans* as opportunistic pathogens. *Med. Mycol.* **36,** 156–165.

Computed Tomography (CT)—Chest (2007). Available from http://www.radiologyinfo.org/en/info.cfm?pg=chestct#part_five.

Corte, A. C., Svoboda, W. K., Navaroo, I. T., Freire, R. L., Malanski, L. S., Shiozawa, M. M., Ludwig, G., Aguiar, L. M., Passos, F. C., Maron, A., Camargo, Z. P., Itano, E. M., *et al.* (2007). Paracoccidioidomycosis in wild monkeys from Parana State, Brazil. *Mycopathologia* **164,** 225–228.

Davies, C., and Troy, G. C. (1996). Deep mycotic infections in cats. *J. Am. Anim. Hosp. Assoc.* **32,** 380–391.

Denning, D. W. (1998). Invasive aspergillosis. *Clin. Infect. Dis.* **26,** 781–803.

Dods, J., Losifidis, E., and Roilides, E. (2007). Central nervous system aspergillosis in children: A systematic review of reported cases. *Int. J. Infect. Dis.* **11,** 381–393.

Elliott, G. S., Whitney, M. S., Reed, W. M., and Tuite, J. F. (1984). Antemortem diagnosis of Paecilomycosis in a cat. *J. Am. Vet. Med. Assoc.* **184,** 93–94.

Erukhimovitch, V., Pavlov, V., Talyshinsky, M., Souprun, Y., and Huleihel, M. (2005). FTIR microscopy as a method for identification of bacterial and fungal infections. *J. Pharm. Biomed. Anal.* **37,** 1105–1108.

Gabbert, N. H., Campbell, T. W., and Beiermann, R. L. (1984). Pancytopenia associated with disseminated Histoplasmosis in a cat. *J. Am. Anim. Hosp. Assoc.* **20,** 119–122.

Geyer, A. S., Fox, L. P., Hussain, S., Della-Latta, P., and Grossman, M. E. (2006). *Acremonium mycetoma* in a heart transplant recipient. *J. Am. Acad. Dermatol.* **55,** 1095–1100.

Greene, R. E., Schlamm, H. T., Oestmann, J. W., Stark, P., Durand, C., Lortholary, O., Wingard, J. R., Herbrecht, R., Ribaud, P., Patterson, T. F., Troke, P. F., Denning, D. W., *et al.* (2007). Imaging findings in acute invasive pulmonary aspergillosis: Clinical significance of the halo sign. *Clin. Infect. Dis.* **44,** 373–379.

Greydanusvanderputten, S. W. M., Klein, W. R., Blankenstein, B., Dehoog, G. S., and Koeman, J. P. (1994). Sporotrichosis in a horse—a case-report. *Tijdschrift Voor Diergeneesk.* **119,** 500–502.

Groll, A. H., and Walsh, T. J. (2001). Uncommon opportunistic fungi: New nosocomial threats. *Clin. Microbiol. Infect.* **7,** 8–24.

Grossgebauer, K. (1980). Detection of *Cryptococcus neoformans* capsule using fluorescent, mucopolysaccharide-binding dyes. *Klinisch. Wochensch.* **58,** 943–945.

Hageage, G., and Harrington, B. (1984). Use of calcofluor white in clinical mycology. *Lab. Med.* **15,** 109–112.

Hoffman, K., Videan, E. N., Fritz, J., and Murphy, J. (2007). Diagnosis and treatment of ocular coccidioidomycosis in a female captive chimpanzee (*Pan troglodytes*)—a case study. *In* "Coccidioidomycosis: Sixth International Symposium," 23–26/08/2006, pp. 404–410. Stanford University, California.

Huggett, J. F., Taylor, M. S., Kocjan, G., Evans, H. E., Morris-Jones, S., Gant, V., Novak, T., Costello, A. M., Zumla, A., and Miller, R. F. (2008). Development and evaluation of a real-time PCR assay for detection of *Pneumocystis jirovecii* DNA in bronchoalveolar lavage fluid of HIV-infected patients. *Thorax* **63,** 154–159.

Janeway, C., Travers, P., Walport, M., and Shlomick, M. J. (2005). Appendix I: Immunologists' toolbox. *In* "Immunobiology: The Immune System in Health and Disease" (E. Lawrence, ed.), pp. 823. Garland Science, New York.

Kumar, J., Kumar, A., Sethy, P., and Gupta, S. (2007). The dot-in-circle sign of mycetoma on MRI. *Diagn. Interv. Radiol.* **13,** 193–195.

Kwon-Chung, K. J., and Bennett, J. E. (1992). The Fungi. *In* "Medical Mycology" (S. Hunsberger, ed.), pp. 866. Lea & Febiger, Pennsylvania.

Lewall, D. B., Ofole, S., and Bendl, B. (1985). Mycetoma. *Skeletal Radiol.* **14,** 257–262.

Lyon, G. M., Bravo, A. V., Espino, A., Lindsley, M. D., Gutierrez, R. E., Rodriguez, I., Corella, A., Carrillo, F., McNeil, M. M., Warnock, D. W., and Hajjeh, R. A. (2004). Histoplasmosis associated with exploring a bat-inhabited cave in Costa Rica, 1998–1999. *Am. J. Trop. Med. Hyg.* **70,** 438–442.

Madigan, M. T., and Martinko, J. M. (2006). Diagnostic microbiology and immunology. *In* "Brock Biology of Microorganisms" (J. Challice, ed.), pp. 1059. Pearson Prentice Hall, Upper Saddle River.

Marr, K. A., Balajee, S. A., McLaughlin, L., Tabouret, M., Bentsen, C., and Walsh, T. J. (2004). Detection of galactomannan antigenemia by enzyme immunoassay for the diagnosis of invasive aspergillosis: Variables that affect performance. *J. Infect. Dis.* **190,** 641–649.

Midgley, G., Clayton, Y. M., and Hay, R. J. (1997). Superficial candidosis. *In* "Diagnosis in Colour: Medical Mycology" (G. Almond, ed.), pp. 155. Mosby-Wolfe, London.

MRI. (2007). Available from http://www.radiologyinfo.org/en/info.cfm?pg=bodymr#part_five.

Murray, P. R., Rosenthal, K. S., and Pfaller, M. A. (2005). Subcutaneous mycoses. *In* "Medical Microbiology" (W. Schmitt and K. Miller, eds.), pp. 963. Elsevier Mosby, Philadelphia.

NHS. CT scan. (2007a). Available from http://www.nhsdirect.nhs.uk/articles/article.aspx?articleId=554.

NHS. MRI scan. (2007b). Available from http://www.nhsdirect.nhs.uk/articles/article.aspx?articleId=556.

Pankajalakshmi, V. V., and Taralakshmi, V. V. (1984). Mycetomas in the tropics. *Indian J. Pathol. Microbiol.* **27,** 223–228.

Perez-Blanco, M., Valles, R. H., Garcia-Humbria, L., and Yegres, F. (2006). Chromoblastomycosis in children and adolescents in the endemic area of the Falcon State, Venezuela. *Med. Mycol.* **44,** 467–471.

Posthaus, H., Krampe, M., Pagan, O., Gueho, E., Suter, C., and Bacciarini, L. (1997). Systemic paecilomycosis in a hawksbill turtle *(Eretmochelys imbricata)*. *J. Mycol. Med.* **7,** 223–226.

Redhead, S., Cushion, M. T., Frenkel, J. K., and Stringer, J. R. (2006). Pneumocystis and *Trypanosoma cruzi*: Nomenclature and typifications. *J. Eukaryot. Microbiol.* **53,** 2–11.

Richardson, M., and Elewski, B. (2000). *In* "Superficial Fungal Infections." pp. 155. Health Press, Oxford.

Santos, A. L. S., Palmeira, V. F., Rozental, S., Kneipp, L. F., Nimrichter, L., Alviano, D. S., Rodrigues, M. L., and Alviano, C. S. (2007). Biology and pathogenesis of *Fonsecaea pedrosoi*, the major etiologic agent of chromoblastomycosis. *FEMS Microbiol. Rev.* **31,** 570–591.

Sarris, I., Berendt, A. R., Athanasous, N., and Ostlere, S. J. (2003). MRI of mycetoma of the foot: Two cases demonstrating the dot-in-circle sign. *Skeletal Radiol.* **32,** 179–183.

Shubitz, L. E. (2007). Comparative aspects of coccidioidomycosis in animals and humans. *In* "Coccidioidomycosis: Sixth International Symposium," 23–26/08/2006, pp. 395–403. Stanford University, California.

Snider, T. A., Joyner, P. H., and Clinkenbeard, K. D. (2008). Disseminated histoplasmosis in an African pygmy hedgehog. *J. Am. Vet. Med. Assoc.* **232,** 74–76.

Sparkes, A. H., Gruffyddjones, T. J., Shaw, S. E., Wright, A. I., and Stokes, C. R. (1993). Epidemiological and diagnostic features of canine and feline dermatophytosis in the United Kingdom from 1956 to 1991. *Vet. Rec.* **133,** 57–61.

Spires, T. (2001). Introduction to fourier transform infrared microscopy; Available fromhttp://mmrc.caltech.edu/FTIR/FTIRintro.pdf.

Stynen, D., Sarfati, J., Goris, A., Prevost, M. C., Lesourd, M., Kamphuis, H., Darras, V., and Latge, J. P. (1992). Rat monoclonal-antibodies against *Aspergillus* galactomannan. *Infect. Immun.* **60,** 2237–2245.

Torres, H. A., Rivero, G. A., Lewis, R. E., Hachem, R., Raad, I. I., and Kontoyiannis, D. P. (2003). Aspergillosis caused by non-fumigatus *Aspergillus* species: Risk factors and *in vitro* susceptibility compared with *Aspergillus fumigatus*. *Diagn. Microbiol. Infect. Dis.* **46,** 25–28.

Toubas, D., Essendoubi, M., Adt, I., Pinon, J. M., Manfait, M., and Sockalingum, G. D. (2007). FTIR spectroscopy in medical mycology: Applications to the differentiation and typing of Candida. *Anal. Bioanal. Chem.* **387,** 1729–1737.

Ueda, Y., Sano, A., Tamura, M., Inomata, T., Kamei, K., Yokoyama, K., Kishi, F., Ito, J., Mikami, Y., Miyajo, M., and Nishimura, K. (2003). Diagnosis of histoplasmosis by detection of the internal transcribed spacer region of fungal rRNA gene from a paraffin-embedded skin sample from a dog in Japan. *Vet. Microbiol.* **94,** 219–224.

Upton, A., Kirby, K. A., Carpenter, P., Boeckh, M., and Marr, K. A. (2007). Invasive aspergillosis following hematopoietic cell transplantation: Outcomes and prognostic factors associated with mortality. *Clin. Infect. Dis.* **44,** 531–540.

Vaneechoutte, M., and VanEldere, J. (1997). The possibilities and limitations of nucleic acid amplification technology in diagnostic microbiology. *J. Med. Microbiol.* **46,** 188–194.

Vansteenhouse, J. L., and Denovo, R. C. (1986). Atypical *Histoplasma capsulatum* infection in a dog. *J. Am. Vet. Med. Assoc.* **188,** 527–528.

Walker, G. M., and White, N. A. (2006). Image section. *In* ''Fungi: Biology and Applications'' (K. Kavanagh, ed.), pp. 1–34. Wiley, Chichester.

Weitzman, I., and Summerbell, R. C. (1995). The dermatophytes. *Clin. Microbiol. Rev.* **8,** 240–259.

Welsh, R. D. (2003). Sporotrichosis. *J. Am. Vet. Med. Assoc.* **223,** 1123–1126.

Wheat, L. J. (2003). Current diagnosis of histoplasmosis. *Trends Microbiol.* **11,** 488–494.

Wheat, L. J., Connolly-Stringfield, P., Kohler, R. B., Frame, P. T., and Gupta, M. R. (1989). *Histoplasma-capsulatum* polysaccharide antigen-detection in diagnosis and management of disseminated histoplasmosis in patients with acquired immunodeficiency syndrome. *Am. J. Med.* **87,** 396–400.

Wheat, L. J., and Kauffman, C. A. (2003). Histoplasmosis. *Infect. Dis. Clin. North Am.* **17,** 1–19.

White, P. L., Barton, R., Guiver, M., Linton, C. J., Wilson, S., Smith, M., Gomez, B. L., Carr, M. J., Kimmitt, P. T., Seaton, S., Rajakumar, K., Holyoake, T., *et al.* (2006). A consensus on fungal polymerase chain reaction diagnosis? A United Kingdom-Ireland evaluation of polymerase chain reaction methods for detection of systemic fungal infections *J. Mol. Diagn.* **8,** 376–384.

CHAPTER 3

Diversity in Bacterial Chemotactic Responses and Niche Adaptation

Lance D. Miller,* Matthew H. Russell,*
and **Gladys Alexandre*,†**

Contents

* Department of Biochemistry, Cellular and Molecular Biology, The University of Tennessee, Knoxville, Tennessee 37996
† Department of Microbiology, The University of Tennessee, Knoxville, Tennessee 37996

Advances in Applied Microbiology, Volume 66
ISSN 0065-2164, DOI: 10.1016/S0065-2164(08)00803-4

Abstract

The ability of microbes to rapidly sense and adapt to environmental changes plays a major role in structuring microbial communities, in affecting microbial activities, as well as in influencing various microbial interactions with the surroundings. The bacterial chemotaxis signal transduction system is the sensory perception system that allows motile cells to respond optimally to changes in environmental conditions by allowing cells to navigate in gradients of diverse physicochemical parameters that can affect their metabolism. The analysis of complete genome sequences from microorganisms that occupy diverse ecological niches reveal the presence of multiple chemotaxis pathways and a great diversity of chemoreceptors with novel sensory specificities. Owing to its role in mediating rapid responses of bacteria to changes in the surroundings, bacterial chemotaxis is a behavior of interest in applied microbiology as it offers a unique opportunity for understanding the environmental cues that contribute to the survival of bacteria. This chapter explores the diversity of bacterial chemotaxis and suggests how gaining further insights into such diversity may potentially impact future drug and pesticides development and could inform bioremediation strategies.

I. INTRODUCTION

The ability of unicellular organisms to survive, to grow, and to compete with other microorganisms in changing environments is dependent on sensing and adapting to changes by modifying their cellular physiology. Bacteria are capable of detecting a wide range of environmental stimuli, including osmolarity, pH, temperature, and chemical ligands of diverse physicochemical properties (Wadhams and Armitage, 2004). Bacteria have evolved several ways of responding to their environment and thereby maintaining optimum physiological states for growth. Sensing and adapting to changes is thus anticipated to play a major role in structuring microbial communities, in affecting dynamic changes in microbial activities, as well as in influencing various microbial interactions with the surroundings. At the cellular level, two-component signal transduction systems mediate specific changes in cellular physiology in response to environmental stimuli. At the molecular level, two-component signal transduction systems typically consist of a sensor histidine kinase that detects specific cues and a cognate response regulator that mediates the response (Lukat and Stock, 1993). One major way bacteria respond to environmental stimuli is by regulating gene expression—either increasing or decreasing expression of a gene or set of genes. Chemotaxis is another way motile bacteria respond to their environment. In fact, most understanding on signal transduction in prokaryotes comes from studying the two-component regulatory system that controls chemotaxis in motile bacteria, such as *Escherichia coli* and *Salmonella typhymurium* (Hazelbauer *et al.*, 2008). Motile bacteria will first detect a cue and then navigate in its direction, if the cue perceived corresponds to an attractant signal or away from it if the cue corresponds to a repellent signal. The bacterial chemotaxis signal transduction system is the sensory perception system that allows motile cells to respond optimally to changes in environmental conditions by facilitating motion towards small-molecule chemoattractants and away from repellents (Wadhams and Armitage, 2004). Thus, chemotaxis provides motile bacteria with an effective means to navigate their environment and to localize in niches that support optimum growth. Such an advantage provided by chemotaxis was observed very early on by Engelmann in the late nineteenth century. Engelmann noticed that bacteria observed under the microscope were actively swimming toward and accumulating near sources of oxygen (algae or air bubbles), but when a harmful chemical was added, the cells slowed down and were no longer capable of actively locating in these niches (Engelmann, 1881). Owing to its role in mediating rapid responses of bacteria to changes in the surroundings, bacterial chemotaxis is a behavior of interest in applied microbiology and it has been previously used to isolate microorganisms with desirable metabolic abilities, especially in bioremediation. This chapter will explore the diversity in bacterial chemotaxis and how such diversity could possibly be used in bioremediation, agriculture, or biosensor applications.

II. MOLECULAR MECHANISMS OF BACTERIAL CHEMOTAXIS

A. Chemotaxis: Control of the motility pattern

1. Biasing the random walk

The ability to swim and navigate the surrounding environment by chemotaxis confers motile bacteria with a competitive advantage in that it allows the cells to maintain and occupy niches that are optimum for growth and survival. In chemotaxis, motile bacteria constantly sample their environment and will detect and process only certain signals as cues. If the cue perceived corresponds to an attractant signal, the cells will move in the direction of the cue, but if the cue provides a repellent signal they will move away from it. The bacterial chemotaxis signal transduction system is the sensory perception system that allows motile cells to respond optimally to changes in environmental conditions by facilitating motion towards small-molecule chemoattractants and away from repellents (Adler and Tso, 1974). Bacteria swim by rotating their flagella and navigate in the environment with a swimming pattern that resembles a three-dimensional walk (Berg, 2000). In a homogeneous chemical environment, cells move along their long axis in relatively straight paths (runs) that are punctuated by brief reorientation events (reversals) where forward motion stops. Reversals are followed by runs in a new and randomly determined direction. In an increasing concentration of a chemical attractant, the cells tend to suppress the reversals and swim longer in the direction of the attractant. Conversely, in an increasing gradient of a repellent, the cells tend to reverse more frequently when swimming towards the repellent and run longer when swimming away from it. Therefore, motile bacteria bias their swimming pattern by modulating the probability of reversals. Such behavior allows motile bacteria to respond efficiently to what are perceived as temporal changes in their environmental conditions. Whether or not a change in a concentration of a stimulus will result in change in swimming pattern is controlled at the level of sensory perception. The fundamental principles of sensory or signal perception and transduction have mostly been studied in model organisms such as *E. coli.* The general principles of signal perception and transduction are conserved in bacteria and archaea despite the diversity of flagellation patterns and mechanisms of flagellar rotation and directional changes (Wadhams and Armitage, 2004).

2. Spatial and temporal sensing

It is often said that bacteria are too small to sense chemical gradients spatially. This may not be true in either theory or experience. The theoretical size limit for spatial sensing has been proposed to be <1 μm (Dusenbery, 1998), and evidence has been found suggesting that a Gram-negative vibriod performs aerotaxis via spatial sensing of steep oxygen gradients (Thar and Kuhl, 2003). Moreover, some cyanobacteria exhibit phototaxis in response to spatial gradients along the length of the cell (Hader, 1987). However, temporal sensing may be more efficient for bacteria, despite the estimated 10% loss in velocity towards increasing attractant gradients due to the biased random walk. Bacteria tend to inhabit chemically heterogeneous environments. What is important for the bacterium may not be to always move up the gradient of the strongest chemoattractant. For instance, a gradient of attractant could coincide with a repellant gradient. In this situation, it would be disadvantageous to merely follow the attractant gradient. Instead, it may be more advantageous for a bacterium to seek a niche with a balance of different chemoattractants optimized for its individual physiology. This is particularly true provided that energy taxis is the dominant behavior of some organisms (Alexandre and Zhulin, 2001; Alexandre *et al.*, 2000, 2004). Instead of sensing chemical gradients spatially, bacteria use a temporal sensing mechanism that compares the instantaneous concentration of an attractant or repellent to that of a few seconds prior. Based on this comparison, a "decision" is made which results in adjusting the probability of random changes in the swimming direction (reversals). Controlling the frequency at which these instances of reorientation occur, cells are able to bias their movement towards environments most suitable for growth and division.

B. Molecular mechanisms of chemotaxis: The *E. coli* paradigm

1. The chemotaxis signal transduction excitation and adaptation pathways

At the center of the protein regulatory network that controls chemotaxis lies a two-component system (TCS). TCS are typically comprised of a membrane bound sensor histidine kinase coupled with a cytoplasmic cognate response regulator (Stock and Da Re, 2000; Zapf *et al.*, 1998). TCS can sense a wide range of environmental stimuli and usually result in a change in gene expression although they can regulate other cellular processes. Until recently, TCS were thought to be the dominant mechanism for prokaryotic signal transduction. However, a recent study has found that so-called one-component systems consisting of a single protein with both input and output domains are more numerous and likely represent the evolutionary precursors to TCS (Ulrich and Zhulin, 2005). In *E. coli*, the TCS at the core of the chemotactic system consists of CheA, the sensor histidine kinase, and CheY, the response regulator (Fig. 3.1). Environmental stimuli are sensed by chemosensory proteins (or chemoreceptors), so-called methyl-accepting chemotaxis proteins or MCPs that are embedded in the cytoplasmic membrane. An N-terminal sensory domain protrudes into the periplasm and receives signals that are then transmitted to the C-terminal signaling domain localized in the cytoplasm. CheW is a linker protein that connects the histidine kinase CheA to the MCPs. When CheA is activated, it autophosphorylates at a conserved histidine residue and in turn phosphorylates its cognate response regulator CheY. Phosphorylated CheY, CheY-P, then freely diffuses through the cytoplasm and binds FliM, the switch protein of the flagellar motor thus effecting a change in the direction of rotation. The signal is terminated by dephosphorylation of CheY-P by CheZ, a phosphatase. The system allowing for the temporal comparison of attractant or repellent concentrations is comprised of CheB and CheR, a methylesterase and methyltransferase, respectively. CheR is constitutively active, constantly adding methyl groups from *S*-adenosylmethionine to the conserved methylation sites on the chemoreceptors. Increasing methylation of MCPs results in a suppression of signaling. CheB becomes active after it is phosphorylated by CheA. Once activated, CheB removes methyl

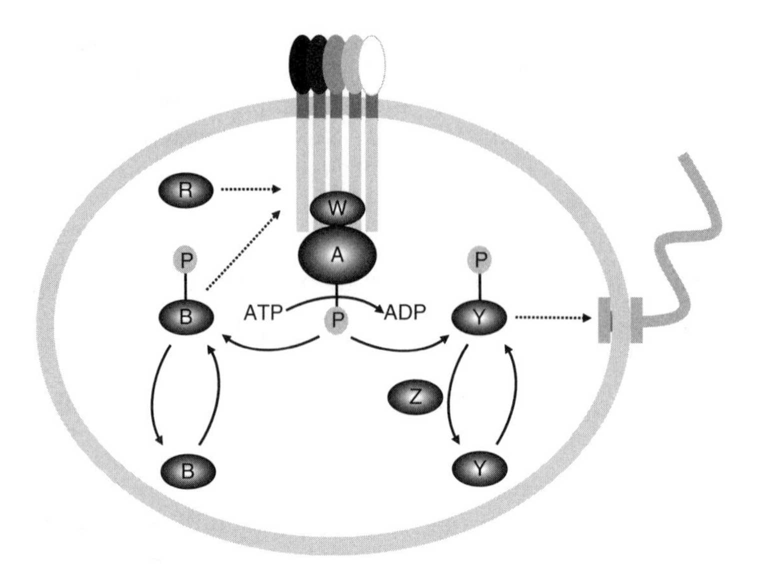

FIGURE 3.1 Schematic diagram of the *E. coli* chemotactic signal transduction pathway. Methyl-accepting chemotaxis proteins, MCPs, transduce signals from chemoeffectors in the periplasmic space thereby modulating the activity of the histidine kinase, CheA, which is linked to the MCPs by CheW. CheA phosphorylates the response regulators CheB and CheY whose phosphorylated forms interact with the MCPs and the flagellar motor, respectively. CheR also interacts with the MCPs whereas the dephosphatase CheZ dephosphorylates CheY.

groups from MCPs thus restoring their ability to signal (Fig. 3.1). CheB and CheR act in concert to allow the cell to adapt to its present level of stimulation. This adaptation system allows a cell to navigate gradients of either attractants or repellents over a wide range of concentrations over several orders of magnitude ranging from the nanomolar to millimolar.

2. Sensing the environment and the chemoreceptors repertoire

The first step in adaptation to the environment is sensing its changes. In motile bacteria, environmental cues are sensed in the extracellular environment by transmembrane receptors (chemoreceptors) that transmit the signal to a cytoplasmic two-component signal transduction pathway controlling the direction of flagellar rotation and chemotaxis (Hazelbauer *et al.*, 2008). Signal perception and transduction are mediated by a series of conformational changes that are transmitted from the N-terminal sensory domain of the chemoreceptor to the C-terminal signaling domain that interacts with CheW and CheA (Ottemann *et al.*, 1998, 1999). In *E. coli*, trimers of dimers of chemoreceptors are clustered at the cell poles in a large lattice that interacts with the cytoplasmic CheA and CheW proteins to form a large array that can be observed under specific conditions by electron microscopy (Zhang *et al.*, 2007). The organization of chemoreceptors into an array and the association of the clustered chemoreceptors' signaling tips with CheA and CheW allow sensory information from multiple chemoreceptors of various sensing capabilities (inputs) to be collected into a single output and trigger coordinated changes in rotational direction of the flagellar motors (Cluzel *et al.*, 2000; Gestwicki and Kiessling, 2002; Jasuja *et al.*, 1999).

The chemoreceptor repertoire is responsible for the specificity of the signal sensed and appears to be quite different in bacteria that occupy different ecological niches. The model organism for chemotaxis, *E. coli* inhabits the mammalian digestive tracts, including human, and it responds chemotactically mainly to amino acids via a metabolism-independent mechanism, where the response is triggered by the amino acid binding to specific transmembrane receptors. *E. coli* possesses four membrane-spanning chemoreceptors (Tsr, Tar, Trg, and Tap) that measure the concentrations of specific chemicals, such as amino acids (serine, leucine, and aspartate), sugars (ribose, galactose, mannose), dipeptides, and also heavy metals, such as cobalt and nickel. It also possesses a fifth chemoreceptor, Aer, that monitors changes in the redox state of the electron transport system and is thus an energy transducer that detects changes in intracellular energy levels (Bespalov *et al.*, 1996; Bibikov *et al.*, 2000; Greer-Philips *et al.*, 2003; Rebbapragada *et al.*, 1997; Repik *et al.*, 2000). Chemotaxis receptors are extremely sensitive and detect chemicals over a wide range of concentrations. For example, the Tar receptor responds to aspartate in a concentration range from 0.1 μM to 100 mM. Many other bacterial species possess a large number of different chemoreceptors (typically 10–20).

C. *Bacillus subtilis*, another model for chemotaxis signal transduction

Similar to *E. coli*, *Bacillus subtilis* possesses a single chemotaxis operon, but it possesses additional accessory proteins that participate in chemotaxis, CheC, CheD, and CheV. CheC functions in adaptation and signal termination. CheD is a deamidase that functions in receptor maturation by deamidating specific glutamine residues of MCPs. This function is carried out by CheB in *E. coli* (Fig. 3.2A). CheV also has a role in adaptation and in coupling CheA to the chemotaxis transducers (Szurmant and Ordal, 2004).

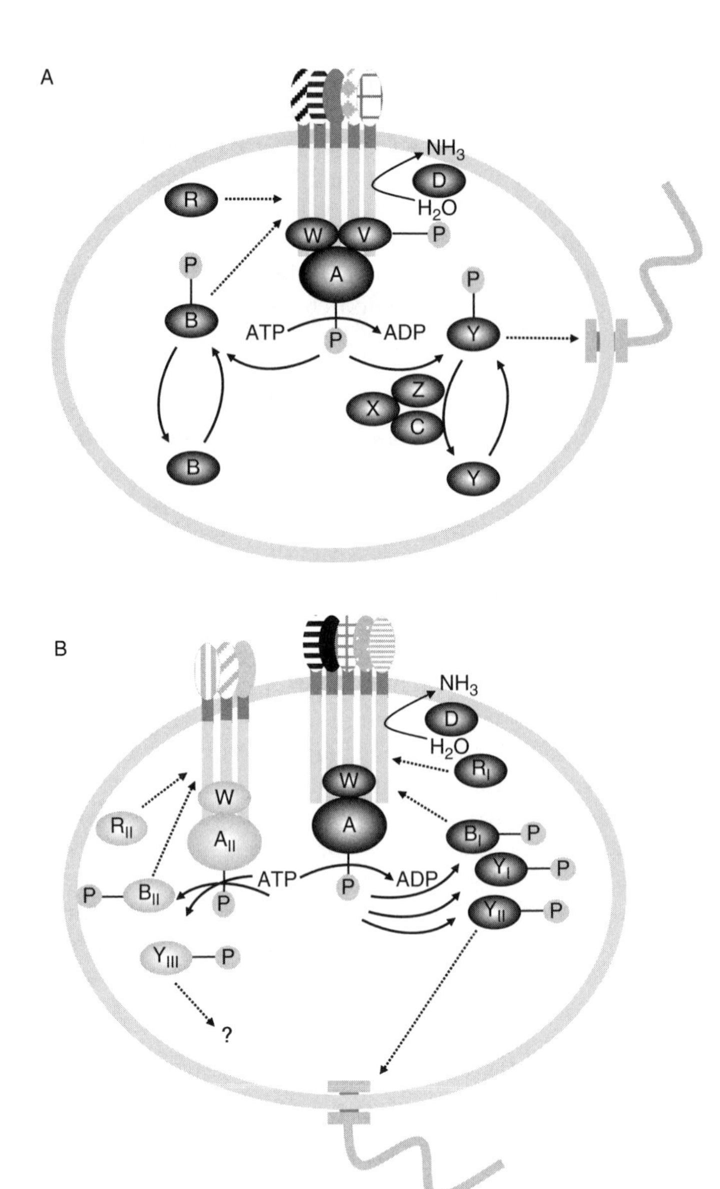

FIGURE 3.2 Examples of diversity in chemotactic signal transduction pathways. (A) Schematic representation of the chemotactic signal transduction pathways in *B. subtilis* and (B) in *R. leguminosarum* bv. *viciae*. The number of chemotaxis-like pathways, the number and diversity of chemoreceptors vary in different bacterial species. Note that the putative sensory domains of chemoreceptors are different the two examples shown to illustrate the diversity in chemosensory systems.

III. DIVERSITY IN CHEMOTAXIS

A. Complete genome sequencing projects and the diversity in chemotaxis

1. Diversity in the number of chemotaxis signal transduction pathways

The majority of bacteria whose genomes have been completely sequenced are motile and possess complex chemotaxis sensory machinery that differ from the model microorganisms studied so far. In addition, these chemotaxis pathways possess additional chemotaxis genes that are not present in the genomes of either *E. coli* or *B. subtilis* and whose functions remain to be determined (Park *et al.*, 2004; Rajagopala *et al.*, 2007) (Fig. 3.2). However, many studies have demonstrated that the principles of signal transduction and perception are conserved among diverse bacteria, although there are some variations on this theme (Szurmant and Ordal, 2004; Wadhams and Armitage, 2004) (Figs. 3.2A and B). The conserved set of CheAWYBRZ proteins and five MCPs comprising the chemotaxis protein network of *E. coli* is a streamlined version resulting from the evolutionary loss of chemotaxis proteins still present in the chemotaxis systems of other organisms (Zhulin, personal communication). The analysis of completely sequenced genomes also revealed that most bacteria possess more than one chemotaxis-like signal transduction pathway (Table 3.1). However, the function of redundant signal transduction pathways cannot be predicted from sequence analysis alone. The chemotaxis operons in these organisms can regulate flagellar and pili-mediated motility as well as perform other cellular functions such as regulation of gene expression and cellular differentiation (Berleman and Bauer, 2005a,b; Bible *et al.*, 2008; Kirby and Zusman, 2003; Vlamakis *et al.*, 2004).

2. Chemoreceptors diversity

Variations in the sensory specificity of the N-terminal domains of the chemoreceptors tailor them to a particular sensory function (Galperin *et al.*, 2001) and the diversity in the sequence of the N-terminal sensory domains reflects the diversity of cues a microorganism can detect and respond to. Therefore, it is not surprising that the greatest diversity in the chemotaxis system is seen in the set of chemoreceptors (Alexandre and Zhulin, 2001). Chemotaxis receptors are extremely sensitive naturally occurring biosensors that allow bacteria to monitor their environments and to detect chemicals over wide concentration ranges. The number of chemoreceptors per microbial genome correlates with the genome size and the metabolic versatility of the bacteria (Alexandre *et al.*, 2004) (Table I).

TABLE 3.1 Number of chemotaxis-like operons and chemoreceptors (MCPs) in completely sequenced genome of selected organisms[a]

Organisms	Number of chemotaxis-like operons	Number of chemoreceptors
Escherichia coli	1	5
Bacillus subtilis	1	10
Helicobacter pylori	1	4
Borrelia burgdorferi	2	5
Shewanella oneidensis	3	27
Rhodobacter sphaeroides	3	11
Rhizobium leguminosarum bv. *viciae*	2	27
Sinorhizobium meliloti	2	10
Ralstonia eutropha JMP134 (pJP4)	3	22
Desulfovibrio vulgaris Hildenborough	3	28
Geobacter metallireducens GS-15	5	17
Dechloromonas aromatica	4	25
Vibrio cholera	3	45
Magnetospirillum magnetotacticum	3	61

[a] The chemotaxis-like operons and chemoreceptors homologs were identified using the MiST database at http://genomics.ornl.gov/mist/ (Ulrich and Zhulin, 2007).

Interestingly, the substrate specificity of bacterial chemoreceptors is known for only a few of them, mainly from enteric bacteria. However, microbial genome sequencing projects reveal the presence of hundreds of different chemoreceptors: some are restricted to few bacteria, whereas others are widespread. The diversity in their sensory abilities is yet to be discovered. Because of their role in bacterial adaptation, establishing the substrate specificity for chemotaxis receptors will lead to the identification of the signals important for the survival of bacteria in different environments and will potentially impact future drug and pesticides development and could inform bioremediation strategies. However, to date, there are few experimental tools available to systematically analyze the sensory specificity of chemoreceptors and methods that usually work well in model organisms may not work well or not at all in other bacteria.

3. Predicting chemotaxis behavior from completely sequenced genomes

The observation that orthologous chemotaxis operons perform different functions in different organisms further complicates the prediction of their role in the biology of the organisms based on sequence comparisons alone.

For example, the dominant chemotaxis operon in the rhizobial species *Agrobacterium tumefaciens* and *Sinorhizobium meliloti* controls chemotaxis and motility, but it does not appear to be a major regulator of chemotaxis under laboratory conditions in *Rhodobacter sphaeroides* (Armitage and Schmitt, 1997; Wright *et al.*, 1998).

In addition, various experimental data indicate that multiple chemotaxis pathways may modulate the motility swimming pattern of organisms that possess several chemotaxis-like operons encoded in their genome. In *R. sphaeroides*, *Rhizobium leguminosarum* bv. *viciae*, and *Azospirillum brasilense*, chemosensory components from at least two operons are required for chemotaxis (Bible *et al.*, 2008; Ferrandez *et al.*, 2002; Miller *et al.*, 2007; Porter and Armitage, 2002, 2004; Stephens *et al.*, 2006).

B. Chemotaxis in bacterial species colonizing diverse niches

1. Ecological advantage provided by chemotaxis

The widespread presence of chemotaxis operons in the complete genomes of bacteria and archaea is evidence of the evolutionary and biological benefits of chemotaxis. Since bacteria inhabit almost every conceivable environmental niche in the biosphere and are capable of undergoing complex cellular differentiation processes in order to live, grow, and divide in different environments, the ecological role of chemotaxis has long been of interest. Studies are beginning to address the question of what role chemotaxis plays in structuring microbial communities and under which environmental conditions chemotaxis provides a competitive advantage (Alexandre *et al.*, 2004). The obvious advantage that chemotaxis provide the cells with is the ability to simultaneously monitor multiple cues and rapidly navigate in the environment. The chemotaxis two-component signal transduction systems are unique compared to other TCSs in that they function with a "molecular memory" provided by the adaptation system which endows cells with the ability to make temporal comparisons about the chemical composition of the environment by modulating sensory sensitivity of the chemoreceptors (Kentner and Sourjik, 2006). As a result of the activity of the adaptation system, cells are able to respond to changes over a wide range of background conditions and they may initiate changes in behavior when thresholds are reached instead of responding to the presence of a dedicated cue as in other TCSs. This level of control is analogous to a lamp controlled by a rheostat compared to a simple on/off switch. The advantage provided by the ability to temporally sample the environment has been recognized by several authors as the main advantages provided by chemotaxis-like signal transduction pathways (Bible *et al.*, 2008; Kirby and Zusman, 2003; Wadhams and Armitage, 2004).

2. Bacterial chemotaxis enhances competitiveness and facilitates adaptation

The ability to sense and navigate towards niches that are optimal for growth is a likely prerequisite for motile bacteria in order to colonize new environments. Therefore, motility and chemotaxis, by allowing the cells to navigate along gradients so that they can localize in optimum environments, have long been proposed to play a major role in the ability to compete for resources (Alexandre *et al.*, 2004). Consistent with this hypothesis, bacterial chemotaxis has repeatedly been shown to be essential for the competitive ability of beneficial, as well as pathogenic, bacteria to colonize their eukaryotic host. For example, chemotaxis was shown to be essential for the beneficial association of *A. brasilense* with wheat (Greer-Phillips *et al.*, 2004), for the symbiotic association of *R. leguminosarum* bv. *viciae* with the common pea plant (Miller *et al.*, 2007), and for the persistent association of *Helicobacter pylori* within a narrow zone of the stomach (Foynes *et al.*, 2000; McGee *et al.*, 2005; Schweinitzer *et al.*, 2008). In contrast, the human pathogen *Vibrio cholera* appears to require motility but not chemotaxis to establish infection (Butler and Camilli, 2004; Gardel and Mekalanos, 1994) and chemotaxis genes are downregulated in stool isolates from infected humans providing further evidence that motility but not chemotaxis is required for infection by *V. cholera* (Butler *et al.*, 2006; Merrell *et al.*, 2002). Such a negative effect of chemotaxis in host colonization has not been identified in any other system studied to date.

Chemotaxis allows bacteria to navigate towards ecological niches that support their metabolism. Chemotaxis could thus be used to "enrich" certain environments with bacteria that possess desirable properties in bioremediation. In fact, such a principle has been applied previously to chemotactically attract certain motile bacteria to environmental pollutants that they can degrade (Harms and Wick, 2006; Lanfranconi *et al.*, 2003; Pandey and Jain, 2002). In a few cases, dedicated chemoreceptors specific to certain pollutants have been identified (Iwaki *et al.*, 2007; Kim *et al.*, 2006).

IV. CHARACTERIZING THE CHEMOTAXIS RESPONSE: QUALITATIVE AND QUANTITATIVE ASSAYS

Insight into chemotaxis diversity and the potential application of this behavior to bioremediation, agriculture, or health-related fields is dependent on the ability to accurately describe the behavior and the development of quantitative methods that enable the determination of parameters that best describe the characteristics of bacterial motion.

Characterization of the motility behavioral responses in nonmodel bacteria requires the development of sensitive qualitative and quantitative methods to measure the chemotaxis response. This was recognized very early on in the study of bacterial chemotaxis since Pfeffer (Pfeffer, 1884) who first used glass capillaries filled with attractant solutions to demonstrate that bacteria accumulated first outside then inside the capillary.

Since then, numerous qualitative and quantitative methods have been developed for chemotaxis and several computerized motion-analysis systems are commercially available (HobsonTracking, Inc., UK; CellTrack, Santa Rosa, CA, etc.) to assist in the characterization and the statistical analysis of bacterial motion parameters.

There are two general types of behavioral assays used to measure bacterial chemotaxis, namely, temporal or spatial gradient assays. In a temporal assay, usually an attractant or repellant is abruptly added to a bacterial suspension to measure the chemotactic response of the motile bacteria. The same compound is then abruptly removed after a given time to measure the bacterial response, usually by examining reversal frequency of cells. However, in spatial chemotaxis assays, cells have to respond to much more shallow gradients, and in the case of the soft-agar plate assay, cells are responding to gradients which they created by metabolizing provided chemicals around the inoculation point (usually a nitrogen or carbon source, or both).

A. Temporal gradient assays

The most common temporal assay measures the probability of changes in the swimming direction (also called cell reversal frequency or reorientation frequency) of free-swimming bacteria when a potential attractant or repellant is added to the suspension. This is the most direct way to assay chemotaxis since one must directly observe the effect of a chemical on the swimming behavior of individual cells. A change in the number of directional changes per second (changes in the reversal frequency) relative to steady-state conditions indicates a chemotactic response to the stimulus. A decrease in reversal frequency (i.e., cells change their swimming direction less frequently) indicates the stimulus is an attractant while an increase in reversal frequency (i.e., cells tend to change swimming directions more frequently) is indicative of a repellant response (Berg and Brown, 1972; Segall *et al.*, 1986). The addition and removal of a stimulus is hardly physiological and would not typically be observed within a bacterium's normal environment but it is a good approach to directly screen the effect of a chemical stimulus on swimming behavior. This assay requires that the bacteria can be tracked and that their mode of changing swimming direction be easily identified, usually with the help of a computerized motion-tracking system. These restrictions limit the use of these assays to only a few bacterial species.

B. Spatial gradient assays

1. Soft agar plate assay

Several spatial behavioral assays are used to screen chemotaxis in motile bacteria (Table 3.2). These assays test for chemotactic responses to supplied chemical stimuli, predominantly nitrogen or carbon to help determine their sensing specificity (Hartmann and Zimmer, 1994). One such assay is the swarm, or soft-agar, plate assay (Fig. 3.3). In this assay, a culture inoculum is placed in the center of a semisolid agar plate, usually 0.2–0.3% agar (w/v), in which the cells can swim (Adler, 1966a; Alexandre *et al.*, 2000). As the cells grow and deplete the agar of the supplied nitrogen and carbon source, they generate a gradient, and if they are motile and chemotactic, they will form typical chemotactic rings at some distance from the inoculation point, indicative of a positive chemotactic response. As the cells continue to metabolize the provided nutrients, the migration outward from the inoculation point continues and the ''chemotactic'' ring expands with time. Therefore, chemotactic rings are observed in this assay if the cells are able to sense and navigate in the chemical gradient formed as a result of metabolism. In this assay, mutants affected in metabolism, motility, or chemotaxis will fail to form a chemotactic ring. Figure 3.3 shows some examples of typical results and their interpretation. In Fig. 3.3A, the bacteria inoculated were able to metabolize the nutrients provided in the plates and grow within the agar and thus they formed a typical sharp chemotactic ring as indicated by the arrow. In contrast, as shown in Fig. 3.3B, the bacteria inoculated were able to grow but did not move outward from the inoculation point. This result indicates that the bacteria are unable to respond chemotactically.

2. Chemical-in-plug assay

The chemical-in-plug assay also utilizes semisolid agar, but unlike the swarm plate assay, cell metabolism is not required to generate a chemical gradient and cells are judged solely on their ability to sense spatial gradients of attractants or repellents (chemotaxis per se). In this assay, a solid (1.5%, w/v) agar plug containing a specific concentration of chemical attractant or repellant in buffer is placed in a semi-soft buffered medium containing a high concentration of motile cells to be tested. A gradient is formed by diffusion of the chemical from the hard agar plug to the surrounding semi-soft agar medium where cells are present. If the cells' sense and respond positively to the chemical stimulus, a ring

TABLE 3.2 Spatial chemotaxis assays and their applications in diverse microorganisms

Methods	Organisms	References
Capillary assay		
	E. coli	Adler (1966a, 1966b, 1973)
	B. subtilis	Ordal and Gibson (1977), Ordal and Goldman (1975)
	Pseudomonas aeruginosa	Kato *et al.* (1992)
	Pseudomonas fluorescens	Singh and Arora (2001)
	B. burgdorferi	Shi *et al.* (1998)
	Ralstonia sp. *SJ98*	Pandey *et al.* (2002)
	Bradyrhizobium japonicum	Chuiko *et al.* (2002), Kurdish *et al.* (2001)
	Acidithiobacillus ferrooxidans	Meyer *et al.* (2002)
	D. Vulgaris	Meyer *et al.* (2002)
Modified capillary assay[a]		
	Pseudomonas putida G7 and Pseduomonas sp. NCIB 9816–4	Grimm and Harwood (1997)
Soft agar (Swarm) plate		
	E. coli	Adler (1966a), Hazelbauer *et al.* (1969)
	R. sphaeroides	Hamblin *et al.* (1997), Porter *et al.* (2002)
	R. leguminosarum bv. *viciae*	Miller *et al.* (2007)
	R. eutropha JMP134 (pJP4)	Hawkins and Harwood (2002)
	Vibrio fischeri	DeLoney-Marino *et al.* (2003)
Chemical-in-plug assay		
	E. coli	Tso and Adler (1974)
	B. burgdorferi	Shi *et al.* (1998)
	Halobacterium salinarum	Hyung and Yu (1997)
Drop assay		
	E. coli	Fahrner *et al.* (1994)
	P. putida G7 and Pseudomonas sp. *NCIB 9816–4*	Grimm and Harwood (1997)

[a] Assay used for chemicals poorly soluble in chemotaxis buffer. These chemicals include petroleum-based products and polyaromatic hydrocarbons such as toluene and naphthalene.

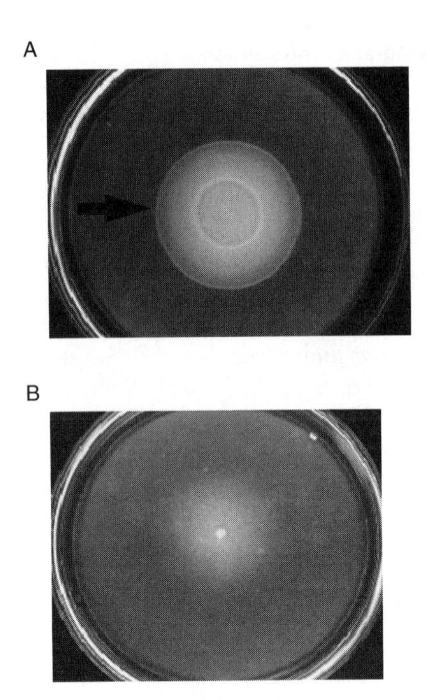

FIGURE 3.3 Examples of typical results obtained using the soft agar plate and chemical-in-plug assays. (A) Soft agar plate result with bacteria positive for chemotaxis under the conditions provided. Arrow indicates location of outer chemotactic ring, true ring specific for the provided nutrients in the plate. (B) Soft agar plate showing growth but no chemotaxis under conditions provided.

will form around the plug within a few hours (Tso and Adler, 1974). If the chemical is a repellant, a zone of clearing, where no bacteria are present, will form around the chemical plugs. Controls should also be used with this assay including a hard plug containing only buffer as a negative control as well as any nonchemotactic or nonmotile strains which may be available for the organism being tested.

3. Capillary assay

The capillary assay as described by Julius Adler (Adler, 1973) is the most widely used method to quantitatively measure chemotaxis in bacteria. Several modifications have been made to Adler's original assay, but it remains the primary assay to measure directed motility in a wide variety of microorganisms (Fig. 3.4). In this method, chemotaxis towards a potential attractant is measured by comparing the number of bacteria from a suspension in buffer that enter a glass capillary tube filled with the potential chemical attractant with the number entering a capillary tube filled with buffer alone. Quantitation is obtained by emptying the bacteria

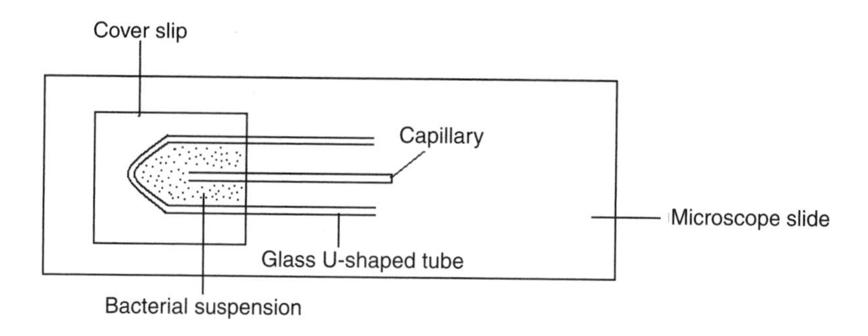

FIGURE 3.4 Setup of Adler's capillary assay. Diagram showing the setup of the capillary assay described by Adler (1973). The bacterial suspension is in the space between cover slip and microscope slide. The glass U-shaped tube is used to keep the suspension within the reservoir created. Prepared solutions of the chemical being tested are loaded within the capillary and place within the bacterial suspension and placed on the microscope for viewing.

that accumulated in the capillary and plating serial dilutions to determine the number of cells that entered the tube. This assay is also valuable for determining the sensitivity of bacteria towards an attractant by finding the threshold concentration required for a chemotactic response. The primary limitation of the capillary assay is the requirement for the test compound to be soluble in an aqueous solution. A modification of this method by Harwood and colleagues allows for a qualitative observation of chemotaxis in bacteria to poorly soluble compounds, including polyaromatic hydrocarbons (PAHs) like toluene or naphthalene (Grimm and Harwood, 1997). In this modified assay, a buffer solution is taken up into a glass capillary followed by pressing the open end of the capillary into a mound of finely granulated crystals of the PAH prior to placing the capillary in the bacterial suspension reservoir. The chemotactic response is judged positive if a visible cloud or mass of cells accumulates near the opening of the capillary. Another useful method for poorly soluble substrates is the drop assay in which a crystal of the compound is added to the center of a viscous cell suspension and chemotaxis is observed as an accumulation of cells encircling the crystal (Fahrner *et al.*, 1994; Grimm and Harwood, 1997).

The capillary aerotaxis assay is useful in testing the cell's ability to sense a spatial oxygen gradient. A flat capillary tube is filled with a suspension of motile cells to be tested. A dense band of cells forms a certain distance from the air–water interface (meniscus) where the cells are essentially trapped. For example, *A. brasilense* is a microaerophilic soil bacterium and seeks an environment with a low optimum oxygen concentration which is suitable for maximum energy production (Zhulin *et al.*, 1996). For

A. brasilense, oxygen is both an attractant and repellant, resulting in a dense aerotactic band of cells at the distance corresponding to its preferred oxygen concentration. Any perturbation in sensing oxygen, by a chemotaxis mutant for instance, results in a change in the distance between the meniscus and the dense cell band (Alexandre *et al.*, 2000).

C. Use of chemotaxis assays to characterize new microbial functions

Chemotaxis assays are not only used to screen for the range of chemicals to which motile bacteria respond. These assays are also useful in determining the sensory specificity of individual chemoreceptors, providing that isogenic mutants are available for the bacterial strain and the chemoreceptor(s) under study. These assays can also be modified to isolate motile bacteria with metabolic abilities desirable for bioremediation or agriculture. For example, the soft agar assay has previously been used to isolate microaerophilic nitrogen-fixing bacteria from the rhizosphere of cereals (Chowdhury *et al.*, 2007; Doroshenko *et al.*, 2007; Tarrand *et al.*, 1978) or to isolate bacteria capable of degrading specific organic contaminants found in a gas oil-contaminated soil (Lanfranconi *et al.*, 2003; Marx and Aitken, 2000a; Parales, 2004). Bacterial chemotaxis is also being considered as a microbially-driven strategy to increase the bioavailability of certain contaminants (Marx and Aitken, 2000b; Pandey and Jain, 2002).

V. CONCLUSIONS AND FUTURE PROSPECTS

Bacterial chemotaxis provides an obvious ecological advantage to motile bacteria since it allows motile cells to rapidly navigate in chemical gradients that may affect their survival and growth. The ecological advantages that chemotaxis provides to motile bacteria is reflected in the observation that the great majority of motile bacteria are chemotactic and that the most metabolically versatile bacterial species that also have the largest genomes possess multiple chemotaxis-like operons (Alexandre *et al.*, 2004; Szurmant and Ordal, 2004; Wadhams and Armitage, 2004). There are few examples where motile flagellated bacteria lack complete chemotaxis signal transduction pathways. However, in these few cases, the gene loss seems to be limited to certain strains within the species. It is possible that motility per se may be required for dispersion while the ability to actively direct the movement (chemotaxis) may not endow the cells with a particular competitive advantage, thereby favoring the loss of this function (Badger *et al.*, 2006; Deckert *et al.*, 1998; Glockner *et al.*, 2003; Moran *et al.*, 2004). Bacterial chemotaxis signal transduction is arguably one of the few biological systems known in great detail and thus it offers a unique opportunity for understanding the environmental cues that contribute to the shaping of bacterial communities. However, several challenges must be overcome, including the characterization of the function of multiple chemotaxis systems in phylogenetically diverse microbial species that have different swimming strategies and inhabit distinct ecological niches. Such analysis would greatly enhance our understanding of the evolution of this signal transduction pathway and it would also facilitate the development of algorithms for the prediction of the function of these pathways in completely sequenced genomes of bacterial species that may not be easily amenable to classical microbial physiology experiments. A particularly intriguing question is what ecological advantage(s) do multiple chemotaxis systems provide the cells with? In addition, the sensory specificity of chemoreceptors is a potential treasure's trove for novel biosensors and for understanding how bacteria couple sensing environmental cues with cellular adaptation. Characterizing the sensory specificity of diverse chemoreceptors remains challenging but it is an attainable objective in the near future owing to the development of sensitive high throughput biochemical methods.

ACKNOWLEDGMENTS

The authors thank NSF (CAREER, MCB-0622777) for support and the University of Tennessee, Knoxville for generous start-up funds.

REFERENCES

Adler, J. (1966a). Chemotaxis in bacteria. *Science* **153,** 708–711.

Adler, J. (1966b). Effect of amino acids and oxygen on chemotaxis in *Escherichia coli*. *J. Bacteriol.* **92,** 121–128.

Adler, J. (1973). Method for measuring chemotaxis and use of method to determine optimum conditions for chemotaxis by *Escherichia coli*. *J. Gen. Microbiol.* **74,** 77–91.

Adler, J., and Tso, W. W. (1974). "Decision"-making in bacteria: Chemotactic response of *Escherichia coli* to conflicting stimuli. *Science.* **184,** 1292–1294.

Alexandre, G., and Zhulin, I. B. (2001). More than one way to sense chemicals. *J. Bacteriol.* **183,** 4681–4686.

Alexandre, G., Greer, S. E., and Zhulin, I. B. (2000). Energy taxis is the dominant behavior in *Azospirillum brasilense*. *J. Bacteriol.* **182,** 6042–6048.

Alexandre, G., Greer-Phillips, S., and Zhulin, I. B. (2004). Ecological role of energy taxis in microorganisms. *FEMS Microbiol. Rev.* **28,** 113–126.

Armitage, J. P., and Schmitt, R. (1997). Bacterial chemotaxis: *Rhodobacter sphaeroides* and *Sinorhizobium meliloti*—variations on a theme? *Microbiology* **143,** 3671–3682.

Badger, J. H., Hoover, T. R., Brun, Y. V., Weiner, R. M., Laub, M. T., Alexandre, G., Mrazek, J., Ren, Q., Paulsen, I. T., Nelson, K. E., Khouri, H. M., Radune, D., *et al.* (2006). Comparative genomic evidence for a close relationship between the dimorphic prosthecate bacteria *Hyphomonas neptunium* and *Caulobacter crescentus*. *J. Bacteriol.* **188,** 6841–6850.

Berg, H. C. (2000). Motile behavior of bacteria. *Phys. Today* **53,** 24–29.

Berg, H. C., and Brown, D. A. (1972). Chemotaxis in *Escherichia coli* analyzed by 3-dimensional tracking. *Nature* **239,** 500–504.

Berleman, J. E., and Bauer, C. E. (2005a). A che-like signal transduction cascade involved in controlling flagella biosynthesis in *Rhodospirillum centenum. Mol. Microbiol.* **55,** 1390–1402.

Berleman, J. E., and Bauer, C. E. (2005b). Involvement of a Che-like signal transduction cascade in regulating cyst cell development in *Rhodospirillum centenum. Mol. Microbiol.* **56,** 1457–1466.

Bespalov, V. A., Zhulin, I. B., and Taylor, B. L. (1996). Behavioral responses of *Escherichia coli* to changes in redox potential. *Proc. Natl. Acad. Sci. USA* **93,** 10084–10089.

Bibikov, S. I., Barnes, L. A., Gitin, Y., and Parkinson, J. S. (2000). Domain organization and flavin adenine dinucleotide-binding determinants in the aerotaxis signal transducer Aer of *Escherichia coli. Proc. Natl. Acad. Sci. USA* **97,** 5830–5835.

Bible, A. N., Stephens, B. B., Ortega, D. R., Xie, Z., and Alexandre, G. (2008). Function of a chemotaxis-like signal transduction pathway in modulating motility, cell clumping and cell length in the alpha-proteobacterium *Azospirillum brasilense. J. Bacteriol.* **190,** 6365–6375.

Butler, S. M., and Camilli, A. (2004). Both chemotaxis and net motility greatly influence the infectivity of *Vibrio cholerae. Proc. Natl. Acad. of Sci. USA* **101,** 5018–5023.

Butler, S. M., Nelson, E. J., Chowdhury, N., Faruque, S. M., Calderwood, S. B., and Camilli, A. (2006). Cholera stool bacteria repress chemotaxis to increase infectivity. *Mol. Microbiol.* **60,** 417–426.

Chowdhury, S. P., Schmid, M., Hartmann, A., and Tripathi, A. K. (2007). Identification of diazotrophs in the culturable bacterial community associated with roots of *Lasiurus sindicus,* a perennial grass of thar desert, India. *Microb. Ecol.* **54,** 82–90.

Chuiko, N. V., Antonyuk, T. S., and Kurdish, I. K. (2002). The chemotactic response of *Bradyrhizobium japonicum* to various organic compounds. *Microbiology* **71,** 391–396.

Cluzel, P., Surette, M., and Leibler, S. (2000). An ultrasensitive bacterial motor revealed by monitoring signaling proteins in single cells. *Science* **287,** 1652–1655.

Deckert, G., Warren, P. V., Gaasterland, T., Young, W. G., Lenox, A. L., Graham, D. E., Overbeek, R., Snead, M. A., Keller, M., Aujay, M., Huber, R., Feldman, R. A., *et al.* (1998). The complete genome of the hyperthermophilic bacterium *Aquifex aeolicus. Nature* **392,** 353–358.

DeLoney-Marino, C. R., Wolfe, A. J., and Visick, K. L. (2003). Chemoattraction of *Vibrio fischeri* to serine, nucleosides, and *N*-acetylneuraminic acid, a component of squid light-organ mucus. *Appl. Environ. Microbiol.* **69,** 7527–7530.

Doroshenko, E. V., Boulygina, E. S., Spiridonova, E. M., Tourova, T. P., and Kravchenko, I. K. (2007). Isolation and characterization of nitrogen-fixing bacteria of the genus *Azospirillum* from the soil of a *Sphagnum* peat bog. *Microbiology* **76,** 93–101.

Dusenbery, D. B. (1998). Spatial sensing of stimulus gradients can be superior to temporal sensing for free-swimming bacteria. *Biophys. J.* **74,** 2272–2277.

Engelmann, T. W. (1881). *Bacterium photometricum*: An article on the comparative physiology of the sense for light and colour. *Arch. Ges. Physiol. Bonn.* **30,** 95–124.

Fahrner, K. A., Block, S. M., Krishnaswamy, S., Parkinson, J. S., and Berg, H. C. (1994). A mutant hook-associated protein (HAP3) facilitates torsionally induced transformations of the flagellar filament of *Escherichia coli. J. Mol. Biol.* **238,** 173–186.

Ferrandez, A., Hawkins, A. C., Summerfield, D. T., and Harwood, C. S. (2002). Cluster II che genes from *Pseudomonas aeruginosa* are required for an optimal chemotactic response. *J. Bacteriol.* **184,** 4374–4383.

Foynes, S., Dorrell, N., Ward, S. J., Stabler, R. A., McColm, A. A., Rycroft, A. N., and Wren, B. W. (2000). *Helicobacter pylori* possesses two CheY response regulators and a histidine kinase sensor, CheA, which are essential for chemotaxis and colonization of the gastric mucosa. *Infect. Immun.* **68,** 2016–2023.

Galperin, M. Y., Nikolskaya, A. N., and Koonin, E. V. (2001). Novel domains of the prokaryotic two-component signal transduction systems. *FEMS Microbiol. Lett.* **203,** 11–21.

Gardel, C. L., and Mekalanos, J. J. (1994). Modus operandi of *Vibrio cholerae*-swim to arrive-stop to kill-the relationship among chemotaxis, motility and virulence. *J. Cell. Biochem.* **56,** 65.

Gestwicki, J. E., and Kiessling, L. L. (2002). Inter-receptor communication through arrays of bacterial chemoreceptors. *Nature* **415,** 81–84.

Glockner, F. O., Kube, M., Bauer, M., Teeling, H., Lombardot, T., Ludwig, W., Gade, D., Beck, A., Borzym, K., Heitmann, K., Rabus, R., Schlesner, H., *et al.* (2003). Complete genome sequence of the marine planctomycete *Pirellula* sp. strain 1. *Proc. Natl. Acad. Sci. USA* **100,** 8298–8303.

Greer-Phillips, S. E., Alexandre, G., Taylor, B. L., and Zhulin, I. B. (2003). Aer and Tsr guide *Escherichia coli* in spatial gradients of oxidizable substrates. *Microbiology* **149,** 2661–2667.

Greer-Phillips, S. E., Stephens, B. B., and Alexandre, G. (2004). An energy taxis transducer promotes root colonization by *Azospirillum brasilense*. *J. Bacteriol.* **186,** 6595–6604.

Grimm, A. C., and Harwood, C. S. (1997). Chemotaxis of *Pseudomonas* spp. to the polyaromatic hydrocarbon naphthalene. *Appl. Environ. Microbiol.* **63,** 4111–4115.

Hader, D. P. (1987). Photosensory behavior in prokaryotes. *Microbiol. Rev.* **51,** 1–21.

Hamblin, P. A., Maguire, B. A., Grishanin, R. N., and Armitage, J. P. (1997). Evidence for two chemosensory pathways in *Rhodobacter sphaeroides*. *Mol. Microbiol.* **26,** 1083–1096.

Harms, H., and Wick, L. Y. (2006). Dispersing pollutant-degrading bacteria in contaminated soil without touching it. *Eng. Life Sci.* **6,** 252–260.

Hartmann, A., and Zimmer, W. (1994). Physiology of *Azospirillum*. *In "Azospirillum*/Plant Associations" (Y. Okon, ed.), pp. 15–41. CRC Press, Boca Raton, FL.

Hawkins, A. C., and Harwood, C. S. (2002). Chemotaxis of *Ralstonia eutropha* JMP134(pJP4) to the Herbicide 2,4-Dichlorophenoxyacetate. *Appl. Environ. Microbiol.* **68,** 968–972.

Hazelbauer, G. L., Falke, J. J., and Parkinson, J. S. (2008). Bacterial chemoreceptors: High-performance signaling in networked arrays. *Trends Biochem. Sci.* **33,** 9–19.

Hyung, S., and Yu, M. A. (1997). An agarose-in-plug bridge method to study chemotaxis in the Archaeon *Halobacterium salinarum*. *FEMS Microbiol. Lett.* **156,** 265–269.

Iwaki, H., Muraki, T., Ishihara, S., Hasegawa, Y., Rankin, K. N., Sulea, T., Boyd, J., and Lau, P. C. K. (2007). Characterization of a pseudomonad 2-nitrobenzoate nitroreductase and its catabolic pathway-associated 2-hydroxylaminobenzoate mutase and a chemoreceptor involved in 2-nitrobenzoate chemotaxis. *J. Bacteriol.* **189,** 3502–3514.

Jasuja, R., Yu, L., Trentham, D. R., and Khan, S. (1999). Response tuning in bacterial chemotaxis. *Proc. Natl. Acad. Sci. USA* **96,** 11346–11351.

Kato, J., Ito, A., Nikata, T., and Ohtake, H. (1992). Phosphate taxis in *Pseudomonas aeruginosa*. *J. Bacteriol.* **174,** 5149–5151.

Kentner, D., and Sourjik, V. (2006). Spatial organization of the bacterial chemotaxis system. *Curr. Opin. Microbiol.* **9,** 619–624.

Kim, H.-E., Shitashiro, M., Kuroda, A., Takiguchi, N., Ohtake, H., and Kato, J. (2006). Identification and characterization of the chemotactic transducer in *Pseudomonas aeruginosa* PAO1 for positive chemotaxis to trichloroethylene. *J. Bacteriol.* **188,** 6700–6702.

Kirby, J. R., and Zusman, D. R. (2003). Chemosensory regulation of developmental gene expression in *Myxococcus xanthus*. *Proc. Natl. Acad. Sci. USA* **100,** 2008–2013.

Kurdish, I. K., Antonyuk, T. S., and Chuiko, N. V. (2001). Influence of environmental factors on the chemotaxis of *Bradyrhizobium japonicum*. *Microbiology* **70,** 91–95.

Lanfranconi, M. P., Alvarez, H. M., and Studdert, C. A. (2003). A strain isolated from gas oil-contaminated soil displays chemotaxis towards gas oil and hexadecane. *Environ. Microbiol.* **5,** 1002–1008.

Lukat, G. S., and Stock, J. B. (1993). Response regulation in bacterial chemotaxis. *J. Cell. Biochem.* **51,** 41–46.

Marx, R. B., and Aitken, M. D. (2000a). A material-balance approach for modeling bacterial chemotaxis to a consumable substrate in the capillary assay. *Biotechnol. Bioeng.* **68,** 308–315.

Marx, R. B., and Aitken, M. D. (2000b). Bacterial chemotaxis enhances naphthalene degradation in a heterogeneous aqueous system. *Environ. Sci. Technol.* **34,** 3379–3383.

McGee, D. J., Langford, M. L., Watson, E. L., Carter, J. E., Chen, Y. T., and Ottemann, K. M. (2005). Colonization and inflammation deficiencies in Mongolian gerbils infected by *Helicobacter pylori* chemotaxis mutants. *Infect. Immun.* **73,** 1820–1827.

Merrell, D. S., Butler, S. M., Qadri, F., Dolganov, N. A., Alam, A., Cohen, M. B., Calderwood, S. B., Schoolnik, G. K., and Camilli, A. (2002). Host-induced epidemic spread of the cholera bacterium. *Nature* **417,** 642–645.

Meyer, G., Schneider-Merck, T., Bohme, S., and Sand, W. (2002). A simple method for investigations on the chemotaxis of *Acidithiobacillus ferrooxidans* and *Desulfovibrio vulgaris*. *Acta Biotechnol.* **22,** 391–399.

Miller, L. D., Yost, C. K., Hynes, M. F., and Alexandre, G. (2007). The major chemotaxis gene cluster of *Rhizobium leguminosarum* bv. *viciae* is essential for competitive nodulation. *Mol. Microbiol.* **63,** 348–362.

Moran, M. A., Buchan, A., Gonzalez, J. M., Heidelberg, J. F., Whitman, W. B., Kiene, R. P., Henriksen, J. R., King, G. M., Belas, R., Fuqua, C., Brinkac, L., Lewis, M., *et al.* (2004). Genome sequence of *Silicibacter pomeroyi* reveals adaptations to the marine environment. *Nature* **432,** 910–913.

Ordal, G. W., and Gibson, K. G. (1977). Chemotaxis toward amino-acids by *Bacillus subtilis*. *J. Bacteriol.* **129,** 151–155.

Ordal, G. W., and Goldman, D. J. (1975). Chemotaxis away from uncouplers of oxidative-phosphorylation in *Bacillus subtilis*. *Science* **189,** 802–805.

Ottemann, K. M., Thorgeirsson, T. E., Kolodziej, A. F., Shin, Y. K., and Koshland, D. E. (1998). Direct measurement of small ligand-induced conformational changes in the aspartate chemoreceptor using EPR. *Biochemistry* **37,** 7062–7069.

Ottemann, K. M., Xiao, W. Z., Shin, Y. K., and Koshland, D. E. (1999). A piston model for transmembrane signaling of the aspartate receptor. *Science* **285,** 1751–1754.

Pandey, G., and Jain, R. K. (2002). Bacterial chemotaxis toward environmental pollutants: Role in bioremediation. *Appl. Environ. Microbiol.* **68,** 5789–5795.

Pandey, G., Chauhan, A., Samanta, S. K., and Jain, R. K. (2002). Chemotaxis of a *Ralstonia* sp. SJ98 toward co-metabolizable nitroaromatic compounds. *Biochem. Biophys. Res. Comm.* **299,** 404–409.

Parales, R. E. (2004). Nitrobenzoates and aminobenzoates are chemoattractants for *Pseudomonas* strains. *Appl. Environ. Microbiol.* **70,** 285–292.

Park, S. Y., Chao, X. J., Gonzalez-Bonet, G., Beel, B. D., Bilwes, A. M., and Crane, B. R. (2004). Structure and function of an unusual family of protein phosphatases: The bacterial chemotaxis proteins CheC and CheX. *Mol. Cell* **16,** 563–574.

Pfeffer, W. (1884). Locomotorische Richtungsbewegungen durch chemische Reize. *Unters Bot Inst Tübingen* **1,** 363.

Porter, S. L., and Armitage, J. P. (2002). Phosphotransfer in *Rhodobacter sphaeroides* chemotaxis. *J. Mol. Biol.* **324,** 35–45.

Porter, S. L., and Armitage, J. P. (2004). Chemotaxis in *Rhodobacter sphaeroides* requires an atypical histidine protein kinase. *J. Biol. Chem.* **279,** 54573–54580.

Rajagopala, S. V., Titz, B., Goll, J., Parrish, J. R., Wohlbold, K., McKevitt, M. T., Palzkill, T., Mori, H., Finley, R. L., and Uetz, P. (2007). The protein network of bacterial motility. *Mol. Syst. Biol.* **3,** 1–13.

Rebbapragada, A., Johnson, M. S., Harding, G. P., Zuccarelli, A. J., Fletcher, H. M., Zhulin, I. B., and Taylor, B. L. (1997). The Aer protein and the serine chemoreceptor Tsr independently sense intracellular energy levels and transducer oxygen, redox, and energy signals for *Escherichia coli* behavior. *Proc. Natl. Acad. Sci. USA* **94,** 10541–10546.

Repik, A., Rebbapragada, A., Johnson, M. S., Haznedar, J. O., Zhulin, I. B., and Taylor, B. L. (2000). PAS domain residues involved in signal transduction by the Aer redox sensor of *Escherichia coli*. *Mol. Microbiol.* **36,** 806–816.

Schweinitzer, T., Mizote, T., Ishikawa, N., Dudnik, A., Inatsu, S., Schreiber, S., Suerbaum, S., Aizawa, S. I., and Josenhans, C. (2008). Functional characterization and mutagenesis of the proposed behavioral sensor TlpD of *Helicobacter pylori*. *J. Bacteriol.* **190,** 3244–3255.

Segall, J. E., Block, S. M., and Berg, H. C. (1986). Temporal comparisons in bacterial chemotaxis. *Proc. Natl. Acad. Sci. USA* **83,** 8987–8991.

Shi, W. Y., Yang, Z. M., Geng, Y. Z., Wolinsky, L. E., and Lovett, M. A. (1998). Chemotaxis in *Borrelia burgdorferi*. *J. Bacteriol.* **180,** 231–235.

Singh, T., and Arora, D. K. (2001). Motility and chemotactic response of *Pseudomonas fluorescens* toward chemoattractants present in the exudate of *Macrophomina phaseolina*. *Microbiol. Res.* **156,** 343–351.

Stephens, B. B., Loar, S. N., and Alexandre, G. (2006). The role of CheB and CheR in a complex chemotaxis and aerotaxis pathway in *Azospirillum brasilense*. *J. Bacteriol.* **188,** 4759–4768.

Stock, J., and Da Re, S. (2000). Signal transduction: Response regulators on and off. *Curr. Biol.* **10,** R420–R424.

Szurmant, L., and Ordal, G. W. (2004). Diversity in chemotaxis mechanisms among the bacteria and archaea. *Microbiol. Mol. Biol. Rev.* **68,** 301–319.

Tarrand, J. J., Krieg, N. R., and Dobereiner, J. (1978). Taxonomic study of *Spirillum lipoferum* group, with descriptions of a new genus, *Azospirillum* gen-NOV and 2 species, *Azospirillum lipoferum* (Beijerinck) comb Nov and *Azospirillum brasilense* sp-NOV. *Can. J. Microbiol.* **24,** 967–980.

Thar, R., and Kuhl, M. (2003). Bacteria are not too small for spatial sensing of chemical gradients: An experimental evidence. *Proc. Natl. Acad. Sci. USA* **100,** 5748–5753.

Tso, W. W., and Adler, J. (1974). Negative chemotaxis in *Escherichia coli*. *J. Bacteriol.* **118,** 560–576.

Ulrich, L. E., and Zhulin, I. B. (2005). Four-helix bundle: a ubiquitous sensory module in prokaryotic signal transduction. *Bioinformatics* **21,** 45–48.

Ulrich, L. E., and Zhulin, I. B. (2007). MiST: A microbial signal transduction database. *Nucleic Acids Res.* **35,** D386–D390.

Vlamakis, H. C., Kirby, J. R., and Zusman, D. R. (2004). The Che4 pathway of *Myxococcus xanthus* regulates type IV pilus-mediated motility. *Mol. Microbiol.* **52,** 1799–1811.

Wadhams, G. H., and Armitage, J. P. (2004). Making sense of it all: Bacterial chemotaxis. *Nat. Rev. Mol. Cell Biol.* **5,** 1024–1037.

Wright, E. L., Deakin, W. J., and Shaw, C. H. (1998). A chemotaxis cluster from *Agrobacterium tumefaciens*. *Gene* **220,** 83–89.

Zapf, J., Madhusudan, C. E., Grimshaw, C. E., Hoch, J. A., Varughese, K. I., and Whiteley, J. M. (1998). Source of response regulator autophosphatase activity: The critical role of a residue adjacent to the SpoOF autophosphorylation active site. *Biochemistry* **37,** 7725–7732.

Zhang, P. J., Weis, R. M., Peters, P. J., and Subramaniam, S. (2007). Electron tomography of bacterial chemotaxis receptor assemblies. *In* "Methods in Cell Biology" (J. R. McIntosh, ed.), Vol. 79, pp. 373–384. Academic Press, San Diego.

Zhulin, I. B., Bespalov, V. A., Johnson, M. S., and Taylor, B. L. (1996). Oxygen taxis and proton motive force in *Azospirillum brasilense*. *J. Bacteriol.* **178,** 5199–5204.

CHAPTER 4

Cutinases: Properties and Industrial Applications

Tatiana Fontes Pio and **Gabriela Alves Macedo**

Contents

Abstract

Cutinases, also known as cutin hydrolases (EC 3.1.1.74) are enzymes first discovered from phytopathogenic fungi that grow on cutin as the sole carbon source. Cutin is a complex biopolymer composed of epoxy and hydroxy fatty acids, and forms the structural component of higher plants cuticle. These enzymes share catalytic properties of lipases and esterases, presenting a unique feature of being active regardless the presence of an oil–water interface, making them interesting as biocatalysts in several industrial processes involving hydrolysis, esterification, and trans-esterification reactions. Cutinases present high stability in organic solvents and

Food Science Department, Faculty of Food Engineering, Campinas State University (UNICAMP), 13083970 Campinas, SP, Brazil

Advances in Applied Microbiology, Volume 66
ISSN 0065-2164, DOI: 10.1016/S0065-2164(08)00804-6

ionic liquids, both free and microencapsulated in reverse micelles. These characteristics allow the enzyme application in different areas such as food industry, cosmetics, fine chemicals, pesticide and insecticide degradation, treatment and laundry of fiber textiles, and polymer chemistry. The present chapter describes the characteristics, potential applications, and new perspectives for these enzymes.

I. INTRODUCTION

Intense research has been carried out in the field of industrial catalysts, both chemical and biological ones, because of the growing demand by production processes that are more effective and environmentally adequate.

Biocatalysts or biological catalysts are proteins that originally catalyze specific chemical reactions in living organisms, and are employed as an alternative to chemical catalysts because of a set of advantages, like high reaction speed, more amenable reaction parameters, good activity in synthetic substrates, catalytic activity in both reaction directions, and also a good degree of selectivity with respect to the type of catalyzed reaction (Campbell, 2000; Faber, 2000; Muderhwa *et al.*, 1988; Paques and Macedo, 2006).

The hydrolytic enzymes (proteases, cellulases, lipases, amylases, and cutinases) are the most employed in organic chemistry. These enzymes are very attractive because of a set of characteristics: low cost, wide availability, substrate specificity, activity without the need for cofactors (Dalla-Vecchia *et al.*, 2004).

In the early 1970s, it was discovered that some phytopathogenic fungi grow on the plant cuticle, using cutin as the sole carbon source, through the production of enzymes that hydrolyze cutin. Since then, several fungal cutinases were purified and characterized. It is well established that cutinases have a pivotal role in the penetration of intact plant surfaces by some fungi. This is supported by a consistent body of knowledge based on molecular biology, monoclonal antibodies, and chemical inhibitors (Fett *et al.*, 1992).

Cutinases are a group of enzymes that catalyze the hydrolysis of cutin, a biopolyester that constitutes the structural component of plant cuticle (Breccia *et al.*, 2003), and are composed by the following: ω-hydroxy fatty acids, dihydroxy palmitic acid, 18-hydroxy-9,10,-epoxy C18 saturated and Δ12 unsaturated fatty acids and 9,10,18-trihydroxy C18 unsaturated, and Δ12 monounsaturated fatty acids.

The specific cutin composition is variable according to the plant species, but as a role, fast growing plants produce cutin mainly containing

16 C fatty acids, predominantly dihydroxypalmitate, while slow-growing plants produce cutinase presenting both 16 C and 18 C fatty acids. Cutinase hydrolyzes cutin into all types of monomers (Kerry and Abbey, 1997).

Cutinases are also able to catalyze reverse reactions, namely etherification, trans-etherification (interesterification, alcoholysis, and radiolysis), and aminolysis. The thermodynamic water activity of the reaction medium is largely responsible for the determination of the predominance of each type of reaction (Pio and Macedo, 2007, Villeneuve *et al.*, 2000).

According to Brenda (2007), cutinase may be classified as EC 3.1.1.3, the recommended name being triacylglycerol lipase, and systematic name being triacylglycerol hydrolyze. The same author also classifies the enzyme as 3.1.1.74, with recommended name of cutinase and systematic name cutin hydrolyze. Therefore, the enzyme is an esterase that hydrolyzes cutin, but also acts and is classified as a lipase.

Cutinases present several interesting properties for application in a wide range of products, from food to detergents (Ferreira *et al.*, 2004). Therefore, intense research has been focused on the structure, function, purification, and applications of cutinases, aiming the development of new processes using these enzymes (Breccia *et al.*, 2003).

The present work provides an overview of the characteristics and potential applications of cutinases in industrial processes.

II. CUTINASE CHARACTERISTICS

Cutinase is one of the smallest members of serine hydrolyzes family, presenting the classical catalytic triad composed by serine, histidine, and a carboxyl group. The subfamily of cutinases is composed by nearly 20 members, based on amino acid sequence similarity. The structure and molecular dynamics are well determined, based on X-ray crystallography and NMR studies. In all serine hydrolyzes, the active site is invariably at the C-terminal end of one β chain (Egmond and De Vlieg, 2000).

The enzyme molecule presents a compact structure composed by 197 residues, that are arranged as a hydrophobic core made by a β central sheet surrounded by 5 α-helices, and a sole triptophan residue (Trp_{69}), located adjacent to a disulfide bond (Cys_{31}–Cys_{109}) (Soares *et al.*, 2003).

According to Borreguero *et al.* (2001), the catalytic site of cutinase is not protected by an amphipatic helicoidal "lid," characteristically present in the lipases. The oxyanion site is preformed in the cutinases, being formed in the lipases during coupling with the substrate. These two characteristics may be responsible for the absence of interfacial activation, detected in the catalytic behavior of cutinases.

Melo *et al.* (2003b) describe, in some cutinases, the presence of a conformation site that opens and closes, indicating that the enzyme is mainly a lipase than an esterase. This movement of active site covering may be responsible for the adaptation of the enzyme in different solvents.

Considering the group of cutinases, the stability is highly dependent on the specific enzyme considered. Petersen *et al.* (1998) observed that the tertiary structure of cutinase remains unchanged in a pH range from 4.0 to 9.0. However, the enzyme retained its activity at pH from 4.0 to 5.0 and also at pH 8.5. This behavior is probably related to the ionization state of His-188, because this residue needs to be deprotonated in order to stabilize the catalytic site. The stability pH of fungal cutinases stays at the basic and acid ranges, while pollen cutinase display stability at neutral pH, the optimal pH value being 6.8 (Kolattukudy, 1984).

Several strategies have been employed in order to increase the thermal and operational stability of cutinases in different reaction media (Carvalho *et al.*, 1997, 1999, 2000; Melo *et al.*, 2003; Ternström *et al.*, 2005). The use of reverse micelles, mainly AOT, is the most employed strategy in the literature, due to the high yields obtained. Reverse micelles are systems composed by amphiphilic molecules with polar groups oriented inwards and apolar groups oriented outwards. Their structure has characteristics that resemble biologic membranes. This feature makes them a useful tool in the study of biologic interactions of bioactive peptides (Melo *et al.*, 2003b).

The use of alcoholic media, mainly using hexanol, increases the retention of cutinase activity in reverse micelles, although the exact mechanism is not fully understood (Sebastião *et al.*, 1993). Gonçalves *et al.* (1999) showed that cutinase remains stable when immobilized in silica.

III. APPLICATIONS OF CUTINASE

Cutinases are a group of versatile enzymes, showing several useful properties for application in products and industrial processes (Table 4.1). In recent years, the stereolytic activity of cutinase has been wide explored. An enzymatic preparation containing cutinase was developed in order to improve the pharmacological effect of agricultural chemicals, taking advantage of its *in vitro* cutinase activity. These enzymes also present great potential in the management of residues from fruits and vegetables. Apple and tomato skin and orange peel are used, through cutin-catalyzed hydrolysis, in the production of important industrial chemicals, for example ricinoleic acid, whose main source is mamone oil (Carvalho *et al.*, 1997).

Cutinases hydrolyze, *in vitro*, a wide array of esters, ranging from soluble synthetic (like *p*-nitrophenyl esther), insoluble long-chain

TABLE 4.1 Reported applications of cutinase in biocatalysis

Publication Number	Publication Date	Author	Name	Reference
WO8809367	01-12-1988	Kolattukudy *et al.*	Methods and compositions regarding the use of cutinase in industrial cleaning processes	European Patent Office
JP3088897	15-04-1991	Eruuseido *et al.*	Method for the utilization of lipases, cutinases, and surfactants in cleaning processes	
NZ337239	28-09-2001	Rainhard and Henrik	Method for the enzymatic degradation of biodegradable polymers	
EP1694903	30-08-2006	Cavaco-Paulo *et al.*	Method for surface modification of polyacrylonitrile and polyamide fibers	
JP2005058228	10-03-2005	Yoshinobu *et al.*	Method for the production of esters in the absence of organic solvents	
CA 2480912	23-10-2003	Salmon *et al.*	Method intended to increase the tensile strength and abrasive ability of cotton fibers	Canadian Patents Database
CA 1262860	14-11-1989	Yuichi *et al.*	Method to enhance the effect of agricultural pesticides	
CA 2060510	10-01-1991	Ayrookaran *et al.*	Method to increase the permeability of fruits and vegetables surface	
CA 2465250	15-05-2003	Shi *et al.*	Method to remove the excess dye in industrial textile dyeing processes	

triglycerides (like triolein and tricaprilin), to emulsified triacylglycerols (Egmond, 2000).

Cutinases can also be used in the synthesis of structured triglycerides, polymers, surfactants, in the production of personal care products, and several processes in agrochemicals and pharmaceutical chemistry compounds containing one or more chiral centers (Macedo and Pio, 2005).

Cutinase has been used as a lipolytic enzyme in the composition of dishwashing and laundry detergents. Cutinase was better than a commercial lipase (Lipolase™) for the removal of triacylglycerols in an industrial washing process, because cutinase is able to hydrolyze fats in calcium-free media (Filipsen *et al.*, 1998).

The environmental impact of a production process is an important aspect, and the use of supercritical fluids is a good strategy in this context. Supercritical fluids present good perspectives in several areas, particularly biocatalysis. It is now well established that many enzymes are able to catalyze reactions in nonaqueous media that are difficult or impossible to occur in aqueous media. Also, some enzymes become more stable and can display an altered selectivity, and some applications of this ''nonaqueous enzymology'' are already in commercial stage (Klibanov, 2001; Krishina, 2002).

Garcia *et al.* (2005) investigated the activity of *Fusarium solani pisi* cutinase in supercritical fluids and organic solvents. The initial rates of transesterification of vinyl butyrate by (*R*,*S*)-2-fenil-1-prophanol were similar in supercritical ethane and in *n*-hexane, and considerably higher than in acetonitrile and in supercritical CO_2 under all conditions tested.

A. Oil and dairy products

Transesterification of fats and oils, as well as stereoselective esterification of alcohols, can be obtained at low water activity using cutinase (Macedo and Pio, 2005).

The technology of oils and fat modifying has raised a considerable interest in the last few years (Clauss, 1996; Gonçalves, 1996; Ahmed, 1995; Lima and Nassu, 1996). The fact that these substances can be obtained from natural, frequently low-cost sources, and turned into important raw materials to be used in food, chemical, and pharmaceutical industries explains such interest. A particularly important issue in this setting is the modification of lipids, in order to fit them for a very particular application. Therefore, the industry has developed several processes to modify the composition of triglycerides (Casey and Macrae, 1992; Gunstone, 1999; Hammond and Glatz, 1988).

The basic structure of oils and fats can be modified by different ways: chemical modification of fatty acids (hydrogenation), breakdown of the ester bond (hydrolysis), and reorganization of fatty acids in the triglyceride main chain (interesterification) (Clauss, 1996).

The biotechnological processes are an interesting option for use in the oil and fat industry, because they have a number of advantages, such as a better process yield, less energy consumption, generation of biodegradable products, cheaper production processes, and smaller residue production. (Hammond and Glatz, 1988; Willis and Maragoni, 1999; Castro *et al.*, 2004).

The milk fat hydrolysis is a typical enzyme-dependent transformation, being intrinsically present in the production processes of several dairy products. Such process can be carried out by enzymes belonging to native microflora (e.g., in the production of cheese from raw milk), or by the intentional use of exogenous enzymes.

According to the extension and specificity of such hydrolysis, the enzymatically modified milk fat will display diverse flavors and smells: from acid-free to butter or cream–cheese. Regado *et al.* (2007) performed an analysis of the milk fat partial lipolysis using 10 microbial lipases and *F. solani pisi* cutinase. The fatty acids liberated in the reaction medium were analyzed using HPLC. The enzymes tested showed an enzymatic modification profile similar to the available commercial enzymes. However, cutinase presented the highest activity against short-chain fatty acids.

B. Flavor compounds

Terpenic esters of short-chain fatty acids are essential oils with a great field of applications as flavor and aroma compounds in food, cosmetic, and pharmaceutical industries. Between terpenic esters, the most important are acetates, propionates, and butyrates of acyclic alcohols (geraniol and citronellol) (Croteau, 1980). These esters are traditionally obtained through chemical synthesis, fermentation, and extraction from natural sources (Welsh *et al.*, 1989).

However, such methods are expensive and present a low reaction yield. Facing the growing demand by natural products, the industry is employing biotechnological technology in order to produce natural flavors, particularly by enzymatic methods (Armstrong *et al.*, 1989). The use of enzymes as industrial catalysts in organic media has important advantages (Klibanov, 1989). In such systems, hydrolytic enzymes can be used to catalyze synthetic reactions, once the reaction chemical balance is altered in the direction of the product of synthesis (John and Abraham, 1991, Ballesteros *et al.*, 1995). Terpenic esters synthesis through direct esterification and transesterification in low water content media were described (Stamatis *et al.*, 1993; Claon and Akoh, 1994; Langrand *et al.*, 1988; Karra-Chaarbouni *et al.*, 1996; Castro *et al.*, 1997).

Ethyl esters of fatty acids are also important flavor substances. Recent advances in biocatalysis employing nonconventional media allowed the utilization of hydrolytic enzymes to catalyze the synthesis of aromatic compounds (Longo and Sanromán, 2006; Vandamme and Soetaert, 2002).

Carvalho, *et al.* (1996) studied the reaction parameters of transesterification of butyl acetate and hexanol, employing *F. solani pisi* recombinant cutinase in reverse micelles of (2-ethyl-1-hexyl) sodium sulfosuccinate (AOT)/isooctane for obtaining hexyl acetate, a short-chain ester with fruit flavor. This type of reaction is potentially important in the synthesis of aromatic esters.

In this study, the enzyme presented high activity at temperatures between 40 and 50 °C. The water:surfactant molar rate had a marked influence on the enzyme activity, with the best results in the range between 5 and 8. The presence of hexanol and the low water content lead to the enzyme stabilization in the interior of the micelles, increasing its thermostability (Carvalho *et al.*, 1999, 2000).

The use of enzymes encapsulated in reverse micelles results in a great interface area, making possible the solubilization of hydrophilic and hydrophobic substrates and frequently increasing the catalytic activity. The separation of reaction products and the enzyme recuperation are easier, and represent an important issue related to micelle research (Larsson *et al.*, 1990).

Barros *et al.* (2007) studied the reaction parameters for the esterification of ethanol and short-chain fatty acids (C_2 to C_6) in organic medium employing cutinase. They describe a 35% increase in molar conversion when the enzyme concentration was changed from 1.4% to 2.1%.

C. Phenolic compounds production

The phenolic acids family is composed by cinnamic (C_6–C_3) and benzoic (C_6–C_1) acids derivatives, as shown in Figs. 4.1 and 4.2. They are characterized by the presence in their structure of a benzenic ring with one or more hydroxyl groups, or a metoxyl group, together with a carboxylic group.

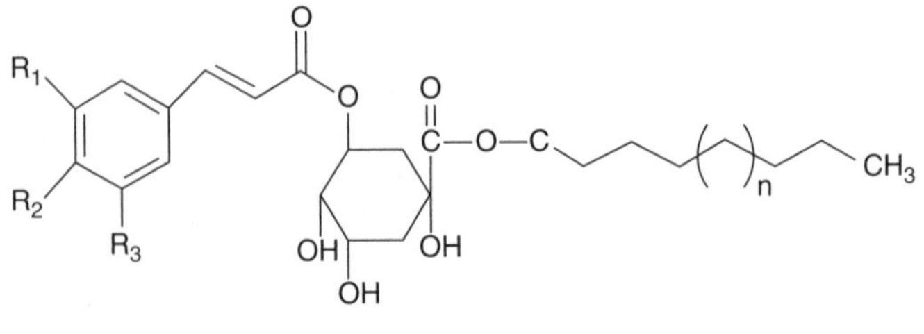

R1
R2
R3
O
O
O
C—O—C
n
CH3
OH
OH
OH

FIGURE 4.1 Chemical structure of cinnamic acid derivatives.

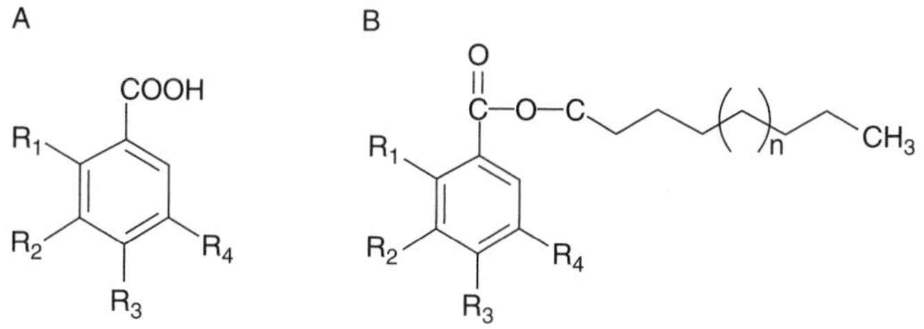

FIGURE 4.2 Chemical structures of benzoic acid derivatives (A) and alcohol esters of benzoic acid (B).

They are naturally occurring hydrophilic antioxidants, ubiquitously present in fruits, vegetables, and aromatic herbs. These substances raise great interest for potential industrial use due to a set of properties such as antioxidants, chelating, free radical scavengers, antiallegic, antiinflammatory, antimicrobial, antiviral, anticarcinogen, and also as a ultraviolet filter (Espinoza and Villeneuve, 2005; Silva *et al.*, 2000).

Because of the low solubility of phenolic acids in aprotic media, their use in lipid-based products is very limited. The esterification of the carboxylic acid with a fatty alcohol greatly increases its hydrophobicity and results in a multifunctional amphiphylic molecule. Such reaction is known as lipophilization, and can be performed chemically or enzymatically.

The chemical lipophilization processes have drawbacks related to the low thermal and pH reaction media stability of these substances. Moreover, the chemical lipophilization processes are not selective, demanding several intermediary reaction steps and purification processes, due to the generation of a great amount of undesirable products (Hills, 2003).

The enzymatic lipophylization of phenolic acids presents advantages such as more selective specificity, less generation of by-products, a greater diversity of pure synthetic substrates, fewer intermediary steps, and more amenable reaction parameters.

The enzymatic lipophylization is studied employing lipases, feruloylesterases, tannases, and cutinases (Soares *et al.*, 2003). Stamatis *et al.* (1999) studied the esterification of ferulic acid with 1-octanol using different enzymes, including *F. solani* cutinase. These enzymes were also able to catalyze the esterification of cinnamic, *p*-coumaric, and *p*-hydroxyphenyl propionic acids at relatively high rates.

D. Insecticide and pesticide degradation

Organophosphate insecticides are widely employed in agricultural plague control and control of cattle parasites. Such chemicals stay for long periods in the environment, mainly in lipids, being easily absorbed by the organisms through various routes, including skin, mucous membranes, lungs, and gastrointestinal tract (Chambers, 1992; Barlas, 1996; Indeerjeet *et al.*, 1997).

From laboratory testing and field experience with these products, the organophosphates have proved to be dangerous to the nervous system of invertebrates, the immune system of vertebrates and adrenal glands and liver of fish. They also produce detectable mutations in blood and lymphatic human cells (Galloway and Handy, 2003).

Kim *et al.* (2006), studied the degradation and detoxification of the organophosphate malathion employing *F. oxysporum* cutinase and yeast esterase. The cutinase degraded 50 and 60% of the initial amount of malathion at 15 and 30 min of reaction, respectively. In contrast, only 35% of the initial amount of malathion was degraded at 48 h of reaction when employing yeast esterase. The authors also analyzed the chemical composition of the final products, as well as its toxicity against bioluminescent bacteria. The final products were highly enzyme-dependent. The use of cutinase degraded malathion mainly to the nontoxic product MDA, while the yeast esterase produced the toxic MMA.

Walz and Schwack (2007) described the development of an enzymatic spectrophotometric assay based on the inhibition of *F. solani pisi* cutinase by organophosphate pesticides. The aim of such assay is the detection of pesticide residues in foods. The cutinase proved to be the most well fitted enzyme for this purpose. The authors also demonstrated that the test can detect the presence of residues even at very low levels.

In a subsequent study (Walz and Schwack, 2008), the authors submitted several organophosphates and carbamates to such test. The oxon organophosphates are detected by the test at very low levels. The thiol organophosphates are weak inhibitors of cutinase, but can be turned into their oxon analogs through the introduction of an oxidative step, previous to the test. The carbamate pesticides were also tested, revealing an efficient cutinase inhibitor effect, though less potent than the organophosphates.

E. Textile industry and laundry

The synthetic fibers are responsible for almost 50% of the textile worldwide demand (Silva *et al.*, 2005), and the production of polyester fibers is greater than cotton fibers. Because of the importance of synthetic fibers for the textile industry, there has been intense research concerning the improvement of production processes, as well as refinements in final product quality.

Synthetic fibers have some characteristics like hydrophobicity, making them uncomfortable in contact with human skin, and low reactivity, which poses some difficulties in the treatment of such fibers with finalizing and inking agents.

The use of strong alkaline agents can improve the hydrophobicity and reactivity of such fibers, but these processes are difficult to control and frequently result in loss of fiber resistance (Carvalho *et al.*, 1998; Soares *et al.*, 2003; Vertommen *et al.*, 2005). Besides, the great amounts of sodium hydroxide and high reaction temperatures required in these processes make them environmentally not attractive.

The best strategy is to modify only the surface during the search for lower hydrophobicity, without changing the fiber interior. Recent research employing enzymes, mainly lipases and cutinases, evidenced the improvement in polyester fiber quality through hydrolysis of ester bonds only at the fiber surface. The reaction parameters are much more bland, presenting low residue amounts, and without the need for complex machinery (Alisch *et al.*, 2004; Fischer-Colbrie *et al.*, 2004; Matamá *et al.*, 2004; Vertommen *et al.*, 2005).

Cutinases and carboxylesterases have shown a similar potential to hydrolyze ester bonds than lipases (Yoon *et al.*, 2002).

Vertommen *et al.* (2005) realized the enzymatic modification of a polyethylene terephtalate surface employing *F. solani pisi* cutinase and lipase A from *Candida antarctica* in an heterogeneous aqueous system. The cutinase displayed a significative hydrolytic activity against the polymer, and no detectable activity was registered in the presence of lipase.

Silva *et al.* (2005) employed cutinase from *F. solani pisi* in the surface modification of synthetic fibers of polyester, polyamide, and acrylic. The cutinase was chosen for use in fiber modification due to its hydrophobic nature and activity against biopolyesters present in plant cuticle.

The activity against polyamide was determined through the production of hexamethylethylenediamine. The activity against polyester was measured by the production of terephtalic acid, and against acrylic by the detection of acetic acid. Most of the amine acid residues near the enzyme active site are hydrophobic. This characteristic probably accounts for the interaction between the enzyme and the fiber surface. The cutinase was able to hydrolyze amide bonds, showing better activity in polyamide than in polyester.

Degani *et al.* (2002) employed cutinase to improve the wetting of cotton fibers. The enzyme was studied alone and in combination with pectine liase. The combination of cutinase and pectine liase presented a synergetic effect. The use of detergents also improved the reaction yield.

F. Polymer chemistry

In the polymer industry, the need for residue management and the search for alternative sources of raw materials are generating new paradigms. Some enzymes, in nonaqueous media, have demonstrated good activity in a wide variety of reactions of polyester and polycarbonate synthesis.

These enzymes are mainly lipases, the most employed being lipase B from *C. antarctica*. Recently, cutinase from *Humicola insolens* (Novozymes®) showed promising activity in polymerization reactions (Hunsen *et al.*, 2007). Most of current research in cutinase-catalyzed biotransformations is focused in polyester degradation and esterification and transesterification of small molecules (Yoon *et al.*, 2002).

More than half a century ago, synthetic polymers started to substitute natural materials in almost all areas, and nowadays plastics are essential to modern lifestyle. The stability and durability of plastics have been improved continuously, making them highly resistant to degradation by the environment, especially microbial. In the last decades, the great amounts of plastic produced resulted in problems with residue management of such materials. Because of the relative short time span of the existence of these materials, the evolution has not still developed new enzymes able to degrade synthetic polymers. The first attempts to produce plastics vulnerable to microbial degradation while retaining their properties started nearly 20 years ago (Mueller, 2006).

It became clear since the beginning that polymers presenting a heteroatom in the main chain, like polyesters, polyeters, polyamides, and polyurethanes can be degraded by microorganisms, and the subsequent developments in biodegradable plastics became focused mainly in polyester-based materials (Bastioli, 1995; Doi, 1990).

The study of hydrolytic enzymes that evolved to degrade highly stable nature polymers such as cellulose, chitin, and cutin may contribute to the knowledge about the degradation of natural and synthetic polymers (Fig. 4.3). This body of knowledge is very relevant in the development of new biodegradable plastic materials.

Plastics based on polyhydroxydecanoates are biodegradable, but their production is very expensive (Mayer and Kaplin, 1994). Polycaprolone is a synthetic polyester prone to degradation by several microorganisms,

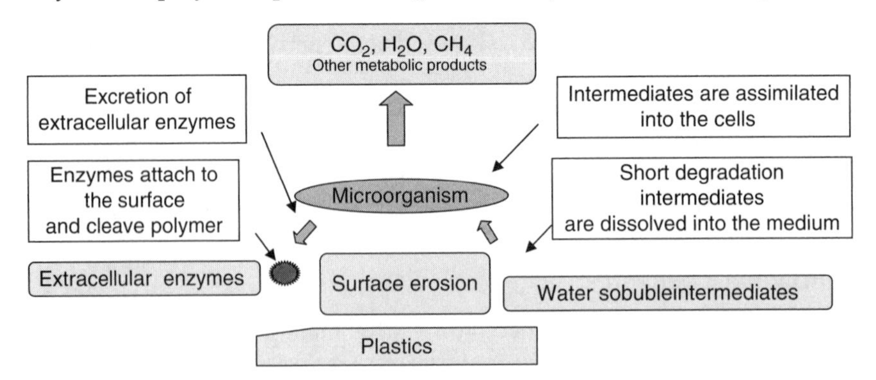

FIGURE 4.3 General mechanism of enzymatic catalyzed hydrolytic polymer degradation.

but its physical properties limit its applications. Murphy *et al.* (1996) showed that *F. solani pisi* cutinase can degrade polycaprolone.

Phtalates are plastifiers employed in polyvinil chloride-based products and in several cosmetics and lacquers (Chang *et al.*, 2004; Sung *et al.*, 2003). Dihexylphtalate (DHP) is one of the most used phthalate esters, employed in the fabrication of toys, gloves, shoes, and food packages. Phtalates, including DHP, are toxic to humans and to the environment. Their use is submitted to a strict control in several countries, but humans can be exposed to such products by several routes, like ingestion, inhalation, and skin contact (Kavlock *et al.*, 2002).

Kim *et al.* (2006) studied the efficacy of *F. oxysporum* cutinase and *Candida cylindracea* esterase in DHP degradation. Cutinase presented better and faster activity than the esterase, with high stability of the hydrolytic activity. The degradation of DHP by cutinase was nearly 70% after 4 h, while 85% of the initial amount of DHP remained intact after 72 h of incubation with the esterase. The toxicity of the reaction products was evaluated by the use of bioluminescent recombinant bacteria. The products of cutinase-catalyzed hydrolysis were less toxic than those employing esterase.

Because of its efficient catalytic activity both in solution and water–lipid interfaces, cutinase is potentially useful for the removal of fats in laundry. However, the unfolding of the enzyme in the presence of anionic surfactants limits its widespread use as an additive in industrial laundry detergents. According to Creveld *et al.* (2001), the stability of *F. solani pisi* cutinase might be increased through mutations designed to avoid the transient formation of hydrophobic groups during protein movement.

G. Enantioselective esterification reactions

The importance of enzymes in biocatalysis is even more evident, because of a set of properties, such as great versatility of catalyzed reactions, bland reaction parameters, and their regio, chemical, and enantioselectivity (Carvalho *et al.*, 2005). Because of their enantioselectivity, some cutinases have been employed in processes involving hydrolysis in aqueous media or synthesis in organic media, as shown in Fig. 4.4.

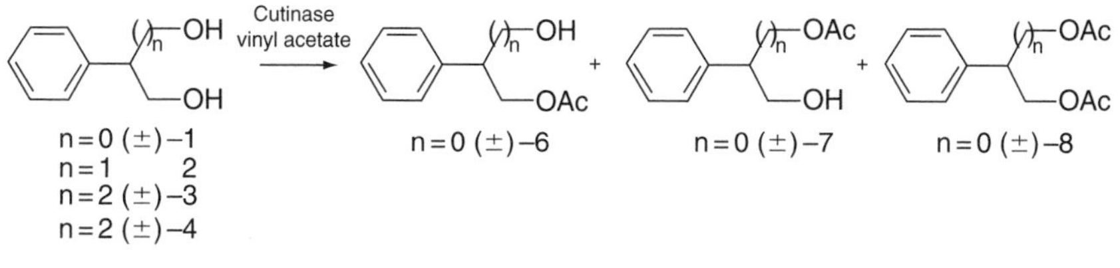

Cutinase
vinyl acetate
OH OH n=0 (±)–1 n=1 2 n=2 (±)–3 n=2 (±)–4
OH OAc n=0 (±)–6 + OAc OH n=0 (±)–7 + OAc OAc n=0 (±)–8

FIGURE 4.4 Cutinase catalyzed resolution of diols 1–4.

Mannesse *et al.* (1995) studied the influence of the 1, 2, and 3 positions of the triglyceride chain length in *F. solani pisi* cutinase activity and stereoselectivity, through the synthesis of different chain length triglycerides with R and S enantiomers. The enzyme showed a preference for the R enantiomer, but this preference was strongly related to the acyl chain length, presenting an R:S activity of 30:1. The enantioselectivity was evidenced in three different systems.

Borreguero *et al.* (2001) studied the regio and enantioselectivity of a recombinant cutinase from *F. solani pisi* in three racemics and one prochiral phenyl alcanodiol through nonreversible transesterification with vinyl acetate in organic media. The increase in the stereoselectivity of the primary hydroxyl group acylation was obtained through the preincubation of the enzyme in the presence of the substrate diol 1; however, there was no correlation with the incubation time. The presence of vinyl acetate had a stabilizing effect on the enzyme activity, similar to the effect of hexanol in reverse micelles.

H. Food industry

The most important fatty acids that take part of marine oils composition are the eicosapentanoic (EPA) and the docosahexanoic (DHA). These fatty acids present therapeutic properties in the field of autoimmune and cardiovascular diseases, and cannot be obtained by conventional heating processes without a substantial degradation (Masson *et al.*, 2000). In the same way, γ-linolenic acid, an important nutrient found in oleaginous seeds, may be obtained by enzymatic hydrolysis under mild temperature conditions (Mukherjee, 1990).

The use of hydrolytic enzymes has also been tested in the fields of biologic degradation and removal of industrial wastes from food industries (Gandhi, 1997; Lie and Molin, 1991; Pandey *et al.*, 1999). These industries, especially those handling meat and dairy products, generate a great amount of foul smelling residues, which damage intrinsically and extrinsically the industrial units. The use of hydrolytic enzymes in the early treatment of such wastes decreases the lipid content, the diameter of the fat droplets and reduces in 60% the time of permanence of the effluent in the treatment tanks (Leal *et al.*, 2002; Masse *et al.*, 2001).

In the production of sugar syrup from wheat starch, the main residue is hemicellulose, which is originated from the cell wall and remains undegraded through the entire production process. The accumulation of hemicellulose in the sieves of the industrial lines impairs the filtration process. Hemicellulose is a heteropolysaccharide containing xylan, which is esterified with arabinofuranoside and ferulic acid. This modified xylan is difficult to hydrolyze by use of xylanase alone. When xylanase is combined with arabinofuranosidase and ferulic acid esterase (FAE), xylan degradation may be improved, thereby removing the impurities and assisting the filtration process. In industrial use, there is a need for a more thermostable xylan degrading enzyme. In this context, Andersen *et al.* (2002) investigated the FAE activity of fungal lipases and cutinases. The cutinases from *Aspergillus oryzae*, *F. solani pisi*, and *Ilumicola insolence* had activity superior to the lipases tested, at pH values of 5.0; 7.0, and 9.0. Cutinase from *I. insolence* showed the best activity. These enzymes display a stability profile that is well-fitted to the industrial process, and because they have a low but significant FAE activity, it might be easier to introduce a high level of FAE activity in cutinases through point mutations.

IV. CONCLUSION

Cutinases are hydrolytic enzymes that share properties of lipases and esterases, and also display the unique characteristic of being active regardless of the presence of an interface. These properties make these enzymes potentially useful as biocatalysts in systems involving hydrolysis, esterification, and transesterification reactions.

These enzymes have been the object of intense and increasing research, with an expressive number of patents already registered, though their widespread industrial use is not yet established.

The field of biotechnology has a well established capacity to improve the original properties of a wide array of raw materials and to submit them to marked transformations, as well as in the treatment of industrial wastes and toxic products degradation. The industrial demand by enzymatic technology is ever growing, and cutinases may become important protagonists in this context.

ACKNOWLEDGEMENTS

The authors thank FAPESP and CNPq for the financial support.

REFERENCES

Ahmed, J. I. (1995). "Trans"-fixed? *Food Sci. Technol. Today* **9,** 228–231.

Alisch, M., Feuerhack, A., Blosfeld, A., Andreaus, J., and Zimmermann, W. (2004). Polyethylene terephthalate fibers by esterases from actinomycete isolate. *Biocatal. Biotransform.* **22,** 347–352.

Andersen, A., Svendsen, A., Vind, J., Lassen, S. F., Hjort, C., Borch, K., and Patkar, S. A. (2002). Studies of ferulic acid esterase activity in fungal lipases and cutinases. *Colloids Surf.* **26,** 47–55.

Ballesteros, A., Bornscheuer, U., Capewell, A., Combes, D., Condoret, J. S., König, K., Kolisis, F. N., Marty, A., Menge, U., Scheper, T., Stamatis, H., and Xenakis, A. (1995). Enzymes in non-conventional phases. *Biocatal. Biotransform.* **13,** 1–42.

Barlas, M. E. (1996). Toxicological assessment of biodegraded malathion in albino mice. *Bull. Environ. Contam. Toxicol.* **57,** 705–712.

Barros, D. P. C., Fonseca, L. P., and Cabral, J. M. S. (2007). Cutinase-catalyzed biosynthesis of short chain alkyl esters. *J. Biotechnol.* **131,** 109–110.

Bastioli, C. (1995). Starch-polymer composites. *In* "Degradable Polymers, Principles and Applications" (G. Scott and D. Gilead, eds.), pp. 112–133. Chapman & Hall, London.

Borreguero, I., Carvalho, C. M. L., Cabral, J. M. S., Sinisterra, J. V., and Alcántara, A. R. (2001). Enantioselective properties of *Fusarium solani pisi* cutinase on transesterification of acyclic diols: Activity and stability evaluation. *J. Mol. Catal. B: Enzym.* **11,** 613–622.

Breccia, J. D., Krook, M., Ohlin, M., and Hatti-Kaul, R. (2003). The search for a peptide ligand targeting the lipolytic enzyme cutinase. *Enzyme Microb. Technol.* **33,** 244.

Brenda: The Comprehensive Enzyme Information System. Assessed online as http://www.brenda-enzymes.info.

Campbell, M. K. (2000). "Bioquímica," 3rd ed. Artmed Editora Ltda, Porto Alegre.

Canadian Patents Database. Assessed online as http://patents.ic.gc.ca.

Carvalho, C. M. L., Aires-Barros, M. R., and Cabral, J. M. S. (1999). Cutinase: From molecular level to bioprocess development. *Biotechnol. Bioeng.* **60,** 17–34.

Carvalho, C. M. L., Aires-Barros, M. R., and Cabral, J. M. S. (2000). Kinetics of cutinase catalyzed transesterification in AOT reversed micelles: Modeling of a batch stirred tank reactor. *J. Biotechnol.* **81,** 1–13.

Carvalho, P. O., Calafatti, S. A., Marassi, M., Silva, D. M., Contesini, F. J., Bizaco, R., and Macedo, G. A. (2005). Potencial de biocatálise enantiosseletiva de lipases microbianas. *Quim. Nova* **28,** 614–621.

Carvalho, C. M. L., Serralheiro, M. L. M., Cabral, J. M. S., and Aires-Barros, M. R. (1997). Application of factorial design to the study of transesterification reactions using cutinase in AOT-reversed micelles. *Enzyme Microb. Technol.* **21,** 117–123.

Casey, J., and Macrae, A. (1992). Biotechnology and the oleochemical industry. *Inform.* **3,** 203–207.

Castro, H. F., Mendes, A. A., Santos, J. C., and Aguiar, C. L. (2004). Modificação de óleos e gorduras por biotransformação. *Quim. Nova.* **27,** 146–156.

Castro, H. F., Oliveira, P. C., and Pereira, E. B. (1997). Evaluation of different approaches for lipase catalyzed synthesis of citronellyl acetate. *Biotechnol Lett.* **19,** 229–232.

Chambers, W. H. (1992). Organophosphorous compounds: An overview. *In* "Organophosphates, Chemistry, Fate, and Effects" (J. E. Chambers and P. E. Levi, eds.), pp. 3–17. Academic Press, San Diego.

Chang, B. V., Yang, C. M., Cheng, C. H., and Yuan, S. Y. (2004). Biodegradation of phthalate esters by two bacteria strains. *Chemosphere* **55,** 533–538.

Claon, P. A., and Akoh, C. C. (1994). Effect of reaction parameters on SP435 lipase-catalyzed synthesis of citronellyl acetate in organic solvent. *Enzyme Microb. Technol.* **16,** 835–838.

Clauss, J. (1996). Interesterificação de Óleo de Palma. *Óleos & Grãos* **5,** 31–37.

Creveld, L. D., Meijberg, W., Berendsen, H. J. C., and Pepermans, H. A. M. (2001). SDS studies of *Fusarium solani pisi* cutinase: Consequences for stability in the presence of surfactants. *Biophys. Chem.* **92,** 61–75.

Croteau, R. (1980). "Fragrance and Flavor Substances." D&PS Verlag, Germany.

Dalla-Vecchia, R., Nascimento, M. G., and Soudi, V. (2004). Aplicações sintéticas de lipases imobilizadas em polímeros. *Quím. Nova.* **27,** 623–630.

Degani, O., Gepstein, S., and Dosoretz, C. G. (2002). Potential use of cutinase in enzymatic scouring cotton fiber cuticle. *Appl. Biochem. Biotechnol.* **102,** 277–289.

Doi, Y. (1990). "Microbial Polyesters." VCH Publishers, New York.

Egmond, M. R., and De Vlieg, J. (2000). *Fusarium solani pisi* Cutinase. *Biochem.* **82,** 1015–1021.
Espinoza, M. C. F., and Villeneuve, P. (2005). Phenolic Acids Enzymatic Lipophilization. *J. Agric. Food Chem.* **53,** 2779–2787.
European Patent Office. Assessed online as http://www.ep.espacenet.com/?locale=en_ep.
Faber, K. (2000). "Biotransformations in Organic Chemistry," 4th ed. Springer-Verlag, New York.
Ferreira, B. S., Calado, C. R. C., Keulen, F., Fonseca, L. P., Cabral, J. M. S., and Fonseca, M. M. R. (2004). Recombinant *Saccharomyces cerevisiae* strain triggers acetate production to fuel biosynthetic pathways. *J. Biotechnol.* **109,** 159–167.
Fett, W. F., Gerard, H. C., Moreau, R. A., Osman, S. F., and Jones, L. E. (1992). Screening of nonfilamentous bacteria for production of cutin-degrading enzymes. *Appl. Environ. Microbiol.* **58,** 2123–2130.
Filipsen, J. A. C., Appel, A. C. M., Van Der Hidjen, H. T. W. M., and Verrips, C. T. (1998). Mechanism of removal of immobilized triacylglycerol by lipolytic enzymes in a sequential laundry wash process. *Enzyme Microb. Technol.* **23,** 274–280.
Fischer-Colbrie, G., Heumann, S., Liebminger, S., Almansa, E., Cavaco-Paulo, A., and Gubitz, G. M. (2004). New enzymes with potential for pet surface modification. *Biocatal. Biotransform.* **22,** 341–346.
Galloway, T., and Handy, R. (2003). Immunotoxicity of organophosphorus pesticides. *Ecotoxicology* **12,** 345–363.
Gandhi, N. N. (1997). Applications of lipases. *J. Am. Oil Chem. Soc.* **74,** 621–633.
Garcia, S., Vidinha, P., Arvana, H., Silva, M. D. R. G., Ferreira, M. O., Cabral, J. M. S., Macedo, E. A., Harper, N., and Barreiros, S. J. (2005). Cutinase activity in supercritical and organic media: Water activity, solvatation and acid-base effects. *Supercrit. Fluids.* **35,** 62–69.
Gonçalves, L. A. G. (1996). *Óleos e Grãos* **5,** 27.
Gonçalves, A. M., Schacht, E., Matthjis, G., Aires-Barros, M. R., Cabral, J. M. S., and Gil, M. H. (1999). Stability studies of a recombinant cutinase immobilized to dextran and derivatized silica supports. *Enzyme Microb. Technol.* **24,** 60–66.
Gunstone, F. D. (1999). What else besides commodity oils and fats? *Fett-Lipid* **101,** 124–130.
Hammond, E. G., and Glatz, B. A. (1988). *In* "Food Biotechnology" (R. D. Kling and P. S. J. Cheetham, eds.), Vol. 2, pp. 173–217. Elsevier, Amsterdam.
Hills, G. (2003). Industrial use of lipases to produce fatty acid esters. *Eur. J. Lipid Sci. Technol.* **105,** 601–607.
Hunsen, M., Azim, A., Mang, H., Wallner, S. R., Ronkvist, A., Xie, W., and Gross, R. (2007). A cutinase with polyester synthesis activity. *Macromolecules* **40,** 148–150.
Indeerjeet, K., Mathur, R. P., Tandon, S. N., and Prem, D. (1997). Identification of metabolites of malathion in plants, water and soil by CG-MS. *Biomed. Chromatogr.* **11,** 352–355.
John, V. T., and Abraham, G. (1991). Lipase catalysis and its applications. *In* "Biocatalysis for Industry" (J. S. Dodrick, ed.), pp. 193–217. Plenum Press, New York.
Karra-Chaabouni, M., Pulvin, S., Touraud, D., and Thomas, D. (1996). Enzymatic synthesis of geraniol esters in a solvent-free system by lipases. *Biotechnol. Lett.* **18,** 1083–1088.
Kavlock, R., Boekelheide, K., Chapin, R., Cunningham, M., Faustman, E., Foster, P., Golub, M., and Henderson, R. (2002). NTP center for the evaluation of risks to human reproduction: phtalates expert panel report on the reproductive and developmental toxicity of di-it *n*-hexyl phthalate. *Reprod. Toxicol.* **16,** 709–719.
Kerry, N. L., and Abbey, M. (1997). Red wine and fractionated phenolic compounds prepared from red wine inhibit low density lipoprotein oxidation *in vitro*. *Atherosclerosis* **135,** 93–102.
Kim, Y. H., Ahn, J. Y., Moom, S. H., and Lee, J. (2005). Biodegradation and detoxification of organophosphate insecticide, malathion by *Fusarium oxysporum* f. sp. *pisi* cutinase. *Chemosphere.* **60,** 1349–1355.

Klibanov, A. M. (1989). Enzymatic catalysis in anhydrous organic solvents. *Trends Biochem. Sci.* **14,** 141–144.

Klibanov, A. M. (2001). Improving enzymes by using them in organic solvents. *Nature* **409,** 241–246.

Kolattukudy, P. E. (1984). Cutinases from fungi and pollen. *In* "Lipases" (B. Borgstrom and T. Brockman, eds.), pp. 471–504. Elsevier Publishing, Amsterdam.

Krishina, S. H. (2002). Developments and trends in enzyme catalysis in nonconventional media. *Biotechnol. Adv.* **20,** 239–267.

Langrand, G., Triantaphylides, C., and Baratti, J. (1989). Lipase catalyzed formation of flavour esters. *Biotechnol Lett.* **10,** 549–554.

Larsson, K. M., Adlercreutz, P., and Mattiasson, B. (1990). Enzymatic catalysis in microemulsions: Enzyme reuse and product recovery. *Biotechnol. Bioeng.* **36,** 135–141.

Leal, M. C. M. R., Cammarota, M. C., Freire, D. M. G., and Sant'Anna, G. L., Jr. (2002). Hydrolytic enzymes as coadjuvants in the anaerobic treatment of dairy wastewaters. *Braz. J. Chem. Eng.* **19,** 175–180.

Lie, E., and Molin, G. (1991). *In* "Bioconversion of Waste Materials to Industrial Products" (A. M. Martin, ed.), Elsevier Applied Science, New York.

Lima, J. R., and Nassu, R. T. (1996). Substitutos de Gorduras em Alimentos: Características e Aplicações. *Quim. Nova.* **19,** 127–134.

Longo, M. A., and Sanromán, M. A. (2006). Production of food aroma compounds: Microbial and enzymatic methodologies. *Food Technol. Biotechnol.* **44,** 335–353.

Macedo, G. A., and Pio, T. F. (2005). A rapid screening method for cutinase producing microorganisms. *Braz. J. Microbiol.* **36,** 388–394.

Mannesse, M. L. M., Cox, R. C., Koops, B. C., Verheij, H. M., Haas, G. H., Egmond, M., Van Der Hijden, H. T. W., and Vlieg, J. (1995). Cutinase from *Fusarium solani pisi* hydrolyzing triglyceride analogues. Effect of acyl chain length and position in the substrate molecule on activity and enantioselectivity. *Biochemistry* **34,** 6400–6407.

Masse, L., Kennedy, K. J., and Chou, S. (2001). Testing of alkaline and enzymatic pretreatment for fat particles in slaughterhouses wastewater. *Bioresour. Technol.* **77,** 145–155.

Masson, W., Loftsson, T., and Haraldsson, G. G. (2000). Marine lipids for products, soft compounds and other pharmaceutical applications. *Pharmacies.* **55,** 172–177.

Matamá, T., Silva, C., O'Neill, A., Casal, M., Soares, C., Gubitz, G.M., and Cavaco-Paulo, A. (2004). Improving synthetic fibers with enzymes. 3rd International Conference on Textile Biotechnology (abstract 5).

Mayer, J. M., and Kaplin, D. L. (1994). Biodegradable materials: Balancing degradability and performance. *Trends Polym. Sci.* **2,** 227–235.

Melo, E. P., Baptista, R. P., and Cabra, J. M. S. (2003a). Improving cutinase stability in aqueous solution and in reverse micelles by media engineering. *J. Mol. Catal. B: Enzymatic.* **22,** 299–306.

Melo, E. P., Costa, S. M. B., Cabral, J. M. S., Fojan, P., and Petersen, S. B. (2003b). Cutinase-AOT interactions in reverse micelles: The effect of 1-hexanol. *Chem. Phys. Lipids* **124,** 37–47.

Muderhwa, J., Pina, M., and Graille, J. (1988). Aptitude à la transesterification de quelques lipases regioselectives 1–3. *J. Oléagineux.* **43,** 385–392.

Mueller, R. J. (2006). Biological degradation of synthetic polyesters—Enzymes as potential catalysists for polyester recycling. *Process Biochem.* **41,** 2124–2128.

Mukherjee, K. D. (1990). Lipase-catalyzed reactions for modification of fats and other lipids. *Biocatalysis.* **3,** 277–293.

Murphy, C. A., Cameron, J. A., Huang, S. J., and Vinopal, R. T. (1996). *Fusarium* polycaprolactone depolimerase is cutinase. *Appl. Environ. Microbiol.* **62,** 456–460.

Pandey, A., Benjamin, S., Soccol, C. R., Nigam, P., Krieger, N., and Soccol, V. T. (1999). The realm of microbial lipases in biotechnology. *Biotechnol. Appl. Biochem.* **29,** 119–131.

Paques, F. W., and Macedo, G. A. (2006). Lipases de látex vegetais: propriedades e aplicações industriais. *Quím. Nova.* **29,** 93–99.

Petersen, S. B., Johnson, P. H., Fojan, P., Petersen, E. I., Petersen, M. T. N., Hansen, S., Ishak, R. J., and Hough, R. J. (1998). Protein engineering the surface of enzymes. *J. Biotechnol.* **66,** 11–26.

Pio, T. F., and Macedo, G. A. (2007). Optimizing the production of cutinase by *Fusarium oxysporum* using response surface methodology. *J. Ind. Microbiol. Biotechnol.* **10,** 101–111.

Regado, M. A., Cristóvão, B. M., Moutinho, C. G., Balcão, V. M., Aires-Barros, R., Ferreira, J. P. M., and Malcata, F. X. (2007). Flavour development *via* lipolysis of milk fat: Changes in free fatty acid pool. *Int. J. Food Sci. Technol.* **42,** 961–968.

Sebastião, M. J., Cabral, J. M. S., and Aires-Barros, M. R. (1993). Synthesis of fatty acid esters by a recombinant cutinase in reversed micelles. *Biotechnol. Bioeng.* **42,** 326–332.

Silva, F. A. M., Borges, F., Guimarães, C., Lima, J. L. F. C., Matos, C., and Reis, S. (2000). Phenolic acids and derivatives: studies on the relationship among structure, radical scavenging activity, and physicochemical parameters. *J. Agric. Food Chem.* **48,** 2122–2126.

Silva, C. M., Carneiro, F., O'Neill, A., Fonseca, L. P., Cabral, J. M. S., Guebitz, G., and Cavaco-Paulo, A. (2005). Cutinase—A new tool for biodegradation of synthetic fibers. *J. Polym. Sci.* **43,** 2448–2450.

Soares, C. M., Teixeira, V. H., and Baptista, A. M. (2003). Protein structure and dynamics in no aqueous solvents: Insights from molecular dynamics simulation studies. *Biophys. J.* **84,** 1628–1641.

Stamatis, H., Kolisis, F. N., and Xenakis, A. (1993). Enantiomeric selectivity of a lipase from *Penicillium simplicissimum* in the esterification of menthol in microemulsions. *Biotechnol. Lett.* **15,** 471–476.

Stamatis, H., Sereti, V., and Kolisis, F. M. (1999). Studies on the enzymatic synthesis of lipophilic derivatives of natural antioxidants. *J. Am. Oil Chem. Soc.* **12,** 1505–1510.

Sung, H. H., Kao, W. Y., and Su, Y. J. (2003). Effects and toxicity of phthalate esters to haemocytes of giant fresh water prawn, *Macrobacillum rosenbergii*. *Aquat. Toxicol.* **64,** 25–37.

Ternström, T., Svendsen, A., Akke, M., and Adlercreutz, P. (2005). Unfolding and inactivation of cutinases by AOT and guanidine hydrochloride. *Biochim. Biophys. Acta.* **1748,** 74–83.

Vandamme, E. J., and Soetaert, W. (2002). Bioflavours and fragrances *via* fermentation and biocatalysis. *J. Chem. Technol. Biotechnol.* **77,** 1323–1332.

Vertommen, M. A. M. E., Nierstrasz, V. A., Van Der Veer, M., and Warmoeskerken, M. M. C. G. (2005). Enzymatic surface modification of poly(ethylene terephtalate). *J. Biotechnol.* **120,** 376–386.

Villeneuve, P., Muderwha, J. M., Graille, J., and Hass, M. J. (2000). Customizing lipases for biocatalysis: A survey of chemical, physical and molecular biologic approach. *J. Mol. Catal. B: Enzym.* **4,** 113–148.

Walz, I., and Schwack, W. (2007). Cutinase inhibition by means of insecticidal organophosphates and Carbamates. *Eur. Food Res. Technol.* **225,** 593–601.

Walz, I., and Schwack, W. (2008). Cutinase inhibition by means of insecticidal organophosphates and Carbamates Part 2: Screening of representative insecticides on cutinase activity. *Eur. Food Res. Technol.* **226,** 1135–1143.

Welsh, W. W., Murray, W. D., and Williams, R. E. (1989). Microbiological and enzymatic production of flavor and fragance chemicals. *Crit. Rev. Biotechnol.* **9,** 105–169.

Willis, W. M., and Maragoni, A. G. (1999). Biotechnology & genetic engineering reviews. *Biotechnol. Genetic Eng. Rev.* **16,** 141–175.

Yoon, M., Kellis, J., and Poulouse, A. J. (2002). Enzymatic modification of polyester. *AATCC* **2,** 33–36.

CHAPTER 5

Microbial Deterioration of Stone Monuments—An Updated Overview

Stefanie Scheerer,* **Otto Ortega-Morales,**[†] and **Christine Gaylarde**[†]

Contents

* Cardiff School of Biosciences, Cardiff University, Cardiff CF10 3TL, United Kingdom
[†] Departamento de Microbiología Ambiental y Biotecnología, Universidad Autónoma de Campeche, Campeche, Campeche, México

Advances in Applied Microbiology, Volume 66
ISSN 0065-2164, DOI: 10.1016/S0065-2164(08)00805-8

Abstract

Cultural heritage monuments may be discolored and degraded by growth and activity of living organisms. Microorganisms form biofilms on surfaces of stone, with resulting aesthetic and structural damage. The organisms involved are bacteria (including actinomycetes and cyanobacteria), fungi, archaea, algae, and lichens. Interactions between these organisms and stone can enhance or retard the overall rate of degradation. Microorganisms within the stone structure (endoliths) also cause damage. They grow in cracks and pores and may bore into rocks. True endoliths, present within the rock, have been detected in calcareous and some siliceous stone monuments and are predominantly bacterial. The taxonomic groups differ from those found epilithically at the same sites. The nature of the stone substrate and the environmental conditions influence the extent of biofilm colonization and the biodeterioration processes. A critical review of work on microbial biofilms on buildings of historic interest, including recent innovations resulting from molecular biology, is presented and microbial activities causing degradation are discussed.

I. INTRODUCTION

A large percentage of the world's tangible cultural heritage is made from stone, and it is slowly but irreversibly disappearing. It has been calculated that, for limestone, an average of 1.5–3 mm of rock will erode away in 100 years in temperate climates, leading to the the disappearance of inscriptions on tombstones in the United Kingdom within 300 years (D. Allsopp, personal communication). The transformation of stone into sand and soil is a natural recycling process, essential to sustain life on earth. However, the deterioration of stone monuments represents a permanent loss of our cultural heritage.

Many different types of stone have been used by artists over the years. The most common are marble and limestone, of the calcareous type, sandstone (which is mainly quartz, feldspar, and iron oxide) and granite (mainly quartz and feldspar), of siliceous type. These differ in hardness, porosity, and alkalinity, properties that affect their susceptibility to biodeterioration.

These stone types are not discrete; there is an overlap between calcareous and siliceous rocks, with types such as calcareous sandstone, or siliceous limestone, existing. In addition, the materials often used to stabilize the building blocks (mortar) and to coat the surface prior to painting (plaster or stucco) could be considered. These are human-made and very variable in composition, sometimes even containing high levels of organic materials; they are generally extremely susceptible to biodeterioration, as is the modern stone substitute, concrete. They will not be included in this review.

Damage to stone caused by microorganisms is often referred to as bioweathering but better called biodeterioration (Gorbushina and Krumbein, 2004); it is the least understood of degradation mechanisms. It was reviewed most recently by Warscheid and Braams (2000) and Gorbushina and Krumbein (2004) in general overviews on biodeterioration of stone, Kumar and Kumar (1999), who reported on biodeterioration of stone in the tropics, and Urzi (2004), who examined these processes in the Mediterranean. Only within the last two decades has it received serious attention from conservators and conservation scientists (Price, 1996; Schnabel, 1991). A thorough understanding of the factors and mechanisms involved in microbial biodeterioration is essential to develop appropriate methods for its control.

II. MICROBIAL ECOLOGY OF OUTDOOR STONE SURFACES

The microflora of external stone surfaces represents a complex ecosystem, which includes not only algae, bacteria, fungi, and lichens, but also protozoa; in addition, small animals, such as mites, may be present and lower and higher plants may develop, once the earlier colonizers have conditioned the surface.

Stone inhabiting microorganisms may grow on the surface (epilithic), in more protected habitats such as crevices and fissures (chasmolithic), or may penetrate some millimetres or even centimetres into the rock pore system (endolithic) (Garcia-Vallès *et al.*, 1997; Golubic *et al.*, 1975; Saiz-Jimenez *et al.*, 1990; Tiano, 2002; Wolf and Krumbein, 1996). They can be found in environments as far apart as the Antarctic (Hirsch *et al.*, 2004) and the (sub)tropics (Althukair and Golubic, 1991; Chacón *et al.*, 2006; Golubic *et al.*, 2005). Endoliths have been classified in more detail by Golubic *et al.* (1981), according to their presence in cracks (chasmoendoliths), pores (cryptoendoliths), or as euendoliths if they show a true boring ability in the stone matrix. The endolithic communities of limestone monuments have been shown, both by culture (Ortega-Morales *et al.*, 2005) and by molecular biology (McNamara *et al.*, 2006) techniques, to be different from those on the surface. These differences may be explained, at least in part, by the protective role of epilithic growth, inorganic matter, and superficial stone layers protecting against incident UV radiation (Cockell *et al.*, 2002), and varying availability of nutrients. It is likely that these microbial communities are also different at the functional level, since increased exposure to UV radiation induces the synthesis of protective pigments (Ehling-Schultz *et al.*, 1997). Warscheid *et al.* (1996) considered that, whereas microorganisms in moderate climates tend to colonize the surface of stones, their tropical and subtropical counterparts prefer to penetrate deeper into the rock profile in order to protect themselves from sunlight and desiccation. Matthes-Sears *et al.* (1997), however, suggested that organisms are driven to become endolithic not for protection, but in the search for increased nutrients and space (lack of competition). As endoliths in natural carbonate rocks are rarely associated with catastrophic failure, their slow growth rates within the rock would lead them to have a relatively stable life for considerable periods (Hoppert *et al.*, 2004).

Microbial colonization generally initiates with a wide variety of phototrophic microorganisms (mainly cyanobacteria and algae). These accumulate biomass, usually embedded in a biofilm enriched with organic and inorganic substances and growth factors (Tiano, 2002; Tomaselli *et al.*, 2000b). Lichens probably follow these on the stone surface (Hoppert *et al.*, 2004). The accumulation of photosynthetic biomass provides an excellent organic nutrient base for the subsequent heterotrophic microflora. Ortega-Morales *et al.* (1999) showed that the C/N ratio of biofilm samples taken under different microclimatic conditions approximated to that of microbial cells (~4), indicating that the main source of organic matter is the biofilm itself. However, the establishment of heterotrophic communities on rocks is possible even without the pioneering participation of phototropic organisms and may in fact facilitate the subsequent growth of photosynthetic populations (Roeselers *et al.*, 2007). In this case, organic substrates from various sources are used, including airborne particles and organic vapors, organic matter naturally present in sedimentary rock (usually between 0.2% and 2%), excreted organic metabolic products and biomass from other organisms, together with synthetic or natural organic substances from previous restoration treatments (Gorbushina *et al.*, 1996; Warscheid and Braams, 2000). Highly degraded stone surfaces, with subsequent alteration of the physical condition of the rock, provide appropriate conditions (a "proto-soil") for the germination of reproductive structures from higher organisms such as cryptogams (mosses and ferns) and higher plants (Tiano, 2002).

Restoration treatments can, indeed, increase microbial colonization when carried out by workers with no microbiological knowledge. Caneva and Nugari (2005) showed that a consolidant made from local plant mucilaginous (carbohydrate-like) extracts (Escobilla), used at the Mayan site of Joya de Ceren, El Salvador, supported the growth of fungi and, particularly, actinomycetes; its use should be critically evaluated. Other treatments, such as simple cleaning with water, have also been shown to exacerbate microbial growth (Young, 1997).

Biodeterioration processes are rarely caused by one distinct group of microorganisms, but are rather an interaction of coexisting groups.

Table 2 shows a list of those that have been detected on stone monuments; the functional microbial groups are discussed later in more detail.

A. Molecular biology in the study of epi- and endo-lithic microorganisms

Our knowledge on the extent of the diversity of the microbial microflora is far from complete, since traditional culture techniques isolate less than 1% of the microbial community (Ward *et al.*, 1990). In recent years, molecular methods have been developed that allow the identification and, to some extent, enumeration of microorganisms in environmental samples (Amann *et al.*, 1995). Techniques such as denaturing gradient gel electrophoresis (DGGE), single strand conformational polymorphism (SSCP), and fluorescent *in situ* hybridization (FISH), point to the possibility that halophilic or alkanophilic eubacteria and archaea are also involved in stone decay (McNamara *et al.*, 2003; Ortega-Morales *et al.*, 2004; Roelleke *et al.*, 1998; Saiz-Jimenez and Laiz, 2000). These extremophiles had not previously been isolated from stone monuments and thus never considered to play a role in their biodeterioration.

Further approaches employing molecular identification techniques have resulted in the identification of previously unknown species of bacteria, including some actinobacteria, and of organisms such as the Acidobacteria, a practically unknown division of bacteria that is widely distributed in a large variety of ecosystems (Heyrman, 2003; McNamara *et al.*, 2006; Saarela *et al.*, 2004; Salazar *et al.*, 2006; Zimmermann *et al.*, 2005). In addition to the Acidobacter group, other rare microorganisms have been detected on historic buildings. Ortega-Morales *et al.* (2004), using SSCP, showed that a pink-stained area of an external wall at the Mayan site of Uxmal contained predominantly bacteria related to the Actinobacteria genus, *Rubrobacter*. These authors also showed, for the first time, putative halophiles of the genera *Halothece* and *Salinibacter*, along with photosynthetic bacteria related to the *Ectothiorhodospiraceae*. The occurrence of this latter group expands our knowledge of the microorganisms that may contribute through their autotrophic metabolism to the fixation of carbon in these terrestrial ecosystems.

Most of the bacteria identified by molecular biology have not been cultured and their role in the ecology of stone surfaces is not understood (Schabereiter-Gurtner *et al.*, 2003). Even less is known about the role of archaea in biodeterioration and conservation, but recent research sheds light on this microbial group. The archaeal species, *Halobacillus trueperi* has been shown to participate in the mineralization of carbonates *in vitro* (Rivadeneyra *et al.*, 2004), and this may be the first indication of the importance of previously uncultured microorganisms in stone deterioration.

There are many practical problems with community analysis using molecular biology methods involving DNA. These include, for example, selective extraction of DNA from different microorganisms, selective amplification in the PCR, lack of amplification of low levels of DNA in a mixture, and interference in the reaction by environmental materials such as polysaccharides or stone constituents. Nevertheless, it has been clearly demonstrated that organisms found by culture and those detected by sequencing methods are not the same. Rölleke *et al.* (1996) identified relatives of the genera *Halomonas*, *Clostridium*, and *Frankia* in an ancient mural painting; these were not detected by culture, which showed the presence of bacteria such as *Bacillus*, *Micrococcus*, and *Arthrobacter*, not detected by the molecular techniques. New strains of the actinomycete genus *Arthrobacter* were detected by a polyphasic study, including molecular analyses, in the internal biofilms on Servilia's tomb, Carmona, Spain, and St. Catherine's chapel, in the Castle of Herberstein, Austria (Heyrman *et al.*, 2005).

Laiz *et al.* (2003) showed that the majority of bacteria detected by culture from artificially inoculated building materials were spore-formers, while a much greater diversity was apparent using the culture-independent technique of DGGE and sequencing. McNamara *et al.* (2006) found a very wide range of bacteria in and on limestone from the Mayan archaeological site of Ek' Balam, Mexico, using total DNA extraction from samples, PCR with 16S rDNA primers and cloning. Although they did not attempt to culture the organisms, comparison with other publications on similar sites indicate that many more, and different, bacterial groups were detected by this method. Using a combined approach of phospholipid fatty acid markers and SSCP profiling, Ortega-Morales *et al.* (2004) determined that the main colonizers in most biofilms at another Mexican Mayan site, Uxmal, were cyanobacteria of the *Pleurocapsales* group, although *Bacillus carboniphilus* was particularly abundant in internal sites (more dense biofilms) and *Rubrobacter*-related bacteria on external surfaces (higher UV radiation). The dangers of relying on only DNA analysis for evaluation of the cyanobacterial microflora were pointed out by Gaylarde *et al.* (2004, 2005) and Chacón *et al.* (2006). Without any doubt, polyphasic detection methods are essential to determine the true nature of epilithic and endolithic communities.

B. Effect of climate and substrate on microflora

Apart from the microorganisms present in the immediate environment, many factors influence the deterioration of stone. Physical, chemical, and biological agents act in associations ranging from synergistic to antagonistic. The physical properties of the stone influence the extent of degradation. For microbial growth, for example, rough surfaces and high porosity favor adhesion and colonization (Caneva *et al.*, 1991; May *et al.*, 2003; Warscheid and Braams, 2000). Environmental pollution, which has increased rapidly within the last century (Wright, 2002), may influence stone degradation directly (e.g., acid rain) or indirectly, by supplying nutrients for microbial growth. It has been shown to enhance detrimental microbial activity on the stone substrate (Herrera and Videla, 2004; Mitchell and Gu, 2000; Sand *et al.*, 2002). Monuments that have survived thousands of years as relicts of extinct cultures have experienced accelerated aging in recent years (Gaylarde and Morton, 2002).

The total properties of a substrate that determine its ability to be colonized by microorganisms have been termed its bioreceptivity (Guillitte, 1995). Although this concept is more used in the engineering field, it could be of interest for heritage conservation, to allow an understanding and assessment of materials to be used in restoration. Prieto and Silva (2005) published a set of simple and well-established methods for assessing bioreceptivity; abrasion pH, bulk density, open porosity, and capillary water. The group is now working on a quantitative method to compare visual observations of surface biogenic color changes on stone (Prieto *et al.*, 2006).

Qualitative and quantitative attributes of the colonizing microflora are strongly influenced by the properties of the stone substrate (Warscheid *et al.*, 1996), and it is well known that different kinds of lichens prefer either calcareous or siliceous rocks (Allsopp *et al.*, 2003). Some empirical evidence for the effect of substrate on microbial colonization comes from studies on natural biofilms on various types of building surfaces (Gaylarde and Gaylarde, 2005). It has also been shown that pollutants deposited on the stone surface from the atmosphere can affect microbial colonization and degradation (Zanardini *et al.*, 2000).

The influence of the chemical composition of stone on general microbial colonization remains unclear; however, the condition of the stone has been said to influence the microflora. May *et al.* (2000) reported that filamentous bacteria were almost never isolated from sound stone in temperate climates, whereas actinomycetes of the genera *Streptomyces*, *Micromonospora*, and *Microphylospora* were the dominant organisms on decayed stone. There is no empirical information on whether these organisms are the cause or the result of the damage, although Mansch and Bock (1998) suggested, with indirect evidence, that colonization of sandstone by nitrifying bacteria is accelerated by chemical weathering.

Published data on the distribution of different taxa of photosynthetic microorganisms do not indicate a clear relationship between the organisms present and stone composition (Tomaselli *et al.*, 2000b) and the major influence is considered to be climate, rather than substrate (Gaylarde and Gaylarde, 2005; Tiano *et al.*, 1995). Tropical and subtropical climates enhance the destructive activity of microorganisms, while in moderate climates air pollution significantly supports microbial biodeterioration.

According to Warscheid (2003), moderate climates with regular rainfall tend to give rise to a mixed consortium of microorganisms on exposed stone surfaces, whereas semiarid climates, with less rain and higher temperatures, support the growth of more specialized microorganisms such as cyanobacteria, black yeasts, and lichens, which tend to dominate the microflora. In arid zones, the detrimental influence of microorganisms is low. Instead a "rock-varnish" is formed, mainly by cyanobacteria and mineral-oxidizing fungi (Gorbushina and Krumbein, 2000; Krumbein and Giele, 1979; Krumbein and Jens, 1981). The highest degree of biodeterioration occurs in the tropics, because of high humidity and temperatures. The stone microflora here is considered to be very aggressive, with a high capacity for "biocorrosion" (more properly called "bioerosion") and biofouling (Warscheid, 2003). These two terms are defined by Warscheid (2003) as: (1) microbially induced or influenced corrosion of materials, altering the structure and stability of the substrate, and (2) the presence of colloidal microbial biofilms on or inside materials, leading to visual impairment and potentially altering the physiochemical characteristics of the substrate. The production of pigments, thick walls, or capsules protects microorganisms from adverse climates; however, their aesthetic damage is severe. Deeply-colored coccoid and filamentous cyanobacteria, which predominate in biofilms on buildings in the hot and humid climates of Latin America (Gaylarde and Gaylarde, 2005), are more frequently present on surfaces of buildings at high altitude in the tropics and subtropics than at lower altitudes (Gaylarde and Englert, 2006; Gaylarde and Gaylarde, 2005; Gaylarde *et al.*, 2004; Fig. 5.1).

Particularly sheltered areas on historic buildings in the United Kingdom have been shown to give rise to rich and homogenous biofilms consisting

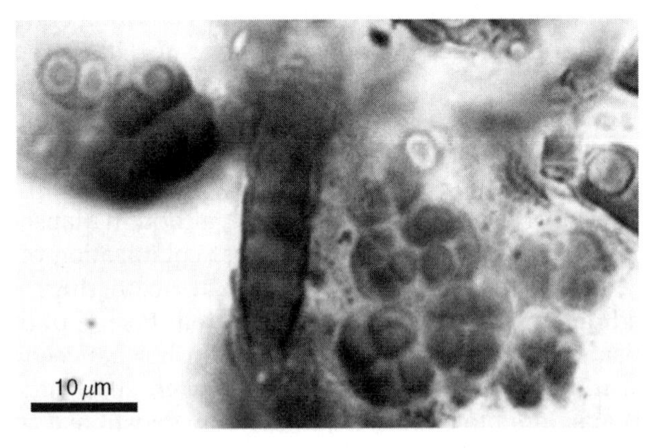

FIGURE 5.1 Mixed coccoid and filamentous cyanobacteria on the external surface of a church in Minas Gerais, Brazil, showing intense pigmentation. This photo is from a rehydrated biofilm. (See Color Plate Section in the back of the book.)

mainly of bacterial rods (May *et al.*, 2003). Biofilms exposed to salt from marine aerosols were of heterogeneous structure with coagulated cells entangling stone particles; whether salting or microbial activity was the main cause of decay is not clear. Exposure to high levels of solar radiation in these temperate climates, with subsequent drying of the substrate, leads to preferential growth of spore forming bacteria, such as *Bacillus* and heat tolerant actinomycetes, over gram-negative bacteria (May *et al.*, 2000). Actinomycetes are frequently found in the more temperate climates of Europe (Palla *et al.*, 2002; Warscheid *et al.*, 1995). Gaylarde and Gaylarde (2005) suggested that they are more common on external surfaces in these milder conditions, whereas in the hot and humid tropics and semitropics, they seem to prefer the interiors of buildings, or to grow as endoliths. However, actinomycetes have also been reported on surfaces in areas of hot climate (Hyvert, 1966; Ortega-Morales *et al.*, 2004), the genus *Geodermatophilus* apparently being common in calcareous stone (Eppard *et al.*, 1996).

Tayler and May (1991) reported seasonal changes in the microbial community of sandstone from ancient monuments in the United Kingdom, with higher bacterial numbers in winter and early spring than in summer and early autumn. Seasonal climate changes tend to result in higher numbers and greater diversity of gram-positive bacteria during summer months in temperate climates. In the warmer Mediterranean climate of Crete, no seasonal changes were observed in heterotrophic bacteria. In this geographical area, the location of the exposed surface (sheltered or not) seemed to play a more significant role than climate change (May *et al.*, 2000). Tomaselli *et al.* (2000a) also reported that there were few seasonal changes in the composition of photosynthetic populations on marble statues from different locations in Italy. Quantitative, rather than qualitative, differences were found, higher numbers of photosynthetic microorganisms being detected during summer months.

Wollenzien *et al.* (1995) showed qualitative differences in fungi occurring on calcareous stone in the Mediterranean. During periods of higher humidity and less sunshine, rapidly growing mycelial fungi, particularly of the genera *Alternaria*, *Aspergillus*, *Cladosporium*, *Phoma*, and *Ulocladium*, were dominant, but they were rarely found during the dry season. Nitrifying bacteria, which are known to be highly dependent on the water regime, have been found to be more abundant in indoor environments and during the rainy season at the archaeological site of Uxmal (Ortega-Morales, 1999).

III. MECHANISMS OF MICROBIAL BIODETERIORATION

The detrimental effects of microorganisms may be aesthetic, biogeochemical, and/or biogeophysical. Microbial cells may contribute directly to the deterioration of stone by using it as a substrate or indirectly by imposing physical stress, serving as nutrients for other organisms, or providing compounds for secondary chemical reactions (Sand, 1996). May (2003) stated that the intimate association of microorganisms with the mineral substrate may reach more than 3 cm deep into the stone, while Wolf and Krumbein (1996) reported microbial contamination in highly degraded, fine grained marble to a depth of 20 cm. These may not have been active boring microorganisms, but such organisms do exist, although their mechanisms of penetration are not fully understood (Salvadori, 2000).

A. Biofilms

Surface biofilms are microbial cells embedded in extracellular polymeric substances (EPS). The simple presence of a biofilm has aesthetic, chemical, and physical effects on the stone. EPS, produced by the cells to allow their adhesion to a given surface, facilitate entrapment of airborne particles, aerosols, minerals, and organic compounds, increasing the dirty appearance of the substrate (Kemmling *et al.*, 2003). Biofilms are areas of high metabolic activity, where digestive enzymes excreted by microorganisms are concentrated. Kemmling *et al.* (2003) found that the EPS in a biofilm from the Market Gate of Miletus (Pergamon Museum) protected cell enzymes against repeated desiccation and rehydration cycles, thus offering the organisms within the biofilm a distinct advantage over nonembedded cells on external surfaces.

Microbial EPS are polymers containing predominantly a range of mainly anionic sugar molecules (but also pigments, proteins, nucleic acids, and lipids) exhibiting several types of functional groups, some of which are capable of binding cations in solution (Moran and Ljungh, 2003). Calcium can be leached from limestone surfaces, or chelated, once solubilized from the matrix, by hexuronic acids, carbonyl, and hydroxyl groups (Ortega-Morales *et al.*, 2001; Perry *et al.*, 2004). Ortega-Morales *et al.* (2001) found higher levels of hexuronic acids in EPS directly extracted from degraded limestone surfaces at Uxmal than on sound stone blocks that were heavily colonized by cyanobacterial biofilms. These molecules may have mediated the deposition of carbonate minerals around coccoid cells, previously demonstrated by SEM (Ortega-Morales *et al.*, 2000).

EPS also act as a physical barrier that protects microorganisms from detrimental substances, such as biocides, and prevents the penetration of conservation materials. The formation of biofilms intensifies microbial attack by weakening the mineral lattice through repeated wetting and drying cycles and subsequent expansion and contraction (Warscheid *et al.*, 1996). They may change the pore size, dry density, water content, surface hardness, and weight of the stone and act as a permeability block for the evaporation of humidity within the stone (May, 2003; Papida *et al.*, 2000).

Biofilms have a lower thermal conductivity than stone, which may lead to uneven heat transfer within the artifact (Dornieden *et al.*, 1997, 2000; Warscheid and Braams, 2000). However, EPS from biofilms have also been reported to have a certain protective nature, owing to a consolidation effect (Kurtz, 2002). Microbial polysaccharides and other naturally occurring biopolymers of various chemical compositions have been shown to inhibit dissolution under certain conditions (Papida *et al.*, 2000; Welch and Vandevivere, 1994).

B. Discoloration

Discoloration is mainly an aesthetic problem (Fig. 5.2). It may be caused by pigments released from, or contained within, the microorganisms (melanins, carotenes, and photosynthetic pigments). Figure 5.3 shows carotene-packed cells of the alga *Trentepohlia umbrina* on a pink-stained limestone surface at the Mayan site of Edzna, Mexico. Crushed samples of stone showed that the calcite crystals had taken up the orange stain (Gaylarde *et al.*, 2006).

Sulfur cycle bacteria can convert limestone into gypsum, common especially in sulfur-polluted environments. This can lead to the formation of dark surface colorations, even when the normally responsible fungi and cyanobacteria are not present. Dark discolorations may also be due to airborne particles trapped in EPS.

FIGURE 5.2 Black biofilms on the surface of the Ambar Fort in Jaipur, India.

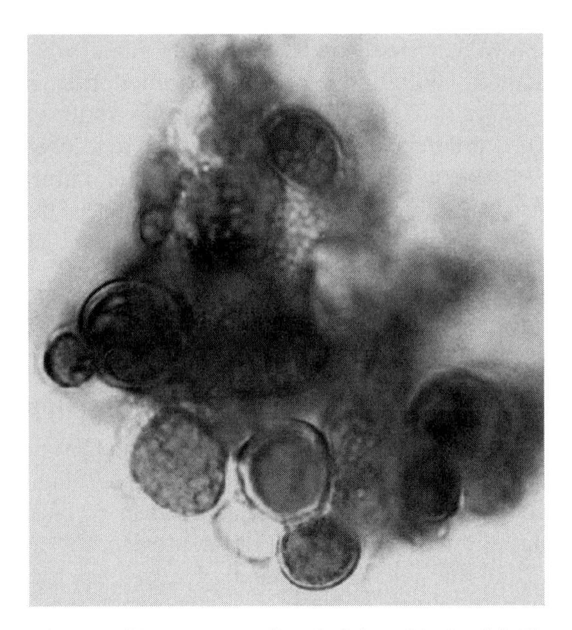

FIGURE 5.3 A pink stained limestone surface in Edzna, Mexico (a). The microscopic image shows bright orange (carotene-packed) cells of *T. umbrina*. (See Color Plate Section in the back of the book.)

Discoloration is not, however, purely aesthetic. Discolored areas may absorb more sunlight, which increases physical stress by expansion and contraction caused by temperature changes (Sand *et al.*, 2002; Warscheid, 2000). Temperatures on darkly stained areas of stone have been shown by Garty (1990) to differ by as much as 8 °C from lighter-colored areas. This effect has been shown experimentally by Carter and Viles (2004), using limestone blocks with and without a lichen covering. Surface temperature change and thermal gradients were greater below the lichen. The darkening of the stone surface decreases its albedo, so that it experiences increased heating/cooling and wetting/drying cycles, causing stresses within the stone (Warke *et al.*, 1996).

A special example of this is the so-called "black crust." Thick (several mm) crusts, such as those found on buildings in Aberdeen, Scotland (Urquart *et al.*, 1996), or on Seville cathedral, Spain (Saiz-Jimenez, 1995), are rich in Ca, Si, and C (Wright, 2002). The carbon component may be composed of hydrocarbons, deposited from vehicle exhausts, and microorganisms within the biofilm may be able to utilize these molecules as nutrients (Saiz-Jimenez, 1995). On the other hand, the more commonly seen thin black crusts (Fig. 5.4) seem to be predominantly cyanobacterial (subtropics/tropics) or fungal (moderate climate) in composition (Gaylarde and Englert, 2006; Gaylarde *et al.*, 2007; Pattanaik and Adhikary, 2002), and are enriched in Si and Fe (Gaylarde and Englert,

FIGURE 5.4 A thin black crust on Campeche cathedral, Mexico. Smaller photo shows close-up. Dark brown branching filamentous cyanobacteria were the main component (authors' unpublished observations).

2006; Wright, 2002). Apart from their aesthetic effects, these crusts also block pores within the stone. This can result in water retention and subsequent spalling of the surface, although it is also possible that they protect the stone by reducing water infiltration (Garcia-Vallès *et al.*, 1997).

C. Salting

Salting, the production of efflorescences, involves secondary minerals produced through reaction of anions from excreted acids with cations from the stone (Fig. 5.5); this mechanism is related to that discussed above. The damage caused by such salts is mainly of a physical nature, leading to blistering, flaking, scaling, and granular disintegration, and this may often be the main mechanism of stone decay (Wright, 2002). Hydration and subsequent swelling of a salt molecule within a small stone pore may cause cracking. During desiccation, the salts crystallize, leading to an increase in volume (Ortega-Morales *et al.*, 2005). At low temperatures, hydration of salts increases the water content and may lead to mechanical damage through ice crystal formation. Salts of biotic or abiotic origin may also increase the production of EPS and biofilm density

FIGURE 5.5 Salting on the internal walls of the tomb of Servilia, Carmona, Spain.

(May, 2003; Papida *et al.*, 2000; Sand, 1996; Sand *et al.*, 2002). Areas of efflorescence present a niche for halophilic/tolerant microorganisms, for example, several Archaea. These specialized microorganisms may have synergistic action with the salts and therefore severely enhance the physical and chemical deterioration processes (May, 2003; Papida *et al.*, 2000).

D. Physical damage

Physical damage may be caused by penetration of filamentous microorganisms (particularly fungal hyphae) into the stone (Hirsch *et al.*, 1995a). Many cyanobacteria, not necessarily filamentous, have also been shown to have this ability. Weakened areas of the stone will be affected first. Danin and Caneva (1990) proposed that calcareous stone is decayed by cyanobacteria by attachment of cyanobacterial cells in small fissures and growth within these fissures. This is followed by water uptake and expansion of cell mass, exerting pressure within the structure, precipitation of carbonates and oxalates around the cells, opening of the fissure with subsequent entry of dust, pollen, grains, etc. and death of some cyanobacterial cells, allowing the establishment of heterotrophic bacteria, fungi, and small animals such as mites. The final increased internal pressure on the superficial layer of the structure leads to its detachment (spalling).

E. Inorganic acids

Inorganic acids, mainly nitric acid (HNO_3) and sulfuric acid (H_2SO_4), but also carbonic acid (H_2CO_3), sulfurous acid (H_2SO_3), and nitrous acid (HNO_2) may dissolve acid-susceptible materials, with the production of substances more soluble in water (calcium sulfate, nitrates, and calcium hydrogen carbonate). This leads to weakening of the stone matrix. This dissolution action proceeds through the depth of the stone (Sand, 1996; Sand *et al.*, 2002). Sulfuric and sulfurous acids are predominantly produced by *Thiobacillus*, as well as by *Thiothrix*, *Beggiatoa*, and some fungi. The involvement of sulfur-oxidizing bacteria in the degradation of sandstone was first proposed in 1904, but later workers, in Germany, did not find these bacteria on natural stone buildings (Bock and Krumbein, 1989), leading to doubts about their role in biodegradation.

Nitric and nitrous acids are produced by ammonia and nitrite oxidizers, heterotrophic nitrifiers, and some fungi. Endolithic nitrifying bacteria were the first microorganisms to be proposed as the cause of stone decay, in 1890 (Muntz, 1890). However, since most natural stone is alkaline in reaction, nitrification will result in the production of nitrate and not nitric acid; thus acid attack is unlikely (Gaylarde and Gaylarde, 2004) and, indeed, Mansch and Bock (1998) suggest that acidic breakdown of stone is required prior to colonization by these bacteria.

Carbonic acid is produced by all living organisms as an end product of energy metabolism after the reaction of CO_2 with water (Sand, 1996). It is a weak acid, however, and unlikely to contribute greatly to stone degradation, especially in calcareous stone, where it will react to form calcium bicarbonate, a weak alkali buffer.

F. Organic acids

Organic acids (e.g., oxalic, citric, acidic, gluconic, malic, succinic acid, but also amino acids, nucleic acids, uronic acids, etc.) may react with the stone, solubilizing it via salt formation and complexation (Sand *et al.*, 2002; Torre de la *et al.*, 1993). Complexation, or chelation, is not, of course, "acid degradation" and pH values may not be reduced. Almost all microorganisms can excrete organic acids, especially when growth is unbalanced (Sand, 1996); perhaps the most frequently mentioned is oxalic acid, produced by fungi and lichens, mainly as the monohydrate form (whewellite), but also the dihydrate (weddelite). Polyfunctional organic acids, such as oxalic, have been shown to enhance the dissolution of siliceous rocks (Bennett *et al.*, 1988), but they may have a protective role in calcareous rock through the formation of calcium oxalate (Di Bonaventura *et al.*, 1999) or malonate (Salinas-Nolasco *et al.*, 2004) films. The brown/yellow oxalate films on stone surfaces are known as "time patinas" and are regarded by many as an attractive feature, enhancing the aged appearance of a monument. They may, or may not, be biogenically produced. However, polyfunctional organic acids such as oxalic have been shown to enhance the dissolution of siliceous rocks (Bennett *et al.*, 1988). The "biomineralization" caused by this metabolic activity has been documented using transmission and scanning electron microscopy in association with electron dispersion spectroscopy (Ascaso *et al.*, 2002; De los Rios and Ascaso, 2005; De los Rios *et al.*, 2004).

G. Osmolytes

Osmolytes are a diverse group of polyol substances (which includes glycerol, other sugars and polysaccharides) produced in response to changes in water activity; they are protectants against freezing, excessive heat and drying, salts, acids, alkalis, and factors such as ethanol. They have been reported in all life forms except protozoa, myxomycetes, and some simple animals. Under alkaline conditions (the majority of stone types), polyols degrade siliceous rock by binding to the crystalline layers, causing expansion, and by enhancing the solubility of organosilicon compounds (Gaylarde and Gaylarde, 2004).

It must be pointed out that much of the above is speculative. There is only little experimental evidence, for example, for the production of acids *in situ* and acidic polysaccharides on the stone surface; most investigators that demonstrated the ability of stone-colonizers to produce acids have tested this in artificial media in the laboratory.

IV. MICROORGANISMS DETECTED ON HISTORIC MONUMENTS

A list of microorganisms present on and in the stone of historic monuments, found in the literature, is given in Table 5.1.

A. Phototrophic microorganisms

Cyanobacteria and algae, as phototrophs, do not require organic material for their growth. They can form biofilms and crusts on stone surfaces, which, depending on the environmental conditions and the predominant strains, can be black, grey, brown, green, or red. Under wet conditions, such biofilms tend to be green, while when dry they are grey or black (Ortega-Morales *et al.*, 2004). This does not mean that the organisms within dry biofilms are dead; indeed, cyanobacterial biofilms have been stored for years in dry and dark conditions and remained viable (our unpublished observations and Gaylarde *et al.*, 2006). It has been shown that certain cyanobacteria, such as *Chroococcidiopsis*, the most dessication-resistant cyanobacterium known (Potts and Friedmann, 1981), regain photosynthetic activity within minutes when rehydrated (Hawes *et al.*, 1992).

Green algae are found mainly in damper areas. Their contribution to biodegradation has not been researched thoroughly and is considered to

TABLE 5.1 Microorganisms detected on stone monuments

Microbial group	Family/genus/species	References
Algae	*Apatococcus, Asterococcus, Cladophora, Chlorella, Chlorococcum, Coccomyxa, Chrysocapsa, Cyanidium, Dimorphococcus, Eustigmatos, Fragilaria, Gongrosira, Heterococcus, Hormidium, Klebsormidium, Muriella, Nanochlorum, Navicula, Nitzschia, Planktospheria, Pleurococcus, Protococcus, Protoderma, Rhizothallus, Stichococcus, Trentepohlia, Ulothrix.*	Crispim *et al.* (2003), Flores *et al.* (1997), Gaylarde *et al.* (2001), Ohba and Tsujimoto (1996), Ortega-Morales *et al.* (2000, 2005), Strzelczyk (1981), Tiano (2002), Tiano *et al.* (1995), Tomaselli *et al.* (2000a,b).
	Apatococcus lobatus, Botryochloris minima, Chlorella vulgaris, C. elipsoidea, Monodus unipapilla, Oocystis parva, O. marssoni, Protococcus viridis, Stichococcus baciliaris, T. umbrina, Ulothrix punctata.	Canela *et al.* (2005), Gaylarde *et al.* (2006), Hoppert *et al.* (2004), Ohba and Tsujimoto (1996), Strzelczyk (1981), Tomaselli *et al.* (2000a).
Cyanobacteria	*Arthrospira, Calothrix, Chlorogloeopsis,* Chroococcales, *Chroococcidiopsis, Chroococcus, Fischerella, Geitlerinema, Gloeocapsa, Gloethece, Hyella, Leptolyngbya, Lyngbya, Mastigocladopsis, Microcoleus, Myxosarcina, Nodularia, Nostoc,* Oscillatoriales, *Phormidium, Plectonema, Pleurocapsa,* Pleurocapsa-group, *Scytonema, Stanieria,* Stigonematales, *Synechococcus, Synechocystis, Tolypothrix, Xenococcus.*	Ascaso *et al.* (2002), Crispim *et al.* (2003), Garcia de Miguel *et al.* (1995), Gaylarde and Morton (2002), Gaylarde *et al.* (2001, 2005), Hoppert *et al.* (2004), McNamara *et al.* (2006), Ortega-Morales *et al.* (2000, 2005), Tiano (2002), Tomaselli *et al.* (2000a,b).
	Acaryochloris marina, Anabaena variabilis, Gloeocapsa helvetica, G. kuetzingiana, G. rupestris, Lyngbya matensiana, L. aerugineocoerulea, Oscillatoria pseudogeminata, O. terebriformis, O. subtilissima,	Caneva *et al.* (2005), Gaylarde and Englert (2006), Gaylarde *et al.* (2005), McNamara *et al.* (2006), Strzelczyk (1981).

(*continued*)

TABLE 5.1 *(continued)*

Microbial group	Family/genus/species	References
	Phormidium lignicola, Stigonema ocellatum, S. hormoides.	
Other photosynthetic bacteria	Chloroflexi.	McNamara *et al.* (2006).
	Rhodoplanes elegans.	McNamara *et al.* (2006).
Archaea	Halophilic bacteria: *Halobacillus, Halobacterium, Halococcus, Halomonas, Natronobacterium*	Heyrman *et al.* (1999), Piñar *et al.* (2001), Rölleke *et al.* (1998), Rölleke *et al.* (1996).
	Methanogenic bacteria, methanotrophic bacteria	Kussmaul *et al.* (1998).
Chemolithotrophic bacteria	Nitrogen cycle: *Nitrobacter, Nitrococcus, Nitrosococcus, Nitrosoglobus, Nitrosomonas, Nitrosospira, Nitrosovibrio, Nitrospira.*	Caneva *et al.* (1991), Gorbushina *et al.* (2002), May (2003), Spieck *et al.* (1992).
	Nitrobacter vulgaris, Nitrosomonas ureae, Nitrospira moscoviensis.	McNamara *et al.* 2006, Pinck and Balzarotti al. (2000).
	Sulfur cycle: *Thiobacillus.*	Caneva *et al.* 1991, Flores *et al.* (1997), Warscheid and Braams, (2000).
	Thiobacillus thiooxidans, T. thiosporus, T. albertis, T. neapolitanus, T. denitrificans.	May (2003), Prieto *et al.* (1995).
Chemoorganotrophic bacteria	Acidobacteria, *Bacillus, Clostridium, Holophaga, Melittangium, Pseudomonas,* sulfate-reducing bacteria.	Flores *et al.* (1997), Gaylarde *et al.* (2001), Gorbushina *et al.* (2002), Heyrman and Swings (2001), Kussmaul *et al.* (1998), McNamara *et al.* (2006), Ortega-Morales and Hernandez-

		Duque, (1998), Ortega-Morales *et al.* (2005), Rölleke *et al.* (1996), Saarela *et al.* (2004).
	Bacillus circulans, B. badius, B. licheni, B. cereus, B. licheniformis, B. barbaricus, B. thuringiensis, B. pumilis, B. megaterium, B. firmus.	Blazquez *et al.* (2000), Gaylarde *et al.* (2001), Heyrman and Swings (2001), McNamara *et al.* (2006), Prieto *et al.* (1997).
Actinomycetes	*Arthrobacter, Aureobacterium, Blastococcus, Brevibacterium, Clavibacter, Geodermatophilus, Micrococcus, Microellobosporium, Micromonospora, Microphylospora, Modestobacter, Nocardia, Nocardiodes, Rhodococcus, Rubrobacter, Streptomyces.*	Aranyanak (1992), Bassi *et al.* (1986), Caneva and Nugari (2005), Caneva *et al.* (1991), Flores *et al.* (1997), Gorbushina *et al.* (2002), Heyrman and Swings (2001), Hyvert (1966), McNamara *et al.* (2006), May *et al.* (2000), May (2003), Ortega-Morales *et al.* (2004, 2005), Rölleke *et al.*, (1996), Saarela *et al.*, (2004), Tiano, (2002), Warscheid and Braams (2000).
	Arthrobacter (Micrococcus) agilis, Geodermatophilus obscurus, Kocuria rosea, Marmoricola aurantiacus, M. lylae, M. roseus, M. varians, M. halobius, M. agilis, Nocardia restricta, Saccharothrix flava.	Blazquez *et al.* (2000), Eppard *et al.* (1996), McNamara *et al.* (2006), May (2003), Prieto *et al.* (1995).
Fungi	*Acremonium (Cephalosporium), Alternaria, Aspergillus, Aureobasidium, Botrytis, Candida, Capnobotryella, Cladosporium, Coniosporium, Cryptococcus, Dictyodesmium, Exophiala, Fusarium, Hortaea, Lichenthelia, Mucor, Nectria, Penicillium,*	Allsopp *et al.* (2003), Blazquez *et al.* (2000), Caneva and Nugari (2005), Gorbushina *et al.* (2002), Hirsch *et al.* (1995b), Monte (2003), Prieto *et al.* (1995), Tiano (2002), Urzi (2004), Urzi

(*continued*)

TABLE 5.1 *(continued)*

Microbial group	Family/genus/species	References
	Phaeococcomyces, Phaeosclera, Phaeotheca, Phoma, Phialostele, Pseudotaeniolina, Rhinocladiella, Rhizopus, Rhodotorula, Sarcinomyces, Sporobolomyces, Sporotrichum, Trichoderma, Trimmatostroma, Ulocladuim.	and De Leo (2001), Urzi *et al.* (2000), Warscheid and Braams (2000).
	Acremonium murorum, A. niger, A. versicolor, A. wentii, Aurobasidium pullulans, Capnobotryella renispora, Chaetomium globosum, Cladosporium cladosporioides, Coniosporium apollinis, C. perforans, C. uncinatus, Exophiala jeanselmei, E. monileae, Hortaea werneckii, Phialophora melinii, Sarcinomyces petricola, Trichoderma viride, Trimmatostroma abietis, Verticillium nigrescens.	Blazquez *et al.* (2000), Caneva and Nugari (2005), Gorbushina *et al.* (2002), Hoppert *et al.* (2004), Urzi (2004).
Lichens	*Aspicilia, Caloplaca, Lecanora, Protoplastenia, Thyrea, Verrucaria, Xanthoria.*	Ascaso *et al.* (1998, 2002), Tiano (2002).
	Caloplaca aurantiaca, C. ceria, C. citrina, C. holocarpa, C. trachyphylla, C. concolor, C. vitellina, Collera crispum, Diploicia canescens, Dirina massiliensis, Lecania rabenhorstii, Lecanora hageni, Ochrolechia parella, Phaeophysica hirsute, Tephronella atra	Ascaso *et al.* (1998), Frey *et al.* (1993), Hoppert *et al.* (2004), Prieto *et al.* (1997, 1999), Seaward (2003).

be mainly to promote the growth of other organisms. This is not so for cyanobacteria, whose role in the deterioration of surfaces of historic buildings has been the subject of several recent studies and reviews (Crispim and Gaylarde, 2004; Ortega-Morales *et al.*, 2001; Tomaselli *et al.*, 2000b). These organisms are generally adapted to resist adverse conditions because of their thick outer envelopes and the presence of protective pigments (Chazal and Smith, 1994; Garcia-Pichel *et al.*, 1992). Cyanobacteria are probably the most resistant of the microflora on monument surfaces (with lichens in second place), if we relate this environment to desert crusts (West, 1990). The Atacama and Namib deserts, the most extreme on earth, have crusts in which cyanobacteria are the only phototrophs (Evenari *et al.*, 1985). The ability of cyanobacteria to survive repeated cycles of dehydration and high levels of UV radiation (Chazal and Smith, 1994; Garcia-Pichel *et al.*, 1992; Potts, 1994) makes them particularly important organisms on outdoor stone surfaces. In spite of their resistance to UV, superficial growth of cyanobacteria and algae is stronger on sheltered indoor surfaces of historic limestone buildings, which have reduced illumination, higher humidity, and more organic nutrients (Gaylarde *et al.*, 2001; Ortega-Morales *et al.*, 2000).

Cyanobacteria have been suggested to be of higher ecological importance as pioneer organisms on exposed stone surfaces of buildings than any other organism (Grant, 1982) and may have the most important influence on weathering of exposed stone (Gaylarde and Morton, 2002). Cyanobacteria have been shown to constitute the major biomass on external surfaces of ancient stone structures in Latin America (Gaylarde *et al.*, 2001; Ortega-Morales *et al.*, 2000), Greece (Anagnostidis *et al.*, 1983), and India (Tripathy *et al.*, 1999). In fact, cyanobacteria and eukaryotic algae had been found, until the recent molecular biology study of McNamara *et al.* (2006), to be the most widespread microorganisms in the endolithic habitat (Sigler *et al.*, 2003). The ability to fix carbon dioxide, and in some species atmospheric dinitrogen (N_2), gives the cyanobacteria an obvious advantage over heterotrophic bacteria. Light quality and intensity are the main factors that control the minimum and maximum depth at which endolithic phototrophic communities grow (Nienow *et al.*, 1988).

Apart from their evident aesthetic deterioration of the stone monument, these phototrophs may cause chemical and physical deterioration by the excretion of chelating agents and stone-dissolving acids (Albertano, 2003; May, 2003; Urzi and Krumbein, 1994), as well as by yet undefined boring activity (Carcia-Pichel, 2006), documented for the Pleurocapsa-group (Mao-Che *et al.*, 1996), *Synechocystis*, *Gloeocapsa*, *Stigonema*, *Schizothrix* (Hoffman, 1989), *Scytonema* (Golubic *et al.*, 2000), and *Mastigocladus* (*Fischerella*) (Boone *et al.*, 2001). Ortega-Morales *et al.* (2000) and Gaylarde and Englert (2006) showed scanning electron micrographs that demonstrate coccoid cyanobacteria sitting in cell-sized depressions in the stone surface, while Gaylarde *et al.* (2006) report the presence of pure colonies of the alga *T. umbrina* within colony-sized pits on limestone. All these images indicate that the cells themselves are causing the degradation.

Photosynthetic organisms deposit $CaCO_3$ in the presence of light and solubilize it at night. The precipitation of calcium salts on cyanobacterial cells growing on limestone suggests the migration of calcium from neighboring sites (Ascaso *et al.*, 1998; de los Rios, 2005; Ortega-Morales *et al.*, 2000; Schultze-Lam and Beveridge, 1994). The external S-layer of *Synechococcus* GL24 binds calcium ions (Schultze-Lam and Beveridge, 1994), which complex with carbonate ions at the pH values (>8.3) produced around the cells (Miller *et al.*, 1990). *Synechococcus* cells can become encrusted with calcite within 8 h in an aqueous environment and must continually shed patches of the mineralized S-layer to remain viable (Douglas and Beveridge, 1998). This mobilization of calcium ions and the trapping of released particles of calcite in the gelatinous sheaths or capsules of cyanobacterial cells (Pentecost, 1987, 1988) are important mechanisms of degradation of calcareous stone. Nitrogen and phosphorus may also be mobilized from the stone and metabolized or stored within the organisms (Albertano, 2003). Warscheid and Braams (2000) also mentioned the possibility that phototrophic organisms take up and accumulate sulfur and calcium into their cells.

Lichens are symbiotic associations between fungi and one or two photobionts, which can be algae or cyanobacteria. They are frequent colonizers of stone monuments and have been mistaken for the remains of ancient rendering when they cover substantial areas of the surface (Seaward, 2003). Lichens are particularly sensitive to air pollution and, indeed, are used as bioindicators of such. It has been suggested that improved air quality has already, or may in the future, lead to an increase in colonization of stone by lichens (Ardron, 2002; Young, 1997). Although many people find lichen growth on stone pleasing, it can be a problem in obscuring fine details of carvings and it is certainly inherently damaging to the structure. Nimis and Monte (1988) reported an interesting effect of lichen growth on the Orvieto duomo (Italy). The alternating dark basalt and light limestone bands have been colonized, respectively, by light and dark, or orange, lichens, completely eliminating the effect desired by the artist on the northern facade.

Lichens cause mechanical damage due to penetration of their rhizines, composed of fungal filaments, and the expansion/contraction of the thallus on wetting/drying, which can lift grains of stone off the surface (De los Rios *et al.*, 2004; Gaylarde and Morton, 2002) (Fig. 5.6). Accumulation of small stone fragments (as small as 5 μm) within the lower thallus have been reported (Gadd, 2007). The depth of penetration depends on the stone substrate and the type of lichen; lichen structures can sometimes be found at least 3 cm below the stone surface (Lee *et al.*, 2003). They also

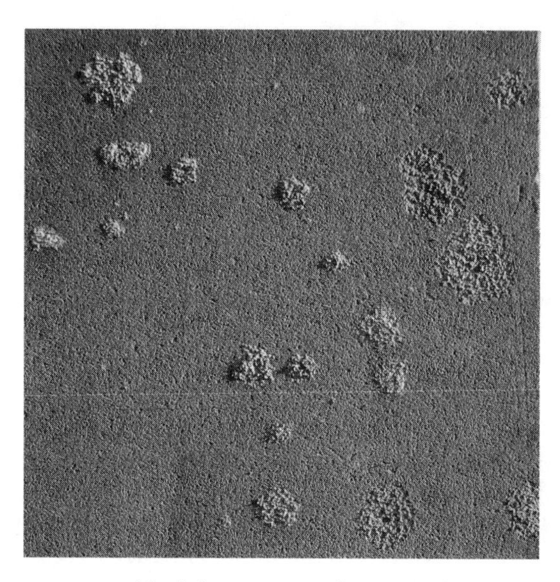

FIGURE 5.6 Damage caused by lichens on a sandstone tomb in Cardiff, Wales.

cause direct chemical attack by the production of significant amounts of acids. "Lichen acids" have been shown to cause damage at the stone/lichen interface (Cameron *et al.*, 1997; Seaward, 2003). The principal acid produced is oxalic, which leads mainly to the formation of calcium oxalate and its different hydrate forms whewellite and wedellite (Gaylarde and Morton, 2002; Tiano, 2002). The lichen thallus has been shown to accumulate from 1% to 50% calcium oxalate, depending on the substrate. Even on siliceous stone, some lichens can accumulate this compound, using calcium from the air, or leachates (Seaward, 2003). Lichens on historic stone buildings have been reviewed recently by Lisci *et al.* (2003).

The CO_2 produced by lichens is transformed within the thallus to carbonic acid (Tiano, 2002), which, although a weak acid, seems to be able to solubilize calcium and magnesium carbonates in calcareous stone. Lichens have been demonstrated to biomobilize certain elements from the stone matrix (De los Rios *et al.*, 2004; Tiano, 2002). The former workers demonstrated magnesium-depleted areas of the stone substrate around the lichen thallus. Saxicolous lichens mobilize magnesium and silicon in rock, causing biochemical weathering (Aghamiri and Schwartzman, 2002). Gordon and Dorn (2005) calculated that a saxicolous lichen increased the weathering rate of basalt by a factor of at least 1.7. The weathering was greatest directly under the lichen colony. About 0.5 mm below the colony, weathering rates fell to those of uncolonized surfaces. Banfield *et al.* (1999) proposed a model, based on high-resolution transmission electron microscopy, of the weathering of silicate rocks by lichen activity. Clear boundaries are shown in the vertical profile, with a direct "biochemical" effect first produced, followed by predominantly biophysical action in the deeper layer of material.

Certain lichens can grow endolithically (Gaylarde and Gaylarde, 2005; Gerrath *et al.*, 1995). They are slow-growing, stress-tolerant organisms, which have been stated to have a similar physiology to epilithic crustose lichens (Tretiach and Pecchiari, 1995), and lead to similar destructive effects.

Under conditions of high abiotic weathering, lichens have been suggested to provide protection for the stone surface from wind and rain through the insoluble oxalate layer (Bungartz *et al.*, 2004; Di Bonaventura *et al.*, 1999; Warscheid and Braams, 2000), or to limit erosion by reducing the level of water within the rock (Garcia-Vallès *et al.*, 2003); their retention of moisture within the thallus reduces thermal stress on a limestone surface (Carter and Viles, 2003). However, they are generally defacing and intrinsically damaging. Even when a protective effect can be shown, subsequent decay of the lichen thallus (which occurs in the centre of the colony of some species) can open this area to further weathering, resulting in cratered mounds on the rock surface (Mottershead and Lucas, 2000). The mechanical removal of crustose lichens is particularly difficult because the thallus forms an intimate association with the substrate. Hence, its removal leads to severe structural damage (Allsopp and Gaylarde, 2004; Gaylarde and Morton, 2002).

Recent work based on molecular approaches has shown that, in addition to algae, lichens, and cyanobacteria, other previously unrecognized phototrophic microorganisms may occur in stone monuments. Ortega-Morales *et al.* (2004) found bacteria related to the Ectothiorhodospiraceae in certain samples at the Mayan site of Uxmal, while McNamara *et al.* (2006) detected *Chloroflexi*-related organisms. This new data, added to the already known complex nature of lithic biofilms on historic monuments, indicates that these organisms may contribute to the carbon pool in autotrophic biofilms. It is likely that their role in stone deterioration, as for algae, is supporting the growth of associated heterotrophs, although the production of osmolytes cannot be ruled out. Interestingly, the halophily of these organisms is congruent with the measured levels of salts in some monuments, where significant amounts of sulfate, chloride, and nitrate have been found (Ortega-Morales, 1999; Ortega-Morales *et al.*, 2004, 2005).

B. Chemoorganotrophic microorganisms

The contribution of heterotrophic microorganisms to stone deterioration, particularly as pioneering colonizers, had long been neglected; however, their degradative role by acid/alkali production and by chelation is now well accepted (Gaylarde and Morton, 2002).

1. Fungi

The effects of fungi are due to physical and chemical actions, which are often synergistic in the degradation of stone. They were recently reviewed by Gadd (2007), and will therefore only be mentioned briefly here. The fungal stone flora consists of filamentous fungi (ubiquitous hyphomycetes and coleomycetes) and microcolonial fungi (black yeasts and yeast-like meristematic fungi) (Gorbushina *et al.*, 2002, 2003; May, 2003; Sterflinger, 2000; Urzi *et al.*, 2000). Meristematic fungi produce swollen, isodiameteric cells with thick, melanin containing cell walls. They remain metabolically active even in low nutrient conditions and have high resistance to desiccation, UV radiation, and osmotic stress (Urzi *et al.*, 2000), thus being well adapted to growth on external walls. Wollenzien *et al.* (1995) suggested that these are the resident fungi in Mediterranean climates; the fast growing, filamentous hyphomycetes being present only in the colder and more humid winter months and therefore considered contamination in this climatic area. Hyphomycetes tend to be the major fungal population in more northerly parts of Europe (Sterflinger, 2002). However, the ubiquitous hyphomycetes can also be found in (sub)tropical climates. Resende *et al.* (1996) identified a wide range of filamentous fungi in soapstone and quartzite in churches in the Brazilian state of Minas Gerais. The most common genera were *Cladosporium* and *Penicillium*. However, it must be emphasized that the detection technique affects the results of such investigations.

Gorbushina *et al.* (2002) detected mainly deuteromycetes, such as *Alternaria*, *Cladosporium*, and *Trichoderma*, on historic marble monuments in St. Petersburg and Moscow. Many of the organisms were obviously derived from the surrounding plants. They applied Koch's postulates to two of the isolates and showed that they could grow on and discolor sterile marble blocks.

Sterflinger (2000) indicated *Aspergillus niger*, *Penicillium simplissimum*, and *Scopulariopsis brevicaulis* as important fungi that attack siliceous stone. These dark pigmented mitosporic fungi ("black fungi") can actively penetrate limestone and marble and produce pits of up to 2 cm diameter on rock surfaces. (Sterflinger and Krumbein, 1997). They are especially important in arid and semiarid environments (hot and cold deserts) because of their ability to resist high temperatures, desiccation, and osmotic stress (Sterflinger, 1998).

In fact, several cryptoendolithic fungi may actively bore into the stone and hence physically disrupt its integrity (Gadd, 2007; Hoffland *et al.*, 2004). Fungi, unlike the phototrophs, do not require light for growth, and so their boring activity can penetrate to greater depths. Golubic *et al.* (2005) discussed such a tunneling activity in carbonate substrates (particularly mollusk shells) in marine environments. Hyphal penetration of materials involves swelling/deflation effects and channeling of water into the substrate. It can form cracks, fissures, and crevices, extend existing ones and lead to the detachment of crystals (Sterflinger, 2000; Urzi *et al.*, 2000). Weaker areas of the stone will be preferably penetrated by thigmotropism (contact guidance on solid surfaces to explore new substrates; Gadd, 2007; Watts *et al.*, 1998).

Biochemical actions of fungi can lead to microtopological alterations through pitting and etching, mineral dislocation and dissolution (Gadd, 2007). They are associated with extracellular mucilaginous substances, which contain, amongst many other metabolites, acidic, and metal-chelating compounds (Burford *et al.*, 2003). Acidic metabolites (oxalic, acetic, citric, and other carbonic acids) deteriorate the stone minerals by a solubilizing and chelating effect (Sterflinger, 2000; Urzi and Krumbein, 1994). Ortega-Morales *et al.* showed that fungi isolated from deteriorated limestone at the Mayan site of Uxmal, Mexico, produced oxalic acid, which reacted with solubilized calcium from the stone to produce crystals of whewellite and weddelite (unpublished results). Fungal oxalic acid had previously been reported to solubilize metals (e.g., iron, aluminum, lithium, manganese) from various other substrates to form oxalates (Devevre *et al.*, 1996; Strasser *et al.*, 1994). Acidolysis and complexolysis, which have been reported to be the primary deteriorative mechanisms of fungi (Gadd, 2007), act on the stone mineral by proton efflux (plasma membrane H^+-ATPase, maintenance of charge balance during nutrient uptake) and siderophores, which mobilize iron (III), or CO_2 production (Gadd, 2007).

Oxidation and reduction of mineral cations are also triggered by fungal activity (Gadd, 2007). Iron and manganese particularly are removed from the stone lattice by redox processes (Warscheid and Braams, 2000) and may be reoxidized at the stone surface, forming "patinas" or "crusts." This biotransfer of metal ions and subsequent formation of patina [called "rock varnish" by Krumbein and Jens (1981) and Krumbein and Giele (1979)], can lead to hardening of the surface layer and exfoliation (Tiano, 2002). However, Urzi *et al.* (2000) emphasized that there is no evidence that meristematic fungi produce acids, oxidize manganese, or are directly responsible for the formation of "rock" or "desert varnish."

Various metabolic substances excreted by fungi are colored, leading to staining of the substrate (Tiano, 2002). The production of melanins by dematiaceous (dark pigmented mitosporic) fungi darkens the stone surface, leading to significant aesthetic alterations, and physical stress.

The literature suggests that fungi are present in low numbers on the surfaces of historic stone buildings. Populations of 10^2–10^5cfu·g^{-1}are common (Gaylarde *et al.*, 2001; Hirsch *et al.*, 1995b; Ortega-Morales *et al.*, 2000; Resende *et al.*, 1992; Urzi, 1993). However, this does not mean that they are unimportant; their activity may be high and erosive! In addition, fungi may be the most important endoliths in built stone, according to De los Rios and Ascaso (2005). They have higher tolerance of low water activity than algae and bacteria and require low nutrient concentrations, as well as having no need for light.

2. Actinomycetes

These filamentous bacteria penetrate their substrate by mechanisms similar to those employed by fungi; they also excrete a wide range of enzymes. They can form a whitish veil on stone or produce various water-soluble dark pigments. Laboratory experiments have demonstrated their ability to utilize nitrites and nitrates and to reduce sulfates (Caneva *et al.*, 1991), and, of course, they are well recognized as degraders of a wide range of different carbon and nitrogen sources. Probably for these reasons, the gram-positive actinomycetes tend to predominate over gram-negative bacteria on exposed stone surfaces (Dornieden *et al.*, 2000; Saarela *et al.*, 2004; Warscheid and Braams, 2000). Their acidic metabolic products can attack calcareous stone, hydrolyze some silicate minerals, and chelate metal ions (Kumar and Kumar, 1999). However, they have been reported to rarely, if ever, produce noteworthy amounts of organic acids and chelates in a rock decay environment (Urzi and Krumbein, 1994). In spite of this, they may cause structural damage by their extensive biofilm formation and penetration of their filaments into the stone substrate.

Actinomycetes have been found as important endoliths in various types of built stone (McNamara *et al.*, 2006; Ortega-Morales *et al.*, 2005), emphasizing their degradative ability in this situation. Ortega-Morales *et al.* (2004), for example, found almost exclusively *Rubrobacter xylanophilus*-related bacteria on external biofilms in Uxmal. Although the genus *Geodermatophilus* has been suggested to be common on and in limestone (Eppard *et al.*, 1996), Urzi *et al.*, 2001 using amplified 16S rDNA analysis (ARDRA) and partial sequencing, found that many of the Geodermatophilaceae family on stone in the Mediterranean belonged to other genera (closest to *Modestobacter multiseptatus*).

There have been a number of publications on the presence of actinomycetes in caves (Laiz *et al.*, 2000; Schabereiter-Gurtner *et al.*, 2004), but, in spite of their obvious importance, there is little in the built cultural heritage literature on this group of microorganisms, apart from mainly superficial comments about their presence. This is an area that demands further attention.

3. Nonfilamentous bacteria

The contribution of heterotrophic bacteria to stone deterioration had long been neglected, as insufficient organic nutrients were assumed to be present on stone surfaces. However, these organisms have been isolated frequently from such surfaces; and it has been found that organic contaminants, such as soil, dust, and dirt, are sufficient to support heterotrophic growth. Furthermore, several of these heterotrophic bacteria are oligotrophic (May, 2003). Chemoorganotrophic bacteria utilize a wide range of nutrients and may serve other microorganisms by the breakdown of poorly degradable compounds (e.g., from atmospheric pollution), which could otherwise not be utilized.

Organisms of the genus *Bacillus* have been very frequently identified on stone buildings (Blazquez *et al.*, 2000; Gaylarde *et al.*, 2001; Heyrman and Swings, 2001; Kiel and Gaylarde, 2006; Laiz *et al.*, 2003; Ortega-Morales *et al.*, 2004; Prieto *et al.*, 1995; Rölleke *et al.*, 1996). This is not unexpected, as they are very common in soil and are able to withstand extreme environments because of their spore-forming ability and ease of culture. Laiz *et al.* (2003), comparing culture and molecular biology techniques, suggested that their proportion of the biofilm on external surfaces of historic buildings is overestimated. However, McNamara *et al.* (2006), using only molecular biology, found that many of the clones were closely related to the low GC Firmicutes and considered that culture techniques may not, in fact, be entirely misleading.

Rather more important than the simple presence of these bacteria in the biofilm, is their potential degradative activity. Kiel and Gaylarde (2006) found that some of their *Bacillus* isolates produced acids and surfactants with autoemulsifying activity in the laboratory, indicating that they had the capacity to accelerate stone degradation. Once again, however, beware extrapolation from laboratory experiments to the real world!

One surprising component of the stone microflora is the group of bacteria producing or utilizing methane. These were isolated from 44 of 225 stone samples from 19 historic buildings in Germany and Italy (Kussmaul *et al.*, 1998). All were Type II methanotrophs, that is, those found at oligotrophic sites under nitrogen limiting conditions. It was suggested that the methane necessary for methanotrophic growth could originate from anthropogenic sources and from endolithic methanogens, which were detected in four of the samples, presumably in anaerobic niches. "Mini-methane producers", such as *Clostridium*, were found in almost half of the 47 samples tested for this activity.

C. Chemolithotrophic microorganisms

The presence of chemolithoautotrophic microorganisms, such as sulfur oxidizers, nitrifying bacteria, and iron and manganese oxidizers, depends on the availability of the specific nutrients supporting their growth (Warscheid and Braams, 2000). Although they were the first group of microorganisms to be implicated in stone decay, their assumed importance has been superceded by later research that suggests the greater role of phototrophs and chemoorganotrophs. Gaylarde and Morton (2002) emphasized that there is little doubt that chemolithotrophic microorganisms have the potential to cause damage to stone; however, their significance to biodeterioration of outdoor stone monuments is still in question.

It appears that sulfur oxidizers and nitrifying bacteria play a more significant role in biodeterioration in humid areas, because of their sensitivity to desiccation (Warscheid and Braams, 2000). In fact, nitrifying bacteria have been suggested to be the most important microbial factors in the decay of sandstone in northern Europe (Bock *et al.*, 1988; Meincke *et al.*, 1988)

1. Sulfur compound oxidizers and reducers

Sulfur-oxidizing bacteria obtain energy by the oxidation of reduced or elemental sulfur to sulfuric acid. Sulfuric acid may react with calcium carbonate to form calcium sulfate (gypsum), which is more soluble in water than the calcium carbonate of the parental rock (Urzi and Krumbein, 1994; Warscheid and Braams, 2000), and thus more readily leached. However, sulfuric acid and calcium sulfate are not always of biogenic origin; they may also derive from atmospheric pollution and acid rain (May, 2003). In fact, Tiano (2002) emphasized that there is as yet no experimental evidence that confirms the direct action of sulfur-oxidizing bacteria in the development of gypsum layers on stone surfaces.

Sulfate-reducing bacteria (which are not chemolithotrophic, but chemoorganotrophs) have been detected in biofilms on limestone (Gaylarde *et al.*, 2001; Ortega-Morales and Hernández-Duque, 1998), but this is apparently rare and no role has been suggested for them in stone decay.

2. Nitrifying bacteria

Ammonia and nitrites on the stone surface are oxidized by chemolithotrophic and, partly, by heterotrophic ammonia and nitrite oxidizers to nitrous and nitric acid, respectively. Ammonia tends to derive from airborne ammonium salts, whereas nitrites may originate from automobiles, industry, and soil (May, 2003). The acids that are produced attack calcium carbonate and other minerals (Urzi and Krumbein, 1994; Warscheid and Braams, 2000). The CO_2 produced can be utilized by the cells to form organic compounds, while calcium cations from the stone matrix form nitrates and nitrites, which are more soluble again than the original mineral phases and thus are leached out of the stone by rain. The characteristic symptom of the activity of nitrifying bacteria is a change in stone properties with no obvious biofilm. It becomes more porous, exfoliation occurs, and fine powder may fall off (Urzi and Krumbein, 1994).

3. Iron- and manganese-oxidizing microorganisms

Iron oxidation is usually rapid and sensitive to pH and oxygen concentrations. Iron-oxidizing bacteria obtain energy by oxidizing ferrous iron in iron-containing minerals to ferric iron, which reacts with oxygen to form iron oxide. The latter process determines the characteristic discoloration and patina formation on stones. Many bacteria and fungi, even algae, are capable of these oxidation steps, causing damaging lesions (Barrionuevo and Gaylarde, 2005; Caneva *et al.*, 1991; Urzi and Krumbein, 1994). It is difficult to distinguish such biooxidation from chemical processes, although evidence for the involvement of living organisms in the formation of a red patina on a dolomite cathedral in Spain has been presented by Valls del Barrio *et al.* (2002).

V. CONTROL OF BIODETERIORATING MICROORGANISMS

This topic deserves a review of its own and will only be mentioned briefly here.

The removal of the microbial community from any given surface is an intervention that must be carefully evaluated. Biocidal treatments may have negative effects on the artifacts (Webster *et al.*, 1992). The removal of the microbial community may give rise to a new succession of microorganisms, which may be more damaging than the old microbial surface populations; and the inhibition of specific groups of microorganisms may favour the growth of others (May, 2003; Warscheid and Braams, 2000). The approach to control biodeterioration must be a polyphasic, interdisciplinary one that considers the history and condition of the artifact as well as physical and chemical damaging factors.

Actions against microbial growth can be divided into four major categories: (1) indirect control by altering environmental conditions; (2) mechanical removal of biodeteriogens; (3) chemicals (biocides); (4) physical eradication methods. Biocides often exhibit detrimental effects on the stone, for example, discoloration, oxidation/reduction of stone compounds, and salt formation, with subsequent crystallization upon drying, leading to exfoliation (Cameron *et al.*, 1997; Caneva *et al.*, 1991; Kumar and Kumar, 1999; Warscheid and Braams, 2000). The ecotoxicity of commercial biocides may make them poor candidates for use in outdoor environments and many countries have prohibited the use of some of the previously most common (and effective) biocides. Furthermore, nitrogen-containing biocides may serve as nutrients for surviving or newly attaching microorganisms (Warscheid and Braams, 2000). Where possible, microbial growth should be prevented by altering growth-supporting conditions (e.g., introduction of a drainage system).

Physical methods such as UV light have long been overlooked in their application on cultural heritage objects, owing to reported long treatment times and low penetration depth (Van der Molen *et al.*, 1980). However, a more recent preliminary study on the removal of lichens by means of a high-intensity pulsed xenon flash lamp gave encouraging results (Leavengood *et al.*, 2000). More research is needed into its effectiveness in controlling a broader range of microorganisms that cause biodeterioration of stone.

There have been a number of publications on the protective effects of microorganisms against biodeterioration of stone. Krumbein (1969) pointed out that after initial destructive processes the microorganisms and the substrate may establish an equilibrium leading to a protective "patina." The influence of lichens has already been discussed and Mottershead *et al.* (2003) suggested that abundant microbial growth on sandstone could protect against salt weathering. They pointed out, however, that this was not the case where colonization is patchy, when degradation was increased. Although a biological control option for stone decay is attractive, it is still some distance in the future.

The human influence of restoration/conservation and its integrated antimicrobial approaches on stone monuments can itself be detrimental, unwittingly improving conditions for microbial colonization and growth. The changes following the discovery of the siliceous stone hieroglyphic stairway at Copan, Honduras, cited by Caneva *et al.* (2005), is a good example. Early removal of the tree cover in the surroundings led to increased exposure to sunlight and better air circulation and resulted in lichen growth. In 1985, a tarpaulin cover was installed to protect the stairway from rain, after various treatments with biocides to remove the growth. By 2005, lichens were much reduced, but there was heavy algal and cyanobacterial growth, interpreted by the authors as the beginning of a new colonization sequence, and doubtless aided by the increased humidity beneath the tarpaulin. Although earlier treatments resulted in a decrease of the more obvious type of biofilms (lichens), the old endolithic population was still present (mainly filamentous fungi and moss protonema were detected), and its removal would be extremely difficult.

VI. CONCLUSIONS

Although it is well established that microorganisms can cause serious damage to stone monuments, knowledge of the precise mechanisms of decay is still fragmentary. This is a field that demands more attention. The development of new identification methods provides us with a broader understanding of the diversity of organisms present on outdoor monuments, and may expand our knowledge of new types of microbial metabolism occurring in these habitats. Most likely, the list of organisms will expand dramatically as further analytical methods for detection and taxonomy are developed. However, very little work has been carried out in studying the general physiology and potential deteriorative activity of the newly identified organisms, using, for example, proteomics. A proteomic and genomic approach would not only shed light on the potential activity of microorganisms, but would also help to design new strategies for isolating and successfully culturing new organisms. Even microorganisms that have long been known to occupy the surface of stone monuments have only rarely been appropriately examined for their actual contribution to stone decay *in vivo*. In order to compare results of different research groups, a standardization of methods for the detection, assessment, and quantification of biodeterioration is necessary.

The possibility of biologically induced stabilization of stone needs to be more thoroughly investigated. Understanding the interactions between microorganisms and with their environment is crucial to determine whether the organism is damaging or protective to the art object. A description of criteria for determining that the decay of a monument is due to microbial action is rare in the literature. Similarly, very few studies aim to quantify biodeterioration processes. In fact, there have been no attempts to define the degree of biodeterioration of an artifact and at what stage antimicrobial actions should be initiated. In order to assess the contribution of microorganisms to the deterioration of cultural heritage objects, as well as the possibilities for their control, interdisciplinary research projects between conservators and scientists, such as microbiologists, geologists, and chemists, are needed.

REFERENCES

Aghamiri, R., and Schwartzman, D. W. (2002). Weathering rates of bedrock by lichens: A mini watershed study. *Chem. Geol.* **188,** 249–259.

Albertano, P. (2003). Methodological approaches to the study of stone alteration caused by cyanobacterial biofilms in hypogean environments. *In* "Art, biology and conservation: Biodeterioration of works of art" (R. J. Koestler, V. H. Koestler, A. E. Charola and F. E. Nieto-Fernandez, Eds.), pp. 302–315. The Metropolitan Museum of Art, New York.

Allsopp, D., and Gaylarde, C. C. (2004). "Heritage Biocare. Training course notes in Biodeterioration for Museums, Libraries, Archives and Cultural Properties." Version 2. Archetype Publications, London (published on CD).

Allsopp, D., Seal, K., and Gaylarde, C. C. (2003). "Introduction to Biodeterioration." Cambridge University Press, Cambridge.

Althukair, A. A., and Golubic, S. (1991). New endolithic cyanobacteria from the Arabian Gulf: 1. Hyella-immanis sp-nov. *J. Phycol.* **27,** 766–780.

Amann, R. I., Ludwig, W., and Schleifer, K. -H. (1995). Phylogenetic identification and *in situ* detection of individual microbial cells without cultivation. *Microbiol. Rev.* **59,** 143–169.

Anagnostidis, K., Economou-Amilli, A., and Roussomoustakaki, M. (1983). Epilithic and chasmolithic microflora (Cyanophyta, Bacillariophyta) from marbles of the Parthenon (Acropolis-Athens, Greece). *Nova. Hedw.* **38,** 227–277.

Aranyanak, C. (1992). Biodeterioration of cultural materials in Thailand. Proceedings of the 2nd International Conference on Biodeterioration of Cultural Property, (K. Toishi, H. Arai, T. Kenjo and K. Yamano, Eds), pp. 23–33. Tokyo: International Communications Specialists, October 5–8,1992, Held at Pacifico Yokohama.

Ardron, P. A. (2002). The increasing diversity of lichens in the South Pennine area and the enigmantic presence of rare species. South Yorkshire Biodiversity and Countryside Conference Part 1: Action for Sustainable Countryside in Urban and Rural South Yorkshire, March, 2002. The South Yorkshire Biodiversity Network. http://www.shu.ac.uk/sybionet/confer/recent/marchconf/bdconf1.htm Accessed on 9th May, 2006.

Ascaso, C., Wierzchos, J., and Castello, R. (1998). Study of the biogenic weathering of calcareous litharenite stones caused by lichen and endolithic microorganisms. *Int. Biodeterior. Biodegrad.* **42,** 29–38.

Ascaso, C., Wierzchos, J., Souza-Egipsy, J., de los Rios, A., and Delgado Rodrigues, J. (2002). *In situ* evaluation of the biodeteriorating action of microorganisms and the effects of biocides on carbonate rock of the Jeronimos Monastery (Lisbon). *Int. Biodeterior. Biodegrad.* **49,** 1–12.

Barrionuevo, M. E. R., and Gaylarde, C. C. (2005). Biodegradation of the sandstone monuments of the Jesuit missions in Argentina. *In* "LABS5. Biodegradation and Biodeterioration in Latin America" (B. O. Ortega-Morales, C. C. Gaylarde, J. A. Narvaez-Zapata and P. M. Gaylarde, eds.), pp. 21–24. Universidad de Campeche, Mexico.

Banfield, J. F., Barker, W. W., Welch, S. A., Taunton, A. (1999). Biological impact on mineral dissolution: Application of the lichen model to understanding mineral weathering in the rhizosphere. Proceedings of the National Academy of Sciences **96,** 3404–3411.

Bassi, M., Ferrari, A., Realini, M., and Sorlini, C. (1986). Red stains on the Certosa of Pavia. A case of biodeterioration. *Int. Biodeterior. Biodegrad.* **22,** 201–205.

Bennett, P. C., Melcer, M. E., Siegel, D. I., and Hassett, J. P. (1988). The dissolution of quartz in dilute aqueous solutions of organic acids at 25 °C. *Geochim. Cosmochim. Act.* **52,** 1521–1530.

Blazquez, A. B., Lorenzo, J., Flores, M., and Gomez-Alarcon, G. (2000). Evaluation of the effects of some biocides against organisms isolated from historic monuments. *Aerobiologia* **16,** 423–428.

Bock, E., and Krumbein, W. E. (1989). Aktivitäten von Organismen und mögliche Folgen für Gestein von Baudenkmälern. Bausubstanzerhaltung in der Denkmalpflege: 2. Statusseminar des Bundesministeriums für Forschung und Technologie (BMFT). "Untersuchung und Eindämmung der Gesteinsverwitterung an Baudenkmälern" 14/15 December 1988. pp. 34–37. Wuppertal.

Bock, E., Sand, W., Meincke, M., Wolters, B., Ahlers, B., Meyer, C., and Sameluck, F. (1988). Biologically induced corrosion of natural stones—strong contamination of monuments with nitrifying organisms. *In* "Biodeterioration" (D. R. Houghton, R. N. Smith and H. O. W. Eggins, Eds.), Vol.7, pp. 436–440. Elsevier Applied Science, London, New York.

Boone, D. R., Castenholz, R. W., and Garrity, G. M. (2001). "Bergey's Manual of Systematic Bacteriology" Vol. 1. Springer, New York.

Bungartz, F., Garvie, L. A. J., and Nash, T. H. (2004). Anatomy of the endolithic Sonoran Desert lichen *Verrucaria rubrocincta* Breuss: Implications for biodeterioration and biomineralization. *Lichenologist* **36,** 55–73.

Burford, E. P., Kierans, M., and Gadd, G. M. (2003). Geomycology: Fungal growth in mineral substrata. *Mycologist* **17,** 98–107.

Caneva, G., and Nugari, M. P. (2005). Evaluation of Escobilla's mucilagino treatments in the archaeological site of Joya de Ceren (El Salvador). *In* "LABS5. Biodegradation and Biodeterioration in Latin America" (B. O. Ortega-Morales, C. C. Gaylarde, J. A. Narvaez-Zapata and P. M. Gaylarde, Eds.), pp. 59–64. Universidad de Campeche, Mexico.

Caneva, G., Nugari, M. P., and Salvadori, O. (1991). "Biology in the Conservation of Works of Art." International Centre for the Study of the Preservation and Restoration of Cultural Property (ICCROM), Rome.

Caneva, G., Salvadori, O., Ricci, S., and Ceschin, S. (2005). Biological analysis for the conservation of the hieroglyphic stairway of Copan (Honduras). *In* "LABS5. Biodegradation and Biodeterioration in Latin America" (B. O. Ortega-Morales, C. C. Gaylarde, J. A. Narvaez-Zapata and P. M. Gaylarde, Eds.), pp. 55–58. Universidad de Campeche, Mexico.

Carcia-Pichel, F. (2006). Plausible mechanisms for the boring on carbonates by microbial phototrophs. *Sediment. Geol.* **185,** 205–213.

Carter, N. E. A., and Viles, H. A. (2003). Experimental investigations into the interactions between moisture, rock surface temperatures and an epilithic lichen cover in the bioprotection of limestone. *Build. Environ.* **38,** 1225–1234.

Carter, N. E. A., and Viles, H. A. (2004). Lichen hotspots: Raised rock temperatures beneath *Verrucaria nigrescens* on limestone. *Geomorphology* **62,** 1–16.

Chacón, E., Berrendero, E., and Garcia Pichel, F. (2006). Biogeological signatures of microboring cyanobacterial communities in marine carbonates from Cabo Rojo, Puerto Rico. *Sediment. Geol.* **185,** 215–228.

Chazal, N. M., and Smith, G. D. (1994). Characterization of a brown *Nostoc* species from Java that is resistant to high light intensity and UV. *Microbiology* **140,** 3183–3189.

Cockell, C. S., Horneck, G., Rettberg, P., Facius, R., and Gugg-Helminger, A. (2002). Influence of snow and ice covers on UV exposure of Antarctic microbial communities–dosimetric studies. *J. Photochem. Photobiol.* **68,** 23–32.

Crispim, C. A., and Gaylarde, C. C. (2004). Cyanobacteria and biodeterioration of cultural heritage: A review. *Microb. Ecol.* **10,** 1007/s00248–003–1052–5.

Crispim, C. A., Gaylarde, P. M., and Gaylarde, C. C. (2003). Algal and cyanobacterial biofilms on calcareous historic buildings. *Curr. Microbiol.* **46,** 79–82.

Danin, A., and Caneva, G. (1990). Deterioration of limestone walls in Jerusalem and marble monuments in Rome caused by cyanobacteria and cyanophilous lichens. *Int. Biodeterior.* **26,** 397–417.

De los Rios, A., and Ascaso, C. (2005). Contributions of *in situ* microscopy to the current understanding of stone biodeterioration. *Int. Microbiol.* **8,** 181–188.

De los Rios, A., Galvan, V., and Ascaso, C. (2004). *In situ* microscopical diagnosis of biodeterioration processes at the convent of Santa Cruz la Real, Segovia, Spain. *Int. Biodeterior. Biodegrad.* **51,** 113–120.

Devevre, O., Garbaye, J., and Botton, B. (1996). Release of complexing organic acids by rhizosphere fungi as a factor in Norway spruce yellowing in acidic soils. *Mycol. Res.* **100,** 1367–1374.

Di Bonaventura, M. P., Gallo, M. D., Cacchio, P., Ercole, C., and Lepidi, A. (1999). Microbial formation of oxalate films on monument surfaces: Bioprotection or Biodeterioration? *Geomicrobiol. J.* **16,** 55–64.

Dornieden, T., Gorbushina, A. A., and Krumbein, W. E. (1997). Änderungen der physikalischen Eigenschaften von Marmor durch Pilzbewuchs. *Int. J. Restoration Buildings Monuments* **3,** 441–456.

Dornieden, T., Gorbushina, A. A., and Krumbein, W. E. (2000). Biodecay of cultural heritage as a space/time-related ecological situation. An evaluation of a series of studies. *Int. Biodeterior. Biodegrad.* **46,** 261–270.

Douglas, S., and Beveridge, T. J. (1998). Mineral formation by bacteria in natural microbial communities. *FEMS Microbiol. Ecol.* **26,** 74–88.

Ehling-Schultz, M., Bilger, W., and Scherer, S. (1997). UV-B-induced synthesis of photoprotective pigments and extracellular polysaccharides in the terrestrial cyanobacterium *Nostoc commume. J. Bacteriol.* **179,** 1940–1945.

Eppard, M., Krumbein, W. E., Koch, C., Rhiel, E., Staley, J. T., and Stackebrandt, E. (1996). Morphological, physiological, and molecular characterization of actinomycetes isolated from dry soil, rocks, and monument surfaces. *Arch. Microbiol.* **166,** 12–22.

Evenari, M., Noy-Meir, I., and Goodall, D. W. (1985). "Hot Deserts and Arid Shrublands. 12A: Ecosystems of the World." Elsevier, Amsterdam.

Ferris, F. G., and Lowson, E. A. (1997). Ultrastructure and geochemistry of endolithic microorganisms in limestone of the Niagara Escarpment. *Can. J. Microbiol.* **43,** 211–219.

Flores, M., Lorenzo, J., and Gomez-Alarcon, G. (1997). Algae and bacteria on historic monuments at Alcala de Henares, Spain. *Int. Biodeterior. Biodegrad.* **40,** 241–246.

Frey, T., von Reis, J., and Barov, Z. (1993). An evaluation of biocides for control of the biodeterioration of artifacts at Hearst Castle. *In* "Preprints of the Triennial Meeting of the International Council of Museums Committee for Conservation (ICOM-CC)" pp. 875–881. Dresden.

Gadd, G. M. (2007). Geomycology: Biogeochemical transformations of rocks, minerals and radionucleotides by fungi, bioweathering and bioremediation. *Mycol. Res.* **111,** 3–49.

Garcia de Miguel, J. M., Sanchez-Castillo, L., Ortega-Calvo, J. J., Gil, J. A., and Saiz-Jimenez, C. (1995). Deterioration of building materials from the Great Jaguar Pyramid at Tikal, Guatemala. *Building Env.* **30,** 591–598.

Garcia-Pichel, F., Sherry, N. D., and Castenholz, R. W. (1992). Evidence for ultraviolet sunscreen role of the extracellular pigment scytonemin in the terrestrial cyanobacterium *Chlorogloeopsis* sp. *Photochem. Photobiol.* **56,** 17–23.

Garcia-Vallès, M., Topal, T., and Vendrell-Saz, M. (2003). Lichenic growth as a factor in the physical deterioration or protection of Cappadocian monuments. *Environ. Geol.* **43,** 776–781.

Garcia-Vallès, M., Vendrell-Saz, M., Molera, J., and Blazquez, F. (1997). Interaction of rock and atmosphere: Patinas on Mediterranean monuments. *Environ. Geol.* **36,** 137–149.

Garty, J. (1990). Influence of epilithic microorganisms on the surface temperature of building walls. *Can. J. Bot.* **68,** 1349–1353.

Gaylarde, P. M., Crispim, C. A., Neilan, B. A., and Gaylarde, C. C. (2005). Cyanobacteria from Brazilian building walls are distant relatives of aquatic genera. *OMICS* **9,** 30–42.

Gaylarde, C. C., and Englert, G. E. (2006). Analysis of surface patina on the church of Nossa Senhora do Rosario, Ouro Preto, Brazil. *In* "Structural Analysis of Historical Constructions" (P. B. Lourenço, P. Roca, C. Modena and S. Agrawal, Eds.), pp. 1708–1715. Maximilian India Ltd.

Gaylarde, P., Englert, G., Ortega-Morales, O., and Gaylarde, C. (2006). Lichen-like colonies of pure *Trentepohlia* on limestone monuments. *Int. Biodeterior. Biodegrad.* **58,** 248–253.

Gaylarde, P., and Gaylarde, C. (2004). Deterioration of siliceous stone monuments in Latin America: Microorganisms and mechanisms. *Corr. Rev.* **22,** 395–415.

Gaylarde, C. C., and Gaylarde, P. M. (2005). A comparative study of the major microbial biomass of biofilms on exteriors of buildings in Europe and Latin America. *Int. Biodeterior. Biodegrad.* **55,** 131–139.

Gaylarde, P. M., Gaylarde, C. C., Guiamet, P. S., Gomez de Saravia, S. G., and Videla, H. A. (2001). Biodeterioration of Mayan Buildings at Uxmal and Tulum, Mexico. *Biofouling* **17,** 41–45.

Gaylarde, C. C., Gaylarde, P. M., Copp, J., and Neilan, B. A. (2004). Polyphasic detection of cyanobacteria in terrestrial biofilms. *Biofouling* **20,** 71–79.

Gaylarde, C., and Morton, G. (2002). Biodeterioration of mineral materials. *In* "Environmental Microbiology" (G. Britton, Ed.), Vol. 1, pp. 516–528. Wiley, New York.

Gaylarde, C. C., Ortega-Morales, B. O., and Bartolo-Perez, P. (2007). Biogenic black crusts on buildings in unpolluted environments. *Curr. Microbiol.* **54,** 162–166.

Gerrath, J. F., Gerrath, J. A., and Larson, D. W. (1995). A preliminary account of the endolithic algae of limestone cliffs of the Niagara Escarpment. *Can. J. Bot.* **73,** 788–793.

Golubic, S., Friedmann, I., and Schneider, J. (1981). The lithobiontic ecological niche, with special reference to microorganisms. *J. Sediment. Res.* **51,** 475–478.
Golubic, S., Perkins, R. D., and Lukas, K. J. (1975). Boring microorganisms and microborings in carbonate substrates. *In* "The Study of Trace Fossils" (R. W. Frey, Ed.), pp. 229–259. Springer-Verlag, Berlin.
Golubic, S., Radtke, G., and Campion-Alsumard, T. (2005). Endolithic fungi in marine ecosystems. *Trends Microbiol.* **13,** 229–235.
Golubic, S., Seong-Joo, L., and Browne, K. M. (2000). Cyanobacteria: Architects of sedimentary structures. *In* "Microbial Sediments" (R. Riding and S. M. Awrami, Eds.), pp. 57–67. Springer, Berlin.
Gorbushina, A. A. (2007). Life on the rocks. *Environ. Microbiol.* **9,** 1613–1631.
Gorbushina, A. A., and Krumbein, W. E. (2000). Subaerial microbial mats and their effects on soil and rock. *In* "Microbial Sediments" (R. E. Riding and S. M. Awramik, Eds.), pp. 161–169. Springer-Verlag, Berlin.
Gorbushina, A. A., and Krumbein, W. E. (2004). Role of organisms in wear down of rocks and minerals. *In* "Micro-organisms in Soils: Roles in Genesis and Functions" (F. Buscot and A. Varma, Eds.), pp. 59–84. Springer, Berlin.
Gorbushina, A., Krumbein, W. E., and Vlasov, D. (1996). Biocarst cycles on monument surfaces. *In* "Preservation and restoration of cultural heritage" (R. Pancella, ed.), pp. 319–332. Proceedings of the 1995. LPC Congress. EPFL, Lausanne.
Gorbushina, A. A., Lyalikova, N. N., Vlasov, D. Y., and Khizhnyak, T. V. (2002). Microbial communities on the monuments of moscow and St. Petersburg: Biodiversity and trophic relations. *Microbiology* **71,** 350–356. (Translated from *Mikrobiologiya* **71,** 409–417).
Gorbushina, A. A., Whitehead, K., Dornieden, T., Nisse, A., Schulte, A., and Hedges, J. L. (2003). Black fungal colonies as units for survival: Hyphal mycosporines synthized by rock-dwelling microcolonial fungi. *Can. J. Bot.* **81,** 131–138.
Gordon, S. J., and Dorn, R. I. (2005). Localized weathering: Implications for theoretical and applied studies. *Prof. Geogr.* **57,** 28–43.
Grant, C. (1982). Fouling of terrestrial substrates by algae and implications for control–A review. *Int. Biodeterior. Bull.* **18,** 57–65.
Guillitte, O. (1995). Bioreceptivity: A new concept for building ecology studies. *Sci. Total Environ.* **167,** 215–220.
Hawes, I., Oward-Williams, C., and Vincent, W. F. (1992). Desiccation and recovery of cyanobacterial mats. *Polar Biol.* **12,** 587–594.
Herrera, L. K., and Videla, H. A. (2004). The importance of atmospheric effects on biodeterioration of cultural heritage constructional materials. *Int. Biodeterior. Biodegrad.* **54,** 125–134.
Heyrman, J. (2003). *Polyphasic characterisation of the bacterial community associated with biodeterioration of mural paintings,* PhD Thesis, Universiteit Gent, Gent:.
Heyrman, J., Mergaert, J., Denys, R., and Swings, J. (1999). The use of fatty acid methyl ester analysis (FAME) for the identification of heterotrophic bacteria present on three mural paintings showing severe damage by microorganisms. *FEMS Microbiol. Lett.* **181,** 55–62.
Heyrman, J., and Swings, J. (2001). 16S rDNA Sequence analysis of bacterial isolates from biodeteriorated mural paintings in the servilia tomb (Necropolis of Carmona, Seville, Spain). *Syst. Appl. Microbiol.* **24,** 417–422.
Heyrman, J., Verbeeren, J., Schumann, P., Swings, J., and De Vos, P. (2005). Six novel *Arthrobacter* species isolated from deteriorated mural paintings. *Int. J. Syst. Evol. Microbiol.* **55,** 1457–64.
Hirsch, P., Eckhardt, F. E. W., and Palmer, R. J., Jr. (1995a). Methods for the study of rock-inhabiting microorganisms–A mini review. *J. Microbiol. Meth.* **23,** 143–167.
Hirsch, P., Eckhardt, F. E. W., and Palmer, R. J., Jr. (1995b). Fungi active in weathering of rock and stone monuments. *Can. J. Bot.* **73,** SUP1, S1384–S1390.

Hirsch, P., Mevs, U., Kroppenstedt, R. M., Schumann, P., and Stackebrandt, E. (2004). Cryptoendolithic actinomycetes from Antarctic rock samples: *Micromonospora endolithica* sp. Nov. and two isolates related to *Micromonospora coerulea* Jensen 1932. *Syst. Appl. Microbiol.* **27,** 166–174.

Hoffman, L. (1989). Algae of terrestrial habitats. *Bot. Rev.* **55,** 77–105.

Hoffland, E., Kuyper, T. W., Wallander, H., Plassard, C., Gorbushina, A. A., Haselwandter, K., Holmström, S., Landeweert, R., Lundström, U. S., Rosling, A., Sen, R., Smits, M. M., *et al.* (2004). The role of fungi in weathering. *Front. Ecol. Environ.* **2,** 258–264.

Hoppert, M., Flies, C., Pohl, W., Günzl, B., and Schneider, J. (2004). Colonization strategies of lithobiontic microorganisms on carbonate rocks. *Environ. Geol.* **46,** 421–428.

Hyvert, G. (1966). Quelques Actinomycetes isolés sur les gres des monuments cambodgiens. *Revue de Mycologie* **31,** 179–86.

Kemmling, A., Kamper, M., Flies, C., Schieweck, O., and Hoppert, M. (2004). Biofilms and extracellular matrices on geomaterials. *Environ. Geol.* **46,** 429–435.

Kiel, G., and Gaylarde, C. (2006). Bacterial diversity in biofilms on external surfaces of historic buildings in Porto Alegre. *World J. Microbiol. Biotechnol.* **22,** 293–297.

Krumbein, W. E. (1969). Über den Einfluss der Mikroflora auf die exogene Dynamik (Verwitterung und Krustenbildung). *Geol. Rundsch.* **58,** 333–363.

Krumbein, W. E., and Giele, C. (1979). Calcification in a coccoid cyanobacterium associated with the formation of desert stromatolites. *Sedimentology* **26,** 593–604.

Krumbein, W. E., and Jens, K. (1981). Biogenic rock varnishes of the Negev desert (Israel)–An ecological study of iron and manganese transformations by cyanobacteria and fungi. *Oecologia* **50,** 25–38.

Kumar, R., and Kumar, A. V. (1999). ''Biodeterioration of stone in tropical environments. An Overview.'' The Getty Conservation Institute, Los Angeles.

Kurtz, H. D., Jr (2002). Endolithic microbial communities as bacterial biofilms: The role of EPS. *In* ''Molecular Ecology of Biofilms'' (R. J. C. McLean and A. W. Decho, Eds.), pp. 105–119. Horizon Scientific Press, Wymondhamchapter 5.

Kussmaul, M., Wilimzig, M., and Bock, E. (1998). Methanotrophs and methanogens in masonry. *Appl. Environ. Microbiol.* **64,** 4530–4532.

Laiz, L., Groth, I., Schumann, P., Zezza, F., Felske, A., Hermosin, B., and Saiz-Jimenez, C. (2000). Microbiology of the stalactites from Grotta dei Cervi, Porto Badisco, Italy. *Int. Microbiol.* **3,** 25–30.

Laiz, L., Piñar, G., Lubitz, W., and Saiz-Jimenez, C. (2003). Monitoring the colonization of monuments by bacteria: Cultivation versus molecular methods. *Environ. Microbiol.* **5,** 72–74.

Leavengood, P., Twilley, J., and Asmus, J. F. (2000). Lichen removal from chinese spirit path figures of marble. *J. Cult. Herit.* **S1,** 71–74.

Lee, C. H., Choi, S. W., and Suh, M. (2003). Natural deterioration and conservation treatment of the granite standing Buddha of Daejosa Temple, Republic of Korea. *Geotech. Geolog. Eng.* **21,** 63–77.

Lisci, L., Monte, M., and Pacini, E. (2003). Lichens and higher plants on stone: A review. *Int. Biodeterior. Biodegrad.* **51,** 1–17.

Mansch, R., and Bock, E. (1998). Biodeterioration of natural stone with special reference to nitrifying bacteria. *Biodegradation* **9,** 47–64.

Mao-Che, L., Le-Campion-Alsumard, T., Boury-Esnault, N., Payri, C., Golubic, S., and Bezac, C. (1996). Biodegradation of shells of the black pearl oyster, *Pinctada margaritifera* var. cumingii, by microborers and sponges of French Polynesia. *Mar. Biol.* **126,** 509–519.

Matthes-Sears, U., Gerrath, J. A., and Larson, D. W. (1997). Abundance, biomass, and productivity of endolithic and epilithic lower plants on the temperate-zone cliffs of the Niagara Escarpment. *Can. Int. J. Plant. Sci.* **158,** 451–460.

May, E. (2003). Microbes on building stone—For good or ill? *Culture* **24,** 5–8.

May, E., Papida, S., and Abdulla, H. (2003). Consequences of microbe-biofilm-salt interactions for stone integrity in monuments. *In* "Art, biology and conservation: Biodeterioration of works of art" (R. J. Koestler, V. H. Koestler, A. E. Charola and F. E. Nieto-Fernandez, Eds.), pp. 452–471. The Metropolitan Museum of Art, New York.

May, E., Papida, S., Hesham, A., Tayler, S., and Dewedar, A. (2000). Comparative studies of microbial communities on stone monuments in temperate and semi-arid climates. *In* "Of microbes and art. The role of microbial communities in the degradation and protection of cultural heritage" (O. Ciferri, P. Tiano and G. Mastromei, Eds.), pp. 49–62. Kluwer Academic/ Plenum Publisher, New York.

McNamara, C. J., Perry, T. D., Bearce, K. A., Hernandez-Duque, G., and Mitchell, R. (2006). Epilithic and endolithic bacterial communities in limestone from a Maya archaeological site. *Microbial. Ecol.* **51,** 51–64.

McNamara, C. J., Perry, T. D., Zinn, M., Breuker, M., and Mitchell, R. (2003). Microbial processes in the deterioration of Mayan archaeological buildings in southern Mexico. *In* "Art, biology and conservation: Biodeterioration of works of art" (R. J. Koestler, V. H. Koestler, A. E. Charola and F. E. Nieto-Fernandez, eds.), pp. 248–265. The Metropolitan Museum of Art, New York.

Meincke, M., Ahlers, B., Krause-Kupsch, T., Krieg, E., Meyer, C., Sameluck, F., Sand, W., Wolters, B., and Bock, E. (1988). Isolation and characterization of endolithic nitrifiers. *In* "Proceedings of the Sixth International Congress on Deterioration and Conservation of Stone" pp. 15–23. Nicolaus Copernicus University, Torun, Poland, 12–14 September 1988.

Miller, A. G., Espie, G. S., and Canvin, D. T. (1990). Physiological aspects of CO_2 and HCO_3^- transport by cyanobacteria—A review. *Can. J. Bot.* **68,** 1291–1302.

Mitchell, R., and Gu, J. -D. (2000). Changes in biofilm microflora of limestone caused by atmospheric pollutants. *Int. Biodeterior. Biodegrad.* **46,** 299–303.

Monte, M. (2003). Oxalate film formation on marble specimens caused by fungus. *J. Cult. Herit.* **4,** 255–258.

Mottershead, D., Gorbushina, A., Lucas, G., and Wright, J. (2003). The influence of marine salts, aspect and microbes in the weathering of sandstone in two historic structures. *Build. Environ.* **38,** 1193–1204.

Mottershead, D., and Lucas, G. (2000). The role of lichens in inhibiting erosion of a soluble rock. *Lichenologist* **32,** 601–609.

Muntz, A. (1890). *Compt. Rend. Acad. Sci.* **110,** 1370.

Nienow, J. A., McKay, C. P., and Friedmann, E. I. (1988). The cryptoendolithic microbial environment in the Ross Desert of Antarctica: Light in the photosynthetically active region. *Microb. Ecol.* **16,** 271–289.

Nimis, P. L., and Monte, M. (1988). Lichens and monuments. *Stud. Geobot.* **8,** 1–133.

Ohba, N., and Tsujimoto, Y. (1996). Soiling of external materials by algae and its prevention I. Situation of soiling and identification of algae. *J. Jpn. Wood Res. Soc.* **42,** 589–595.

Ortega-Morales, B. O. (1999). Approche des communautés microbiennes et leur rôle dans la biodeterioration des monuments Mayas du site archéologique d'Uxmal (Yucatan, Mexique). PhD Thesis. France: University of Brest, Brest.

Ortega-Morales, B. O., Gaylarde, C. C., Englert, G. E., and Gaylarde, P. M. (2005). Analysis of salt-containing biofilms on limestone buildings of the Mayan culture at Edzna, Mexico. *Geomicrobiol. J.* **22,** 261–268.

Ortega-Morales, O., Guezennec, J., Hernandez-Duque, G., Gaylarde, C. C., and Gaylarde, P. M. (2000). Phototrophic biofilms on ancient Mayan buildings in Yucatan, Mexico. *Curr. Microbiol.* **40,** 81–85.

Ortega-Morales, B. O., and Hernandez-Duque, G. (1998). Biodeterioro de monumentos históricos Mayas. Caso de biodegradación microbiana de la edificaciones en Uxmal (Yucatan) y la evaluación de medios de protección y preservación. *Ciencia y Desarrollo* **24,** 48–53.

Ortega-Morales, B. O., Hernández-Duque, G., Borges-Gómez, L., and Guezennec, J. (1999). Characterization of epilithic microbial communities associated with Mayan stone monuments in Yucatan, Mexico. *Geomicrobiol. J.* **16,** 221–232.

Ortega-Morales, B. O., Lopez-Cortes, A., Hernandez-Duque, G., Crassous, P., and Guezennec, J. (2001). Extracellular polymers of microbial communities colonizing ancient limestone monuments. *In* "Methods in Enzymology. Microbial Growth in Biofilms Part A: Developmental and Molecular Biological Aspects" (R. J. Doyle, Ed.), Vol. 336, pp. 331–339. Academic Press, San Diego.

Ortega-Morales, B. O., Narvaez-Zapata, J. A., Schmalenberger, A., Dousa-Lopez, A., and Tebbe, C. C. (2004). Biofilms fouling ancient limestone Mayan monuments in Uxmal, Mexico: A cultivation-independent analysis. *Biofilms* **1,** 79–90.

Palla, F., Federico, C., Russo, R., and Anello, L. (2002). Identification of *Nocardia restricta* in biodegraded sandstone monuments by PCR and nested-PCR DNA amplification. *FEMS Microbiol. Ecol.* **39,** 85–89.

Pattanaik, B., and Adhikary, S. P. (2002). Blue-green algal flora at some archaeological sites and monuments of India. *Feddes Rep.* **113,** 289–300.

Pentecost, A. (1987). Growth and calcification of the fresh-water cyanobacterium *Rivularia haematites. Proc. R. Soc. Lond. B* **232,** 125–136.

Pentecost, A. (1988). Growth and calcification of the cyanobacterium *Homoeothrix crustacea. J. Gen. Microbiol.* **134,** 2665–2671.

Perry, T. D., Duckworth, O. W., McNamara, C. J., Martin, S. T., and Mitchell, R. (2004). Effects of the biologically produced polymer alginic acid on macroscopic and microscopic calcite dissolution rates. *Environ. Sci. Technol.* **38,** 3040–3046.

Piñar, G., Ramos, C., Rölleke, S., Schabereiter-Gurtner, C., Vybiral, D., Lubitz, W., and Denner, E. B. M. (2001). Detection of indigenous *Halobacillus* populations in damaged ancient wall paintings and building materials: Molecular monitoring and cultivation. *Appl. Environ. Microbiol.* **67,** 4891–4895.

Pinck, C., and Balzarotti-Kämmlein, R. (2000). Biocidal efficacy of Algophase against nitrifying bacteria. Poster presented at: Protection and Conservation of the Cultural Heritage of the Mediterranean Cities, *5th International Symposium on the Conservation of Monuments in the Mediterranean Basin,* 5th–8th April, 2000, Seville, Spain. Available at http://www.sci.port.ac.uk/ec/Poster%20Seville%20CSPU.htm, accessed 17/5/06.

Potts, M. (1994). Desiccation tolerance of prokaryotes. *Microbiol. Rev.* **58,** 755–805.

Potts, M., and Friedmann, E. I. (1981). Effects of water stress on cryptoendolithic cyanobacteria from hot desert rocks. *Arch. Microbiol.* **130,** 267–271.

Price, C. A. (1996). "Stone Conservation. An overview of current research." Research in Conservation Series. The Getty Conservation Institute, Santa Monica.

Prieto, B., Seaward, M. R., Edwards, H. G., Rivas, T., and Silva, B. (1999). Biodeterioration of granite monuments by *Ochrolechia parella* (L.) mass: An FT Raman spectroscopic study. *Biospectroscopy* **5,** 53–59.

Prieto, B., and Silva, B. (2005). Estimation of the potential bioreceptivity of granitic rocks from their intrinsic properties. *Int. Biodeterior. Biodegrad.* **56,** 206–215.

Prieto, B., Silva, B., Aira, N., and Álvarez, L. (2006). Towards a definition of a bioreceptivity index for granitic rocks. Perception of the change in appearance of the rock. *Int. Biodeterior. Biodegrad.* **58,** 150–154.

Prieto, B., Silva, B., Rivas, T., Wierzchos, J., and Ascaso, C. (1997). Mineralogical transformation and neoformation in granite caused by the lichens *Tephronella atra* and *Ochorelechia parella. Int. Biodeterior. Biodegrad.* **40,** 191–199.

Resende, M. A., Leite, C. A., Warscheid, T., Becker, T. W., and Krumbein, W. E. (1992). Microbiological investigations on quartzite and soapstone of historical monuments in Minas Gerais, Brazil. *In* "FIRST LABS, Proc. First Latin American Biodeter. Symp., Campos do Jordao, Brazil" (W. C. Latorre and C. C. Gaylarde, Eds.), pp. 17–22. TecArt Editora, Sao Paulo.

Resende, M. A., Rezende, G. C., Viana, E. V., Becker, T. W., and Warscheid, T. (1996). Acid production by fungi isolated from historic monuments in the Brazilian state of Minas Gerais. *In* "Biodegradation & Biodeterioration in Latin America, Mircen/UNEP/ UNESCO/ICRO-FEPAGRO/UFRGS" (C. C. Gaylarde, E. L. S. de Sá and P. M. Gaylarde, Eds.), pp. 65–67. Porto Alegre.

Rivadeneyra, M. A., Párraga, J., Delgado, R., Ramos-Cormenzana, A., and Delgado, G. (2004). Biomineralization of carbonates by *Halobacillus trueperi* in solid and liquid media with different salinities. *FEMS Microbiol. Ecol.* **48,** 39–46.

Roeselers, G., van Loosdrecht, M. C. M., and Muyzer, G. (2007). Heterotrophic pioneers facilitate phototrophic biofilm development. *Microbiol. Ecol.* **54,** 578–585.

Rölleke, S., Muyzer, G., Wawer, C., Wanner, G., and Lubitz, W. (1996). Identification of bacteria in a biodegraded wall painting by denaturing gradient gel electrophoresis of PCR-amplified gene fragments coding for 16S rRNA. *Appl. Environ. Microbiol.* **62,** 2059–2065.

Roelleke, S., Witte, A., Wanner, G., and Lubitz, W. (1998). Medieval wall painting–A habitat for Archaea: Identification of Archaea by denaturing gradient gel electrophoresis (DGGE) of PCR amplified gene fragments coding 16S rRNA in a medieval wall painting. *Int. Biodeterior. Biodegrad.* **41,** 85–92.

Saarela, M., Alakomi, H. -L., Suihko, M. -L., Maunuksela, L., Raaska, L., and Mattila-Sandholm, T. (2004). Heterotrophic microorganisms in air and biofilm samples from Roman catacombs, with special emphasis on actinobacteria and fungi. *Int. Biodeterior. Biodegrad.* **54,** 27–37.

Saiz-Jimenez, C. (1995). Deposition of anthropogenic compounds on monuments and their effect on airborne microorganisms. *Aerobiologia* **11,** 161–175.

Saiz-Jimenez, C., Garcia-Rowe, J., Garcia del Cura, M. A., Ortega-Calvo, J. J., Roekens, E., and Van Grieken, R. (1990). Endolithic cyanobacteria in Maastricht limestone. *Sci. Total Environ.* **94,** 209–220.

Saiz-Jimenez, C., and Laiz, L. (2000). Occurrence of halotolerant/halophilic bacterial communities in deteriorated monuments. *Int. Biodeterior. Biodegrad.* **46,** 319–326.

Salazar, O., Valverde, A., and Genilloud, O. (2006). Real-time PCR for the detection and quantification of Geodermatophilaceae from stone samples and identification of new members of the genus *Blastococcus*. *Appl. Environ. Microbiol.* **72,** 346–352.

Salinas-Nolasco, M. F., Méndez-Vivar, J., Lara, V. A., and Bosch, P. (2004). Passivation of the calcite surface with malonate ion. *J. Colloid Interface Sci.* **286,** 16–24.

Salvadori, O. (2000). Characterisation of the endolithic communities of stone monuments and natural outcrops. *In* "Of microbes and art. The role of microbial communities in the degradation and protection of cultural heritage" (O. Ciferri, P. Tiano and G. Mastromei, Eds.), pp. 89–101. Kluwer Academic/ Plenum Publisher, New York.

Sand, W. (1996). Microbial mechanisms. *In* "Microbially influenced corrosion of materials" (E. Heitz, H. -C. Flemming and W. Sand, Eds.), pp. 15–25. Springer Verlag, Berlin, Heidelberg.

Sand, W., Jozsa, P. -G., and Mansch, R. (2002). Weathering, Microbiol. *In* "Environmental Microbiology" (G. Britton, Ed.), Vol. 6, pp. 3364–3375. Wiley, New York.

Schabereiter-Gurtner, C., Piñar, G., Lubitz, W., Roelleke, S., and Saiz-Jimenez, C. (2003). Acidobacteria in Paleolithic painting caves. *In* "Molecular biology and cultural heritage" (C. Saiz-Jimenez, Ed.), pp. 15–22. Swets & Zeitlinger B. V., Lisse, The Netherlands.

Schabereiter-Gurtner, C., Saiz-Jimenez, C., Pinar, G., Lubitz, W., and Rolleke, S. (2004). Phylogenetic diversity of bacteria associated with Paleolithic paintings and surrounding rock walls in two Spanish caves (Llonin and La Garma). *FEMS Microbiol. Ecol.* **47,** 235–247.

Schnabel, L. (1991). The treatment of biological growths on stone: A conservator's viewpoint. *Int. Biodeterior.* **28,** 125–131.

Schultze-Lam, S., and Beveridge, T. J. (1994). Physicochemical characteristics of the mineral-forming S-layer from the cyanobacterium *Synechococcus* strain GL24. *Appl. Environ. Microbiol.* **60,** 447–453.

Seaward, M. R. D. (2003). Lichens, agents of monumental Destruction. *Microbiol. Today* **30,** 110–112.

Seeler, J. -S., and Golubic, S. (1991). *Iyengariella endolithica* sp. nova, a carbonate boring stigonematalean cyanobacterium from a warm spring-fed lake: Nature to culture. *Algol. Stud.* **64,** 399–410.

Sigler, W. V., Bachofen, R., and Zeyer, J. (2003). Molecular characterization of endolithic cyanobacteria inhabiting exposed dolomite in central Switzerland. *Environ. Microbiol.* **5,** 618–627.

Spieck, E., Meinecke, M., and Bock, E. (1992). Taxonomic diversity of *Nitrosovibrio* strains isolated from building sandstones. *FEMS Microbiol. Ecol.* **102,** 21–26.

Sterflinger, K. (1998). Temperature and NaCl-tolerance of rock-inhabiting meristematic fungi. *A. Van Leeuw. J. Microb.* **74,** 271–281.

Sterflinger, K. (2000). Fungi as geologic agents. *Geomicrobiol. J.* **17,** 97–124.

Sterflinger, K. (2002). Fungi on stone monuments: State of the art. *Coalition* **4,** 3–5.

Sterflinger, K., and Krumbein, W. E. (1997). Dematiaceous fungi as a major agent for biopitting on mediterranean marbles and limestones. *Geomicrobiol. J.* **14,** 219–230.

Strasser, H., Burgstaller, W., and Schinner, F. (1994). High yield production of oxalic acid for met al leaching purposes by *Aspergillus niger*. *FEMS Microbiol. Lett.* **119,** 365–370.

Strzelczyk, A. B. (1981). Microbial biodeterioration: Stone. *In* "Economic microbiology" (A. H. Rose, Ed.), Vol. 6, pp. 62–80. Academic Press, London.

Tayler, S., and May, E. (1991). The seasonality of heterotrophic bacteria on sandstones on ancient monuments. *Int. Biodeterior.* **28,** 49–64.

Tiano, P. (2002). Biodegradation of cultural heritage: Decay,mechanisms and control methods. Seminar article held at the Universidade Nova de Lisboa, Departamento de Concervação e Restauro, 7 - 12. January, 2002. http://www.arcchip.cz/w09/w09_tiano.pdf, accessed: 22/08/2005

Tiano, P., Accolla, P., and Tomaselli, L. (1995). Phototrophic biodeteriogens on lithoid surfaces—An ecological study. *Microbiol Ecol.* **29,** 299–309.

Tomaselli, L., Lamenti, G., Bosco, M., and Tiano, P. (2000b). Biodiversity of photosynthetic microorganisms dwelling on stone monuments. *Int. Biodeterior. Biodegrad.* **46,** 251–258.

Tomaselli, L., Tiano, P., and Lamenti, G. (2000a). Occurrence and fluctuations in photosynthetic biocoenoses dwelling on stone monuments. *In* "Of microbes and art. The role of microbial communities in the degradation and protection of cultural heritage" (O. Ciferri, P. Tiano and G. Mastromei, Eds.), pp. 49–62. Kluwer Academic/ Plenum Publisher, New York.

Torre de la, A., Gomez-Alarcon, G., Vizcaino, C., Garcia, T., and de la Torre, M. A. (1993). Biochemical mechanisms of stone alteration carried out by filamentous fungi living in monuments. *Biogeochemistry* **19,** 129–147.

Tretiach, M., and Pecchiari, M. (1995). Gas exchange rates and chlorophyll content of epi- and endolithic lichens from the Trieste Karst (NE Italy). *New Phytol.* **130,** 585–592.

Tripathy, P., Roy, A., Anand, N., and Adhikary, S. P. (1999). Blue-green algal flora on the rock surface of temples and monuments of India. *Feddes Repert.* **110,** 133–144.

Urquart, D. C. M., Young, M. E., MacDonald, J., Jones, M. S., and Nicholson, K. A. (1996). Aberdeen granite buildings: A study of soiling and decay. *In* "Processes of urban stone decay" (P. J. Smith and P. A. Warke, Eds.), pp. 66–77. Donhead Publishing, London.

Urzi, C. (1993). Interactions of some microbial communities in the biodeterioration of marble and limestone. *In* "Trends in microbiology" (R. Guerrero and C. Pedros-Alio, Eds.), pp. 667–672. Spanish Society for Microbiology.

Urzi, C. (2004). Microbial deterioration of rocks and marble monuments in Mediterranean Basin: A review. *Corros. Rev.* **22,** 441–457.

Urzi, C., Brusetti, L., Salamone, P., Sorlini, C., Stackebrandt, E., and Daffonchio, D. (2001). Biodiversity of Geodermatophilaceae isolated from altered stones and monuments in the Mediterranean basin. *Environ. Microbiol.* **3,** 471–479.

Urzi, C., and De Leo, F. (2001). Sampling with adhesive tape strips: An easy and rapid method to monitor microbial colonization on monument surfaces. *J. Microbiol. Meth.* **44,** 1–11.

Urzi, C., de Leo, F., de How, S., and Sterflinger, K. (2000). Recent advances in the molecular biology and ecophysiology of meristematic stone-inhabiting fungi. *In* "Of microbes and art. The role of microbial communities in the degradation and protection of cultural heritage" (O. Ciferri, P. Tiano and G. Mastromei, Eds.), pp. 3–21. Kluwer Academic/Plenum Publisher, New York.

Urzi, C., and Krumbein, W. E. (1994). Microbiological impacts on cultural heritage. *In* "Durability and change. The science, responsibility, and cost of sustaining cultural heritage" (W. E. Krumbein, Ed.), pp. 343–368. Wiley, Chichester, Report of the Dahlem Workshop 6—11 December 1992. Environmental science research report 15.

Valls del Barrio, S., Garcia-Vallès, M., Pradell, T., and Vendrell-Saz, M. (2002). The red–orange patina developed on a monumental dolostone. *Eng. Geol.* **63,** 31–38.

Van der Molen, J. M., Garty, J., Aardema, B. W., and Krumbein, W. E. (1980). Growth control of algae and cyanobacteria on historical monuments by a mobile UV unit (MUVU). *Stud. Conserv.* **25,** 71–77.

Ward, D. M., Weller, R., and Bateson, M. M. (1990). 16S rRNA sequences reveal numerous uncultured microorganisms in a natural community. *Nat. Lond.* **345,** 63–65.

Warke, P. A., Smith, B. J., and Magee, R. W. (1996). Thermal response characteristics of stone: Implications for weathering of soiled surfaces in urban environments. *Earth Surf. Proc. Land.* **21,** 295–306.

Warscheid, T. (2000). Integrated concepts for the protection of cultural artefacts against biodeterioration. *In* "Of Microbes and Art. The role of microbial communities in the degradation and protection of cultural heritage" (O. Ciferri, P. Tiano and G. Mastromei, Eds.), pp. 185–202. Kluwer Academic/Plenum Publisher, New York.

Warscheid, T. (2003). The evaluation of biodeterioration processes on cultural objects and approaches for their effective control. *In* "Art, biology and conservation: Biodeterioration of works of art" (R. J. Koestler, V. H. Koestler, A. E. Charola and F. E. Nieto-Fernandez, Eds.), pp. 14–27. The Metropolitan Museum of Art, New York.

Warscheid, T., and Braams, J. (2000). Biodeterioration of stone: A review. *Int. Biodeterior. Biodegrad.* **46,** 343–368.

Warscheid, T., Becker, T. W., and Resende, M. A. (1996). Biodeterioration of Stone: A comparison of (sub-) tropical and moderate climate zones. *In* "Biodegradation & biodeterioration in Latin America" (C. C. Gaylarde, E. L. S. de Sá, and P. M. Gaylarde, Eds.), pp. 63–64. Mircen/UNEP/UNESCO/ICRO-FEPAGRO/UFRGS, Porto Alegre.

Warscheid, T., Oelting, M., and Krumbein, W. E. (1991). Physico-chemical aspects of biodeterioration processes on rocks with special regard to organic pollutants. *Int. Biodeterior.* **28,** 37–48.

Warscheid, T., Petersen, K., and Krumbein, W. E. (1995). Effect of cleaning on the distribution of microorganisms on rock surfaces. *In* "Biodeterioration and biodegradation 9" (A. Bousher, M. Chandra and R. Edyvean, Eds.), pp. 455–460. Institute of Chemical Engineering, Rugby, UK.

Watts, H. J., Very, A. A., Perera, T. H. S., Davies, J. M., and Gow, N. A. R. (1998). Thigmotropism and stretch-activated channels in the pathogenic fungus *Candida albicans*. *Microbiology* **144,** 689–695.

Weber, B., Wessels, D. C. J., and Budel, B. (1996). Biology and ecology of cryptoendolithic cyanobacteria of a sandstone outcrop in the Northern Province, South Africa. *Algol. Stud.* **83,** 565–579.

Webster, R. G. M., Andrew, C. A., Baxter, S., MacDonald, J., Rocha, M., Thomson, B. W., Tonge, K. H., Urquhart, D. C. M., and Young, M. E. (1992). Stonecleaning in Scotland - Research Report to Historic Scotland and Scottish Enterprise by Masonry Conservation Research Group. Gilcomston Litho, Aberdeen; http://www2.rgu.ac.uk/schools/mcrg/misst.htm, accessed: June 2005.

Welch, S. A., and Vandeviviere, P. (1994). Effect of microbial and other naturally occurring polymers on mineral dissolution. *Geomicrobiol. J.* **12,** 227–238.

West, N. E. (1990). Structure and function of microphytic soil crusts in wild wind ecosystems of arid to semi-arid regions. *Adv. Ecol. Res.* **20,** 179–223.

Wolf, B., and Krumbein, W. E. (1996). Microbial colonization on the surface and at the depth of marble elements of the "Freundschaftstempel" at Snassouci (Potsdam) *In* "Proceedings of the VIII th international congress on deterioration and conservation of stone" (J. Riederer, Ed.), pp. 637–642. Möller Druck und Verlag, Berlin.

Wollenzien, U., De Hoog, G. S., Krumbein, W. E., and Urzi, C. (1995). On the isolation of colonial fungi occurring on and in marble and other calcareous rocks. *Sci. Total Environ.* **167,** 287–294.

Wright, J. S. (2002). Geomorphology and stone conservation: Sandstone decay in Stoke-on-Trent. *Struct surv* **20,** 50–61.

Young, M. E. (1997). Biological growths and their relationship to the physical and chemical characteristics of sandstones before and after cleaning. PhD Thesis. The Robert Gordon University Aberdeen, UK.

Zimmermann, J., Gonzalez, J. M., Ludwig, W., and Saiz-Jimenez, C. (2005). Detection and phylogenetic relationships of a highly diverse uncultured acidobacterial community on paleolithic paintings in Altamira Cave using 23S rRNA sequence analyses. *Geomicrobiol. J.* **22,** 379–388.

CHAPTER 6

Microbial Processes in Oil Fields: Culprits, Problems, and Opportunities

Noha Youssef, Mostafa S. Elshahed, and **Michael J. McInerney**[1]

Contents

Department of Microbiology and Molecular Genetics, Oklahoma State University, 1110 S Innovation Way, Stillwater, Oklahoma 74074
[1] Corresponding author: Department of Botany and Microbiology, University of Oklahoma, 770 Van Vleet Oval, Norman, Oklahoma 73019

Advances in Applied Microbiology, Volume 66
ISSN 0065-2164, DOI: 10.1016/S0065-2164(08)00806-X

Abstract

Our understanding of the phylogenetic diversity, metabolic capabilities, ecological roles, and community dynamics of oil reservoir microbial communities is far from complete. The lack of appreciation of the microbiology of oil reservoirs can lead to detrimental consequences such as souring or plugging. In contrast, knowledge of the microbiology of oil reservoirs can be used to enhance productivity and recovery efficiency. It is clear that (1) nitrate and/or nitrite addition controls H_2S production, (2) oxygen injection stimulates hydrocarbon metabolism and helps mobilize crude oil, (3) injection of fermentative bacteria and carbohydrates generates large amounts of acids, gases, and solvents that increases oil recovery particularly in carbonate formations, and (4) nutrient injection stimulates microbial growth preferentially in high permeability zones and improves volumetric sweep efficiency and oil recovery. Biosurfactants significantly lower the interfacial tension between oil and water and large amounts of biosurfactant can be made *in situ*. However, it is still uncertain whether *in situ* biosurfactant production can be induced on the scale needed for economic oil recovery. Commercial microbial paraffin control technologies slow the rate of decline in oil production and extend the operational life of marginal oil fields. Microbial technologies are often applied in marginal fields where the risk of implementation is low. However, more quantitative assessments of the efficacy of microbial oil recovery will be needed before microbial oil recovery gains widespread acceptance.

Key Words: Petroleum microbiology, Biosurfactants, Sulfate reducers, Souring, Plugging, Oil recovery.

I. INTRODUCTION

World population is projected to increase by nearly 45% over the next 4 decades and by the middle of the century there will be more than 9 billion people (U.S. Census Bureau, International Data Base, August 2006;

http://www.census.gov/ipc/www/idb/worldpopinfo.html). The per capita energy consumption is a good predictor of the standard of living, which means that the demand for energy will continue to increase with the world's population and its desire to improve living standards (Hall *et al.*, 2003). The economic prosperity and security of nations will depend on how societies manage their energy resources and needs. An important question is how will we meet the future demand for more energy. Historically, the combustion of fossil fuels–oil, coal, and natural gas–has supplied more than 85% of world energy needs (Energy Information Agency, 2006). The reliance on fossil fuel energy has increased CO_2 emissions and fostered global climate change. For these reasons, it is advantageous to diversify our energy sources with the inclusion of more carbon-free or carbon-neutral fuels. However, even the most optimistic projections suggest that renewable energy sources will comprise less than 10% of the world's requirements through 2030 (Energy Information Agency, 2006). The most critical energy need is in the transportation sector. The use of nonpetroleum sources such as ethanol and of unconventional oil sources (shale oil, gas-to-liquids, and coal-to-liquids) will increase substantially, but each will only account for <10% of the demand by 2030 (Energy Information Agency, 2007). Thus, crude oil will likely continue as the dominant source of transportation fuels in the near future.

Current technologies recover only about one-third to one-half of the oil contained in the reservoirs. Globally, about 1 trillion barrels (0.16 Tm^3) of oil have been recovered, but about 2–4 trillion barrels (0.3–0.6 Tm^3) remain in oil reservoirs and are the target of enhanced oil recovery (EOR) technologies (Hall *et al.*, 2003). In the United States, more than 300 billion barrels (47.6 Gm^3) of oil remain unrecoverable from U.S. reservoirs, after conventional technologies reach their economic limit (Lundquist *et al.*, 2001). A critical feature of U.S. oil production is the importance of marginal wells whose production is <1.6 m^3 of oil or <1600 m^3 of natural gas per day. Currently, 27% of the oil (about the same amount imported from Saudi Arabia) and 8% of the natural gas produced in the U.S. onshore (excluding Alaska) is produced from marginal wells. These wells are at risk of being prematurely abandoned and it is estimated that about 17.5 million m^3 of oil was lost because of the plugging and abandonment of marginal wells between 1994 and 2003. New technologies to recover entrapped oil and to slow the decline in oil production in marginal wells are needed to increase oil reserves.

The amount of oil recovered by EOR technologies is not large, about 0.3 million m^3 per day (Anonymous, 2006), even though a number of economic incentives have been used to stimulate the development and application of EOR processes. Chemical-flooding technologies such as micellar or alkaline-surfactant-polymer flooding displace tertiary oil efficiently, but they have been marginally economic because of high chemical costs. Chemical losses because of adsorption, phase partitioning, trapping, and bypassing when mobility control is not maintained can be severe (Green and Willhite, 1998; Strand *et al.*, 2003; Weihong *et al.*, 2003). The only way to compensate for these losses is by increasing the volume of the surfactant solutions (Green and Willhite, 1998). Further, the implementation of these processes is complicated by reservoir heterogeneity and the need for large capital investment. All of these factors make chemical flooding a high-risk venture.

Microbially EOR (MEOR) processes have several unique characteristics that may provide an economic advantage. Microbial processes do not consume large amounts of energy as do thermal processes, nor do they depend on the price of crude oil as many chemical processes do. Because microbial growth occurs at exponential rates, it should be possible to produce large amounts of useful products rapidly from inexpensive and renewable resources. The main question is whether microbial processes do, in fact, generate useful products or activities in amounts and at rates needed for significant oil recovery (Bryant and Lockhart, 2002).

In this chapter, we will review what is known about the microbiology of oil fields from cultivation-dependent and cultivation-independent approaches. The technical feasibility of MEOR processes will be assessed by analyzing laboratory and field data to determine (1) if microbial products or activities mobilize entrapped oil from laboratory model systems, (2) if it is possible to produce the needed products or to stimulate the appropriate activity in the reservoir, and (3) if oil production coincides with the *in situ* product formation or activity stimulation. The recent book on petroleum microbiology provides an excellent resource (Ollivier and Magot, 2005). A number of compilations of the results of microbial field trials are available that provide detailed information on the characteristics of the reservoirs and the microorganisms and the nutrients used (Bass and Lappin-Scott, 1997; Hitzman, 1983, 1988; Lazar, 1991). Metabolism of hydrocarbons has been recently reviewed (Van Hamme *et al.*, 2003) and will only be covered here in its relationship to EOR. Other reviews and summaries of MEOR are available (Finnerty and Singer, 1983; Islam and Gianetto, 1993; Jack, 1988; McInerney *et al.*, 2005a). Methodologies for MEOR have been reviewed (McInerney and Sublette, 1997; McInerney *et al.*, 2007). Two other reviews provide an excellent summary of the early history of MEOR (Davis and Updegraff, 1954; Updegraff, 1990).

II. FACTORS GOVERNING OIL RECOVERY

An understanding of the multiphase flow properties of reservoir rock and the mechanisms that entrap oil is important for the success of any EOR project, including those involving microorganisms. When a well is drilled into an oil reservoir, oil and water are pushed to the surface by the natural pressure within the reservoir. As this pressure dissipates, pumps are placed on the well to assist in bringing the fluids to the surface. This stage of oil production is called primary production (Planckaert, 2005). Eventually, additional energy must be added to the reservoir to continue to recover oil. Often surface water, seawater, or brine from a subterranean formation is injected into the formation to push the oil to recovery (production) wells. This stage of oil production is called secondary oil production. After extensive water flooding, a large amount of oil still remains entrapped in the reservoir (called residual oil) and is the target of EOR (or the tertiary stage) processes.

The capillary pressure within the vicinity of the well governs the rates of oil and water production (Craig, 1980; Donaldson, 1985). The relative fluid saturations of water and oil in this region are functions of the capillary pressure between these two fluids, which is determined by the pore entrance size distribution of the rock. If the pore entrance size distribution decreases, the capillary pressure shifts and causes oil production to cease at an oil saturation, which otherwise would normally allow oil production. The accumulation of small particles, scale, paraffins, and asphaltenes precipitates, and the compaction of the sand plug drainage channels for oil and cause changes in capillary pressure. Mobile oil may be available only a short distance from the well, if the appropriate drainage patterns can be reestablished. Technologies that remove particulates, scales, paraffin, and asphaltene deposits can restore the drainage pattern or alter fluid saturations making oil mobile and increasing the rate of oil production from the well (Donaldson, 1985).

The goal of EOR, MEOR, when microorganisms are involved, is to increase the ultimate amount of oil recovered from a reservoir and not just to increase the productivity of individual wells. The efficiency of oil recovery is defined by the following equation (Craig, 1980; Green and Willhite, 1998):

$$E_r = E_d \times E_v \tag{6.1}$$

where E_r is the recovery efficiency expressed as a fraction of the original oil-in-place, E_d is the microscopic oil displacement efficiency expressed as the fraction of the total volume of oil displaced from a unit segment of rock, and E_v is the volumetric sweep efficiency expressed as the fraction of the total reservoir that is contacted by the recovery fluid. The microscopic displacement efficiency is a measure of the amount of oil that remains in small pores or dead-end pores after a recovery process. The viscosity and the capillary forces that hold the oil in place are expressed as a ratio called the capillary number (N_{ca}) (Tabor, 1969):

$$N_{ca} = (\mu_w v_w)/(s_{ow}) \tag{6.2}$$

where μ_w is the viscosity, v_w is the volumetric fluid flux, and s_{ow} is the oil–water interfacial tension (IFT). Chemical methods such as surfactant, micellar–polymer, or caustic or polymer flooding increase the capillary number by reducing interfacial tension or increasing water viscosity, respectively. Thermal methods reduce oil viscosity. Significant oil recovery requires a 100- to 1000-fold increase in capillary number (Reed and Healy, 1977). Some biosurfactants generate the very low interfacial tensions needed for significant oil recovery (Lin *et al.*, 1994; McInerney *et al.*, 1990; Nguyen *et al.*, 2008; Youssef, *et al.*, 2007a).

The volumetric sweep efficiency often dominates the recovery process when large variations between the viscosity of the recovery fluid and the oil or between the permeability of different zones of the formation exist (Craig, 1980). Large differences between the oil and aqueous phase viscosities will result in irregular movement of these fluids with water moving more rapidly than oil and reaching the production well first. The relative mobility of the two phases is expressed in the mobility ratio (Craig, 1980):

$$M = (k_w \mu_o)/(k_o \mu_w) \tag{6.3}$$

where M is the mobility ratio, k_w is the relative permeability of water in the waterflooded zone, k_o is the relative permeability of oil in the oil saturated zone, μ_o is the viscosity of the oil, and μ_w is the viscosity of water. Mobility ratios less than 1 are favorable and result in a uniform displacement of oil. Mobility ratios much greater than 1 are unfavorable and result in water channeling through the oil. The addition of polymers, such as xanthan gum, increases the viscosity of the water phase, resulting in favorable mobility ratios.

Poor sweep efficiency also occurs in reservoirs that have large permeability variations (Craig, 1980; Hutchinson, 1959). Because most oil reservoirs are composed of heterogeneous layers of rock, permeability variation is often an important factor controlling the sweep efficiency and the ultimate recovery of oil. Water preferentially flows through high permeability layers and little or no flow occurs in low permeability layers. Waterfloods will push the oil out of high permeability layers but the oil in low permeability layers remains unrecovered. Bacteria will preferentially plug high permeability layers, which will lead to more uniform water movement through the reservoir and improved oil displacement (Crawford, 1962, 1983; Raiders *et al.*, 1986a).

III. MICROBIAL ECOLOGY OF OIL RESERVOIRS

Multiple groups of microorganisms with diverse physiological and metabolic abilities and phylogenetic affiliations have routinely been recovered from oil reservoirs. The ability of microorganisms to sustain an underground deep biosphere, which is independent of above ground primary productivity (Chapelle *et al.*, 2002; Krumholz *et al.*, 1997; Lin *et al.*, 2007), coupled to the proved abilities of anaerobic microorganisms to utilize multiple oil components (Heider *et al.*, 1998) attest to the presence of indigenous microbial communities in oil reservoirs, and currently, it is a well-established scientific fact that oil reservoirs harbor and sustain diverse bacterial and archaeal communities. This section will summarize past efforts to characterize the microbial communities in oil reservoirs by culture-dependent and culture-independent approaches and highlight the effect of prevalent geochemical *in situ* conditions on the microbial communities' compositions.

A. Origins of microorganisms recovered from oil reservoirs

Determining whether a microorganism is autochthonous (indigenous) or allochthonous (foreign or transient) to an oil reservoir is essential before any conclusions can be made, regarding its role in the ecosystem. Contamination of oil-reservoir materials obtained during sampling is a thorny issue that constantly concerns petroleum microbiologists due to the number of possible sources of contamination upon sampling. Another important point is the effect of water-flooding procedure on the native-microbial community (Vance and Thrasher, 2005). Reinjecting produced water after being exposed to surface conditions will result in the reinoculation of the reservoir with surface microorganisms. Water, brine, or seawater injection, besides introducing exogenous microorganisms, could alter the geochemistry of the formation temporally or permanently, such as with the introduction of sulfate or oxygen, that could result in changes to the indigenous-microbial community structure. Further, some microorganisms could possess exceptional survival abilities and could be detected by various culture-dependent and culture-independent efforts long after being introduced to the formation.

The issue of contamination during sampling oil reservoirs has previously been debated in great detail (Magot *et al.*, 2000). Although more elaborate and expensive sampling procedures have been used to sample the deep terrestrial subsurface (Griffin *et al.*, 1997; Krumholz *et al.*, 1997; Lin *et al.*, 2007), they have often not been used while sampling oil reservoirs due to cost. Therefore, personal judgment becomes a critical factor in determining the origin (indigenous or nonindigenous) of isolates and 16S rRNA gene sequences encountered in oil reservoirs. Magot (2005) suggests two main criteria to determine the indigenous nature of microbial strains obtained from reservoir fluids: (1) comparing the isolate's growth optima to the *in situ* conditions in the oil reservoir, and (2) comparing the global distribution of the strain's phylotype in oil reservoir samples worldwide.

While useful, both approaches could be judged as harsh or exclusive. The optimum temperature of a microorganism is not necessarily an accurate reflection of the *in situ* temperature and could also be governed by other physiological and ecological considerations. Thermophilic isolates with much lower temperature optima than their environment (Vetriani *et al.*, 2004), and thermotolerant isolates with a low temperature optima have been reported from high-temperature ecosystems (Takai *et al.*, 2004). Similarly, some halophilic and halotolerant microorganisms recovered from salt crystals have a relatively low salt tolerance (Mormile *et al.*, 2003; Vreeland *et al.*, 2002, 2007). A more reasonable approach may be considering range (minimum and maximum growth limits) or the ability to survive for prolonged periods of time at the *in situ* reservoir condition. However, care should be taken while assessing the growth limits of slow-growing isolates, as false negatives may result because of the extended incubation time. The global presence of specific microbial lineages in geographically isolated oil reservoirs is indeed a good indication of their indigenous nature. However, this criterion could theoretically exclude novel groups that are indigenous to a specific oil reservoir where specialized niche exists.

B. Microorganisms isolated from oil reservoirs

In general, oil reservoirs have low redox potentials and hence harbor mainly anaerobic and facultative microorganisms. Electron donors in oil reservoirs include hydrogen, volatile fatty acids (VFAs) such as acetate, propionate, and benzoate (Fisher, 1987), petroleum hydrocarbons (aromatic hydrocarbons of various ring numbers and aliphatic hydrocarbons of various chain lengths), and inorganic electron donors (e.g., sulfide). Sulfate and carbonate minerals are important electron acceptors in many oil reservoirs. Some oil field isolates use iron (III) as an electron acceptor, but it is unclear how prevalent iron (III) is in oil reservoirs. Nitrate and oxygen are limiting in most oil reservoirs unless added with injected fluids.

In addition to redox potential and the availability of electron donors and acceptors, temperature and salinity appear to be the most important environmental factors that shape oil reservoir microbial communities. Below, we will summarize the various types of microorganisms that have been isolated from oil reservoirs and highlight the effect of prevalent environmental conditions on the phylogenetic diversity of isolated species. It is important to note that it is difficult to determine whether a microorganism is indigenous or not to an oil reservoir and the reader should consult the original manuscript if this decision is critical.

1. Methanogens

Methanogens metabolize hydrogen and CO_2, acetate, methylamines, and dimethylsulfides with the concurrent production of methane. Currently, methanogens are distributed among five orders (Methanomicrobiales, Methanobacteriales, Methanosarcinales, Methanococcales, and Methanopyrales) within the subkingdom Euryarchaeota, domain Archaea (Euzeby, 2008). Oil reservoir methanogenic isolates capable of metabolizing all the above-mentioned substrates that have been described, and collectively belong to four out of the five recognized methanogenic orders. No oil reservoir Methanopyrales have been described so far.

a. Hydrogenotrophic methanogens Mesophilic, hydrogenotrophic methanogens isolated from low-salinity oil reservoir include members of the genus *Methanobacterium* within the order Methanobacteriales (Belyaev *et al.*, 1986; Davydova-Charakhch'yan *et al.*, 1992a), and the genus *Methanoplanus* within the order Methanomicrobiales (Ollivier *et al.*, 1997). At higher temperatures, hydrogenotrophic methanogens include members of the genera *Methanobacterium, Methanothermobacter* (Davydova-Charakhch'yan *et al.*, 1992a; Jeanthon *et al.*, 2005; Ng *et al.*, 1989; Orphan *et al.*, 2000), *Methanoculleus* (order Methanomicrobiales (Orphan *et al.*, 2000)), and *Methanococcus* and *Methanothermococcus* (order Methanococcales (Nilsen and Torsvik, 1996a; Orphan *et al.*, 2000)). Halotolerant hydrogenotrophic methanogens (e.g., *Methanocalculus halotolerans*) has also been recovered from oil reservoirs with elevated salinities (Ollivier *et al.*, 1998).

b. Methylotrophic methanogens Methylotrophic methanogens have also been isolated from oil fields. Most of these isolates are mesophiles, such as *Methanosarcina siciliae* (Ni and Boone, 1991), *Methanosarcina mazei* (which can also utilize acetate (Obraztsova *et al.*, 1987)), in addition to the non-thermophilic halophile *Methanhalophiluus euhalobius* (Obrazstova *et al.*, 1988). However, a recent thermophilic methylotrophic isolate (*Methermicoccus shengliensis*) that belongs to the order Methanosarcinales and represents a novel family (Methermicoccaceae) within this order has been reported (Cheng *et al.*, 2007).

c. Aceticlastic methanogens Acetate-utilizing methanogens belong to the order Methanosarcinales. A single Methanosarcinales-affiliated isolate (*Methanosarcina mazei*) has been reported from oil fields (Obraztsova *et al.*, 1987). However, active mesophilic (Belyaev and Ivanov, 1983; Grabowski *et al.*, 2005b,c) and thermophilic (Bonch-Osmolovskaya *et al.*, 2003; Orphan *et al.*, 2000, 2003) aceticlastic enrichments derived from oil reservoir materials have frequently been reported. The scarcity of aceticlastic methanogenic isolates from oil reservoirs is probably a reflection of the difficulty of isolating this group of microorganisms rather than a reflection of their rarity in oil fields.

2. Sulfate-reducing bacteria

Sulfate-reducing bacteria (SRB) were the first microorganisms recovered from oil fields (Bastin *et al.*, 1926). Sulfate-reducing capability is currently identified in four different bacterial phyla (Proteobacteria, Firmicutes, Nitrospira, and *Thermodesulfobacterium*), as well as in the phyla Euryarchaeota and Crenarchaeota within the Archaea. Sulfate-reducing oil-reservoir isolates belonging to three out of the four bacterial phyla, as well as the Euryarchaeota within the archaeal domain have been reported.

a. Proteobacteria Sulfate-reducing Proteobacteria belong to the class δ-Proteobacteria. Within the δ-Proteobacteria, eight orders are currently described, five of which are predominantly composed of sulfate-reducing microorganisms (Euzeby, 2008). Two of these six orders (Desulfarculales and Desulfurellales) currently contain very few recognized species, and no isolates belonging to these orders have been encountered in oil reservoirs. SRB isolated from oil reservoirs are members of the orders Desulfovibrionales, Desulfobacterales, and Syntrophobacterales.

i. Desulfovibrionales Desulfovibrionales isolates recovered from the oil field are predominantly mesophiles. The majority are members of the genera *Desulfovibrio* and *Desulfomicrobium* (Birkeland, 2005; Leu *et al.*, 1999; Magot *et al.*, 1992, 2004; Miranda-Tello *et al.*, 2003; Nga *et al.*, 1996; Rozanova *et al.*, 1988; Tardy-Jacquenod *et al.*, 1996). Members of both genera use H_2, lactate, and pyruvate as electron donors. The apparent ubiquity of these two genera in oil reservoirs suggests their pivotal role in hydrogen metabolism in sulfidogenic oil fields, although this might be a reflection of the relative ease of their isolation. In addition, to the previously mentioned genera, a novel isolate that represents a new Desulfovibrionales genus (*Desulfovermiculus halophilus* gen. nov., sp. nov.) has recently been recovered from a Russian oil-field (Beliakova *et al.*, 2006) and is capable of completely degrading several organic compounds (malate, fumarate, succinate, propionate, butyrate, crotonate, ethanol, alanine, formate, etc.), in addition to using hydrogen, lactate, and pyruvate. More interestingly, this novel microorganism can grow at NaCl concentrations up to 23%.

ii. Desulfobacterales Collectively, members of the order Desulfobacterales are capable of degrading H_2, organic acids, ethanol, as well as various small molecular weight petroleum hydrocarbons. Most Desulfobacterales genera are known for their ability to completely oxidize substrates to CO_2, while others (e.g., members of the genus *Desulfobulbus*) are incomplete oxidizers, metabolizing substrates only to the point of acetate. Oil field isolates belonging to the order Desulfobacterales include members of the genera *Desulfobacter* (Lien and Beeder, 1997), *Dsulfobulbus* (Lien *et al.*, 1998b), *Desulfotignum* (Ommedal and Torsvik, 2007), and *Desulfobacterium* (Galushko and Rozanova, 1991). Most oil field Desulfobacterales isolates are VFA degraders. However, *Desulfobacterium cetonicum* has been shown to metabolize *m*- and *p*-cresol (Muller *et al.*, 1999, 2001). In addition, *Desulfotignum toluenicum* (Ommedal and Torsvik, 2007) and strains oXyS1 and mXyS1 (Harms *et al.*, 1999) were isolated from oil reservoir model columns and oil separators, respectively, and are capable of metabolizing toluene, *o*- and *m*-xylene anaerobically.

iii. Syntrophobacterales Oil-well sulfate-reducing Syntrophobacterales described so far are members of the exclusively thermophilic genera *Desulfacanium* (Rees *et al.*, 1995; Rozanova *et al.*, 2001b) and *Thermosulforhabdus* (Beeder *et al.*, 1995). However, it is important to note that the order Syntrophobacterales encompasses, in addition to thermophilic sulfate-reducers mentioned above, mesophilic sulfate-reducers as well and microorganisms that are capable of syntrophic degradation of organic compounds. Hydrocarbon-degrading capabilities have also been reported by mesophilic sulfate-reducing isolates belonging to this order (Cravo-Laureau *et al.*, 2004; Davidova *et al.*, 2006), none of which however, originated from oil-reservoirs.

b. Firmicutes Within the gram-positive Firmicutes, multiple genera with sulfate-reduction abilities have been reported, all of which belong to the order Clostridiales, for example, *Desulfotomaculum*, *Desulfurispora*, *Desulfovirgula*, *Desulfosporosinus*, and *Thermodesulfobium* (Euzeby, 2008). The genus *Desulfotomaculum* is the most ubiquitous amongst gram-positive SRBs, and the only Firmicutes that has been encountered in oil fields so far. Mesophilic oil reservoir *Desulfotomaculum* isolates have been reported (Tardy-Jacquenod *et al.*, 1998), but the majority of *Desulfotomaculum* isolates from oil reservoir are thermophiles (Nazina *et al.*, 1988; Nilsen *et al.*, 1996b; Rosnes *et al.*, 1991). Hydrocarbon-degrading capabilities have been reported within the members of this genus (Londry *et al.*, 1999; Tasaki *et al.*, 1991) but not in any of the oil-reservoir isolates so far.

c. Thermodesulfobacteria *Thermodesulfobacteria* represents a distinct bacterial phylum with only two genera (*Thermodesulfobacterium* and *Thermodesulfatator*) both of which are thermophilic sulfate-reducers. Oil reservoir isolates belonging to the genus *Thermodesulfobacter* have been obtained from thermophilic terrestrial and marine oil reservoirs (Christensen *et al.*, 1992; L'Haridon *et al.*, 1995).

d. Archaea Only Euryarchaeota sulfate-reducing microorganisms, but no sulfate-reducing Crenarchaeota (members of the genus *Caldivirga*), have been isolated from oil fields so far. Sulfate-reducing Archaea recovered from oil reservoirs are members of the genus *Archaeoglobus* and have been recovered mainly in various North Sea oil wells (Beeder *et al.*, 1994; Stetter *et al.*, 1993). The indigenous nature of this hyperthermophilic sulfate-reducer has been a matter of intense debate (see below).

3. Fermentative microorganisms

A number of fermentative microorganisms have been isolated from high-temperature and low-temperature oil reservoirs. It is important to note that many microorganisms in this group possess dual fermentative and respiratory metabolic abilities (e.g., sulfur and thiosulfate reduction) and could theoretically utilize both strategies for their *in situ* growth and survival.

A large fraction of thermophilic fermentative microorganisms recovered from oil fields are either members of the phylum Thermotogae or members of the order Thermoanaerobacteriales within the class Clostridia, phylum Firmicutes. Thermotogae is a phylum exclusive for thermophilic, anaerobic fermenters, although recent phylogenetic and metagenomic evidence for the presence of low-temperature "Mesotoga" have recently been reported (Nesbo *et al.*, 2006). Thermotogae isolates have consistently been shown to be members of high-temperature oil reservoirs, suggesting an indigenous nature of these microorganisms to oil reservoirs (Davey *et al.*, 1993; L'Haridon *et al.*, 2001, 2002; Lien *et al.*, 1998a; Miranda-Tello *et al.*, 2004, 2007; Takahata *et al.*, 2001). Members of the Thermotogae can grow on complex substrates, as well as sugars, with acetate and hydrogen being the final end products. Oil field isolates have been identified in four out of the six currently recognized genera in this phylum (*Thermotoga*, *Thermosipho*, *Geotoga*, and *Petrotoga*) with members of the last two genera being exclusively recovered from oil reservoirs. Sulfur and/or thiosulfate reduction is widely distributed among members of this phylum.

Members of the order Thermoanaerobacteriales within the Firmicutes are also commonly encountered in oil fields, and include isolates belonging to the genera *Thermoanaerobacter* (Cayol *et al.*, 1995; L'Haridon *et al.*, 1995), *Thermoanaerobacterium* (Grassia *et al.*, 1996), *Caldanaerobacter* (Fardeau *et al.*, 2004; Grassia *et al.*, 1996), and *Mahella* (Bonilla Salinas *et al.*, 2004), all of which are thermophilic sugar fermenters. In addition to Thermoanaerobacteriales isolates, other thermophilic, Firmicutes-affiliated oil reservoir microorganisms have been identified, including the organic acid fermenter *Anaerobaculum thermoterrenum* (Rees *et al.*, 1997) and the amino acid degrader *Thermovirga lienii* (Dahle and Birkeland, 2006), both of which are members of the order Clostridiales.

Few studies have focused on isolating fermentative microorganisms from oil reservoirs with elevated salinities. Fermentative halophilic oil reservoir isolates exclusively belong to the genus *Haloanaerobium* (Order Haloanaerobiales) such as *H. acetethylicum*, *H. salsuginis*, *H. congolese*, and *H. kushneri* (Bhupathiraju *et al.*, 1994, 1999; Ravot *et al.*, 1997). Most *Haloanaerobium* spp. are saccharolytic and proteolytic, and produce H_2, acetate, and CO_2 as end products of fermentation.

In spite of the general metabolic and phylogenetic diversity of mesophilic fermentative microorganisms, few isolates that belong to this metabolic group have been recovered from oil fields. In general, studies on the isolation of mesophilic fermentative microorganisms from oil fields are extremely sparse and are far from adequate to identify any global distribution patterns or link members of any of the identified phylogenetic groups to a specific ecological role in oil reservoirs. Fermentative, mesophilic Firmicutes-affiliated isolates include *Fusibacter paucivorans*, a new genus belonging to the order Clostridiales that utilizes a limited number of carbohydrates and was isolated from an offshore oil field in Congo (Ravot *et al.*, 1999), *Dethiosulfovibrio peptidovorans* a proteolytic microorganism that grows on peptones and individual amino acids but not sugars (Magot *et al.*, 1997b), and "*Acetobacterium romashkovi*," an acetogenic microorganism that is also capable of fermenting sugars and amino acids (Davydova-Charakhch'yan *et al.*, 1992b). In addition to Firmicutes, a novel species within the genus *Spirochaeta* has been identified from oil reservoirs (Magot *et al.*, 1997a). A novel genus (*Petrimonas*) within the phylum Bacteroidetes has recently been isolated from oil reservoirs as well (Grabowski *et al.*, 2005c), making it the first member of this phylum to be isolated from oil reservoirs. A recent study has enriched for a wide range of fermentative mesophilic microorganisms from a low-temperature oil reservoir in Canada, and 16S rRNA gene-based analysis of bacterial enrichments identified various members of putatively fermentative members of the Clostridiales, genus *Bacteroides*, and genus *Spirochetes* (Grabowski *et al.*, 2005b).

4. Other microbial isolates

a. Hyperthermophiles Although the presence and recovery of hyperthermophiles from oil fields has previously been demonstrated, the indigenous nature of these microorganisms remains in question (L'Haridon *et al.*, 1995; Stetter *et al.*, 1993). Geological studies correlating *in situ* biodegradation of oil components to temperatures in oil reservoirs (Head *et al.*, 2003) and those correlating VFA levels to temperatures (Fisher, 1987) suggest an upper limit of 80–90 °C for *in situ* biological activities in oil reservoirs, which is far from being the upper growth limit within the microbial world (Kashefi and Lovley, 2003). In addition, it has been observed that these hyperthermophilic microorganisms recovered in high-temperature reservoirs could remain viable for long periods of time at seawater temperatures, increasing the possibility that they are immigrants to the ecosystem introduced during well manipulations. Hyperthermophiles recovered from oil reservoirs include members of the archaeal genera *Archaeoglobus*, *Pyrococcus*, and *Thermococcus* (Grassia *et al.*, 1996; L'Haridon *et al.*, 1995; Stetter *et al.*, 1993; Takahata *et al.*, 2000).

b. Syntrophic microorganisms Syntrophic microorganisms are responsible for the degradation of a wide range of organic compounds in association with hydrogen- and acetate-utilizing methanogens. A previous report demonstrated that glycerol fermentation by two oil reservoir-derived species of the genus *Halanaerobium* is greatly facilitated when grown in a coculture with the hydrogen-scavenging sulfate-reducer *Desulfohalobium retbaense* (Cayol *et al.*, 2002). To our knowledge, no pure isolate capable of degrading compounds that obligatory-require syntrophic interactions under methanogenic conditions have been retrieved from oil reservoirs. The notoriously fastidious nature of syntrophic microorganisms could partly be responsible for the lack of syntrophic oil well isolates. Thermodynamic considerations argue for the involvement of syntrophic microorganisms in the anaerobic degradation of most aromatic and aliphatic components of oil under methanogenic conditions. It follows that documenting the degradation of oil components *in situ* or in methanogenic laboratory enrichments derived from oil reservoirs could be regarded as an indirect evidence for the presence of syntrophic microorganisms in oil fields. Recent work demonstrating active methanogenic hydrocarbon metabolism in oil field reservoirs clearly attests to the presence of a native syntrophic population within the oil fields (Atiken *et al.*, 2004; Jones *et al.*, 2008). Indeed, methanogenic alkane-degrading enrichments derived from North Sea oil reservoirs have been reported (Jones *et al.*, 2008) and 16S rRNA gene clone libraries derived from these enrichments indicated the presence of a clone affiliated with the genus *Syntrophus*, all members of which are capable of syntrophic metabolism. Finally, members of the genus *Syntrophus* were also identified in enrichments degrading heptadecane under methanogenic conditions that was derived from Pelikan lake oil field in western Canada (Grabowski *et al.*, 2005a).

c. Autotrophs Respiratory microorganisms capable of utilizing hydrogen as a sole electron donor coupled to sulfate (Birkeland, 2005; Leu *et al.*, 1999; Magot *et al.*, 1992, 2004; Miranda-Tello *et al.*, 2003; Nga *et al.*, 1996; Rozanova *et al.*, 1988; Tardy-Jacquenod *et al.*, 1996) or nitrate and/or iron (III) (Greene *et al.*, 1997; Myhr and Torsvik, 2000; Nazina *et al.*, 1995b; Slobodkin *et al.*, 1999) as electron acceptors have frequently been isolated from oil fields. Acetogenic microorganisms utilizing H_2 and CO_2 to produce acetate have also been reported. "*A. romashkovii*," a member of the order Lactobacillales within the Firmicutes, has been isolated from an oil field in Siberia (Davydova-Charakhch'yan *et al.*, 1992b). Also, an acetogenic enrichment containing a microorganism is highly similar to a known acetogen (*Acetobacterium carbinolicum*) (99% 16S rRNA gene sequence similarity) has been reported (Grabowski *et al.*, 2005b). Finally, chemolithoautotrophic sulfide-oxidizing, nitrate-reducing bacteria (NRB) have been isolated from oil production fluids in a Canadian oil field (Gevertz *et al.*, 2000).

d. Nitrate, iron, and manganese reducers Various nitrate-reducing microorganisms with autotrophic (Myhr and Torsvik, 2000), heterotrophic (Nazina *et al.*, 1995b), and chemolithotrophic (Gevertz *et al.*, 2000) abilities have been isolated. Many of the NRB, including some that were recovered from oil fields, are facultative and metabolically versatile, raising the question of their contribution to nitrate reduction in oil fields (Huu *et al.*, 1999; Nazina *et al.*, 1995b). Two thermophilic respiratory genera that appear to be especially abundant in oil reservoirs are *Geobacillus* and *Deferribacter*. The genus *Geobacillus* belongs to the order Bacillales within the Firmicutes, and oil reservoir-derived *Geobacillus* isolates are thermophilic microaerophiles that degrade alkanes only under aerobic conditions and some could reduce nitrate anaerobically (Nazina *et al.*, 2001). The genus *Deferribacter* contains three species, all of which are thermophiles and are capable of anaerobic respiration using multiple electron acceptors. *Deferribacter thermophilus* was isolated from Beatrice oil field in the North Sea and is capable of coupling oxidation of hydrogen, organic acids, and complex substrates to manganese and iron reduction (Greene *et al.*, 1997).

C. Culture-independent analysis of microbial communities in oil reservoirs

Isolation efforts have identified numerous bacterial and archaeal species that are capable of mediating various metabolic processes occurring in oil fields. Nevertheless, culture-dependent characterization of microbial communities is governed by several factors that limit its utility to describe the ecosystem's community completely. Isolation of a single microorganism mediating a specific metabolic process could hardly be a representative of the entire community mediating this process *in situ* because microorganisms that are easiest to obtain in pure cultures are not necessarily the most numerically abundant and/or metabolically active *in situ*. The media routinely used for isolation of environmental microorganisms (including those in oil reservoirs) are often carbon and nitrogen-rich compared to the prevailing environmental condition. Culturing from samples collected at a specific time and production stage of oil reservoir does not capture the changes in microbial community that occur throughtout the entire exploration and exploitations stages. In addition, the relatively low number of isolates obtained from oil fields, especially from low-temperature oil reservoirs, severely limits our ability to deduce distribution patterns or unifying ecological themes regarding oil-reservoir community. For example, while the recent isolation of a spirochete (Magot *et al.*, 1997a) and a member of the phylum Bacteroidetes (Grabowski *et al.*, 2005a) establishes the presence of members of these phyla in oil reservoirs, these isolations hardly capture the intra-phylum diversity, abundance, and distribution of members of these phyla in various oil fields. Finally, the inability of microbiologists to isolate a large fraction of microorganisms present in nature is a well-established phenomenon (Rappe and Giovannoni, 2003; Zengler, 2006; Zengler *et al.*, 2002).

In comparison, culture-independent 16S rRNA gene-based surveys are extremely valuable in providing an overall view of the community composition in a specific ecosystem, regardless of the metabolic abilities of the community members. These studies also provide preliminary information on the relative abundance of different groups within the ecosystem and could be used for monitoring temporal and spatial changes in an ecosystem. Thousands of culture-independent 16S rRNA gene-sequencing surveys have already been reported in almost all accessible ecosystems on Earth (Keller and Zengler, 2004; Pace, 1997). The collective conclusion from these studies clearly enforces the notion that the scope of microbial diversity is much broader than implied by culture-dependent studies (Janssen, 2006; Rappe and Giovannoni, 2003). The discovery of novel phyla and subphyla as the most numerically abundant members of several habitats drastically changed our view of the community compositions of multiple globally relevant ecosystems, including soil (Janssen, 2006) and pelagic marine environments (DeLong, 2005; Rappe and Giovannoni, 2003).

Several culture-independent sequencing surveys have been conducted in high and low temperatures, and in marine and terrestrial oil reservoirs. Compared to the majority of previously studied ecosystems, the information (number of studies and number of sequences analyzed per study) currently available is very sparse. Surprisingly, 16S rRNA gene-based analysis of oil reservoir communities have not lead to any significant, paradigm-shifting discoveries and have not drastically altered our view of the oil reservoir communities. To our knowledge, this is one of the few environments in which culture-independent analysis did not dramatically alter our view of its microbial community composition. Culture-independent studies have rather been confirmatory of the results previously obtained by culture-dependent surveys, in spite of the inherent limitations outlined above. The lack of significant discoveries does not necessarily mean that novel, yet-uncultured microbial groups are not present in oil reservoirs, because the absence of novel phyla could also be attributed to (1) the small number of studies conducted so far, (2) the relatively small number of sequences that have been analyzed in these studies, and (3) the fact that most studies, so far, have been conducted in thermophilic oil reservoirs where extreme conditions could limit the overall microbial diversity within the ecosystem (Lozupone and Knight, 2007).

In addition, a review of currently available culture-independent sequencing surveys show that these studies, similar to isolation-based efforts, are effected by the issues of contamination and reinoculation during water flooding described above, which confers uncertainty regarding the indigenous nature of identified oil-reservoir populations. Therefore, while many of the culture-independent studies generated fairly large clone libraries (Li *et al.*, 2006, 2007a), the detection of perceived contaminants and their exclusion from further analysis severely diminished the number of ''relevant'' sequences recovered. This issue deprives many of these studies of one of the most important strengths of culture-independent surveys, for example, the ability to identify large number of microorganisms in a single experiment.

Culture-independent surveys of high-temperature oil reservoirs have been conducted in oil fields in continental and offshore California (Orphan *et al.*, 2000, 2003), an offshore oil field in Qinghuang unit, China (Li *et al.*, 2006), Huabei oil field in continental China (Li *et al.*, 2006, 2007a), and in Troll oil formation in the North sea (Dahle *et al.*, 2008). Collectively, the results highlight the problem of sample contamination that plagues the work. While thermophilic enrichment and isolation studies could partially alleviate this problem by setting enrichments targeting thermophiles, the indiscriminatory nature of culture-independent surveys does not selectively detect indigenous oil field thermophiles. As a result, a large number of sequences that belong to marine and terrestrial mesophilic lineages (e.g., *Pseudomonas, Marinobacter, Sinorhizobium*) often represent a majority of the clones in a specific library and microorganisms perceived as native, for example, *Thermotogales,* while readily enriched from the same samples and often represent the majority of the population based on microscopic observation, are often the minority (Li *et al.*, 2006), or entirely absent (Orphan *et al.*, 2000) from these clone libraries.

The main characteristics of these libraries (site, reservoir conditions, target group of microorganisms, library size, and salient findings) are listed in Table 6.1. Clones that we subjectively judge as indigenous that were identified in these clone libraries mainly belong to lineages previously identified as inhabitants of high-temperature oil reservoirs by using culture-based approaches (e.g., *Thermotogales, Thermoanaerobacteriales, Thermococcus, Methanomicrobiales, Methanosarcinales*, and δ-Proteobacteria). Novel microorganisms/lineages identified in these studies that could potentially be indigenous, but have not previously been cultured, include

TABLE 6.1 Culture independent 16S rRNA gene sequencing surveys conducted in oil reservoirs

Study site	Reservoir conditions	Primers target	Clones /OTUs	Lineages detected	Novel lineages previously unidentified by culturing studies, general comments	Reference
Troll oil formation, North sea	70 °C, low-salinity, not water flooded	Bacteria, Archaea	88/29 (Bacteria), 22/3 (Archaea)	Firmicutes, γ, δ-Proteobacteria, *Thermotogales*, *Spirochetes*, *Bacteroidetes*, *Methanococcus*, *Methanolobus*, *Thermococcus*	Detection of moderately thermophilic members of the *Bacteroidetes* (genus *Anaerophaga*) that have not been previously isolated from oil reservoirs. Large fraction of clones belongs to nonthermophilic lineages, and appears to be contaminants	Dahle *et al.* (2008)
Multiple oil fields, California	Multiple high-temperature, low salinity reservoirs at different stages of flooding	Unviersal, Archaea	118/41 (Bacteria), 168/11 (Archaea). Archaeal sequences identified in both clone libraries	α, β, γ, δ-Proteobacteria, *Bacteroidetes*, Firmicutes, *Methanomicrobiales*, *Methanosarcinales*, *Thermococcales*	Majority of Proteobacteria, *Bacteroidetes*-affiliated clones belong to nonthermophilic lineages and appear to be contaminants. No *Thermotogales* clones detected in spite of being identified in a parallel enrichment effort	Orphan *et al.* (2000)

Hubei oil field, China	75 °C, low salinity (1.6%), water flooded	Bacteria	337/74	α, β, γ, ε-Proteobacteria Firmicutes, *Actinobacteria*, *Thermotogales*, Nitrospira	Majority of γ and β-Proteobacteria and *Actinomycetes*, as well as some Firmicutes clones belong to nonthermophilic lineages and appear to be contaminants. Detection of clones affiliated with sulfate-reducing genus *Thermodesulfovibrio* (Phylum Nitrospira) that has not been previously isolated from oil reservoirs	Li *et al.* (2006)
Hubei oil field, China	75 °C, low salininty (1.6%), water flooded	Archaea	237/28	*Methanobacteriales*, *Methanococcales*, *Methanomicrobiales*, *Methanosarcinales*	First identification of *Methanocorpusculum* clones (order *Methanomicrobiales*) in high-temperature oil reservoirs, origin still uncertain.	Li *et al.* (2007a)

(continued)

TABLE 6.1 *(continued)*

Study site	Reservoir conditions	Primers target	Clones /OTUs	Lineages detected	Novel lineages previously unidentified by culturing studies, general comments	Reference
Qinghuang offshore oil field, China	65 °C, low salinity, water flooded	Bacteria, Archaea	338/60 (Bacteria) 220/28 (Archaea)	Firmicutes, Nitrospira, *Thermotogae*, α, β, γ, ε-Proteobacteria, *Methanobacteriales*, *Methanococcales*, *Crenarchaeota*	Detection of clones affiliated with sulfate-reducing genus *Thermodesulfovibrio* (Phylum Nitrospira) that has not been previously isolated from oil reservoirs. Most α, β, γ-Proteobacteria and *Crenarchaeota* clones belong to nonthermophilic lineages and appear to be contaminants	Li *et al.* (2007b)
Pelikan lake oil field	Low-temperature (18–20 °C), low salinity	Bacteria, *Archaea*	151/1, 192/12	ε-Proteobacteria, *Methanomicrobiales*, *Methanosarcinales*	Detection of *Methanocorpusculum*, and *Methanosaeta* clones both of which have not been previously isolated from oil reservoirs. Only one OTU identified in the entire bacterial clone library	Grabowski *et al.* (2005b)

Western Canadian oil fields	Low-temperature (25 °C), low salinity	Bacteria	36/ND*	δ-Proteobacteria, *Deferribacteres*, Firmicutes	Most clones appear to be indigenous, minimum amount of contaminants clones. Detection of *Synergistes*-like microorganisms (Phylum *Deferribacteres)* that have not been previously isolated from oil reservoirs	Voordouw *et al.* (1996)

* ND; not determined.

thermophilc *Bacteroidetes* (thermophilic members of the genus *Anaerophaga*) (Dahle *et al.*, 2008), clones affiliated with members of the genus *Methanocorpusculum* (order Methanomicrobiales) (Li *et al.*, 2007a) and clones affiliated with members of the sulfate-reducing genus *Thermodesulfovibrio* (Phylum Nitrospira) (Li *et al.*, 2006, 2007a).

Few 16S-based analysis of the microbial community in low-temperature reservoirs has been reported. Compared to surveys of thermophilic oil reservoirs, judging the indigenous nature of encountered microorganisms is more problematic because there is no litmus test (i.e., affiliation with thermophilic lineages) available to attest to the indigenous nature of the community. Grabowski *et al.* (2005b) investigated the bacterial diversity in a low-temperature, nonwater flooded oil reservoir (Pelikan lake oil field) in western Canada and reported a bacterial community with extremely low diversity; only one phylotype related to the genus *Arcobacter* (ε-Proteobacteria) was encountered in the bacterial clone library. Voodrouw *et al.* (1996) presented an analysis of 36 16S rRNA gene clones from a low-temperature oil reservoir in western Canada. The study remains one of the few culture-independent surveys of oil reservoirs in which contaminants' clones do not appear to represent a significant fraction of the total number of clones in the library and hence plausible ecological functions could be assigned to the majority of the sequenced clones. Several potential metabolic groups of microorganisms were identified: fermentative and/or acetogenic microorganisms, sulfide-oxidizers, and sulfate-reducers (Voordouw *et al.*, 1996).

Besides 16S rRNA gene-based analysis, a few hybridization-based studies on the oil reservoir microbial communities have been reported. Voordouw *et al.* used a reverse sample genome probing approach, in which labeled environmental DNA is hybridized to genomes of target microorganisms, to detect and quantify sulfate-reducing microorganisms in multiple oil reservoir-derived samples and enrichments (Voordouw *et al.*, 1991, 1992, 1993). A more recent study used an oligonucleotide microchip that targets key genera of thermophilic *Bacteria* and *Archaea* (17 probes with varying degrees of specificity) to identify communities in a high-temperature reservoir in Western Siberia (Bonch-Osmolovskaya *et al.*, 2003). The study identified several microorganisms (members of the phylum Aquificales, genus *Thermus* within the phylum Deinococcus-Thermus, and members of the genus *Desulfurococcus* within the Crenarchaeota, Domain *Archaea*) that have escaped a parallel cultivation effort and have not been previously detected in oil reservoirs members. This significant contribution with a relatively limited number of probes highlights the potential of hybridization-based community investigations using newly available broad range microchips that are capable of reliably detecting thousands of microbial species with a high degree of sensitivity simultaneously.

IV. DELETERIOUS MICROBIAL ACTIVITIES: HYDROGEN SULFIDE PRODUCTION (OR SOURING)

The onset of hydrogen sulfide production (or souring) often occurs when reservoirs are flooded with brine or seawater that contains high levels of sulfate (McInerney *et al.*, 1991; Sunde and Torsvik, 2005). The presence of sulfate and nitrogen and phosphorous sources in the injection water, the reduction in reservoir temperature due to the injection of cooler displacement fluids, and the electron donors present in the reservoir (organic acids and hydrocarbons) create conditions favorable for the growth of SRB and the production of hydrogen sulfide (Sunde and Torsvik, 2005; Vance and Thrasher, 2005). The increase of H_2S (known as souring) is associated with corrosion of pipelines, platform structures, and other equipment; increases refining costs of oil and gas; plugs reservoirs by the accumulation of sulfides minerals; and increases health risks due to the toxicity of H_2S (Chen *et al.*, 1994; Davidova *et al.*, 2001; Eckford and Fedorak, 2002; Myhr *et al.*, 2002; Nemati *et al.*, 2001a,b).

A. Current souring control approaches

The detrimental consequences of souring have caused the oil industry to invest heavily in strategies for souring control. Biocides such as bronopol, formaldehyde, glutaraldehyde, benzalkonium chloride, cocodiamine, and tetrakishydroxymethyl phosphonium sulfate (THPS) are commonly applied to injection waters and production facilities to reduce H_2S concentrations below the threshold levels defined by NACE international standards (Vance and Thrasher, 2005). The problem with biocides is the need for high concentrations and frequent treatments to achieve the desired results, especially when dealing with biofilms (Burger, 1998; Kjellerup *et al.*, 2005; Vance and Thrasher, 2005). Some biocides may also pose a health risk to operators. If the reservoir was sour prior to water flooding, H_2S was probably not of recent biological origin and liquid- or solid-phase H_2S scavengers (e.g., triazines, sodium hydroxide, aldehydes, metal oxides, and iron and zinc oxide-based biocides) could be used. Another approach to control souring is to remove sulfate or significantly lower sulfate concentrations below 50 mg/l from the injection waters. Nanofiltration technology has been applied to water injection facilities to achieve desulfation (Rizk *et al.*, 1998; Vance and Thrasher, 2005). Manipulation of the injection water's salinity could potentially inhibit H_2S production, if NaCl concentrations above 12% can be achieved (Vance and Thrasher, 2005).

B. Microbial control of souring

Recently, the stimulation of NRB by addition of nitrate, nitrite, or nitrate/molybdate mixtures with or without the addition of NRB has been used to control souring (Hubert *et al.*, 2003; Sunde and Torsvik, 2005). The idea of using nitrate to abate hydrogen sulfide production is not new and has been used to control odors in sewage (Carpenter, 1932; Heukelelekian, 1943). There are several mechanisms by which nitrate addition can control souring: (1) competition for electron donors between NRB and SRB, (2) sulfide oxidation by NRB, (3) increase in redox potential and subsequent inhibition of SRB, and (4) the inhibition of SRB by nitrite, or other nitrogen oxides (e.g., nitrous oxide) (Hubert *et al.*, 2003, 2004; Jenneman *et al.*, 1986; Montgomery *et al.*, 1990; Nemati *et al.*, 2001b; Reinsel *et al.*, 1996; Sunde and Torsvik, 2005). Two physiological types of NRBs are involved in the control of SRB activity; heterotrophic, NRB (hNRB) and sulfide-oxidizing, NRB (SO-NRB) (Eckford and Fedorak, 2002; Sunde and Torsvik, 2005). hNRB compete with SRB for common electron donors. The reduction of nitrate or nitrite is energetically more favorable than sulfate reduction. Thus, hNRB have higher molar growth yields and faster growth rates than SRB so that hNRB effectively outcompete SRB for common electron donors (Sunde and Torsvik, 2005). If hNRB are present, the prediction is that they will prevent the growth of SRB (Kjellerup *et al.*, 2005). On the other hand, SO-NRB oxidize H_2S to sulfur or sulfate with nitrate or nitrite as the electron acceptor, but do not inhibit the growth of SRB (Kjellerup et al., 2005; Montgomery *et al.*, 1990). Some microorganisms such as *Sulfurospirilum* spp. are capable of both hNRB, and SO-NRB types of metabolism (Hubert and Voordouw, 2007).

Most of the laboratory experiments indicate that sulfide oxidation by SO-NRB is an important mechanism for sulfide inhibition (Montgomery *et al.*, 1990; Myhr *et al.*, 2002; Nemati *et al.*, 2001a). In presence of pure cultures of SO-NRB and nitrate, sulfide concentrations decreased (McInerney *et al.*, 1992, 1996; Montgomery *et al.*, 1990; Nemati *et al.*, 2001a). The addition of nitrate or small amounts of nitrite to an up-flow packed-bed bioreactor inoculated with produced water from an oil field inhibited sulfide accumulation and resulted in an increase in SO-NRB populations but the SRB population was not affected (Hubert *et al.*, 2003). Similarly, the addition of nitrate inhibited sulfide production and increased the NRB populations (Davidova *et al.*, 2001). Because the SRB population was not affected after treatment, it is likely that SO-NRB were responsible for the observed decrease in sulfide concentration.

One study involving a coculture of *Sulfurospirilum* sp. and *Desulfovibrio* sp. implicates the accumulation of nitrite as an important mechanism for inhibiting SRB activity (Haveman *et al.*, 2004). In an up-flow packed-bed bioreactor inoculated with produced water from an oil field and amended with nitrate, *Sulfurospirilum* spp. were dominant. The authors argued that the metabolic versatility of *Sulfurospirilum* spp. (the ability to perform heterotrophic or sulfide-oxidizing nitrate reduction) coupled with the ability to produce inhibitory concentrations of nitrite were the main reasons that SRB activity was inhibited. Similarly, nitrite accumulation concomitent with the inhibition of sulfide was observed following nitrate addition to porous columns inoculated with brine from North Slope of Alaska reservoirs (Reinsel *et al.*, 1996).

Competitive exclusion of SRB by hNRB is also possible. An increase in the population of hNRB was observed when the concentration of NO_2^- was increased in an up-flow packed-bed bioreactor inoculated with produced water from an oil field (Hubert *et al.*, 2003). Eckford and Fedorak (Eckford and Fedorak, 2002) found both hNRB and SO-NRB populations increased in batch experiments when sulfide-containing produced water was amended with nitrate. Finally, some studies show that nitrite and molybdate act synergistically to inhibit sulfide production by pure cultures of SRB or by sulfate-reducing enrichments from produced waters (Kjellerup *et al.*, 2005; Nemati *et al.*, 2001b; Reinsel *et al.*, 1996).

Several field trials show that nitrate injection is effective in inhibiting or reducing reservoir souring (Table 6.2). The injection of about 58 kg per day of ammonium nitrate for 45 days into a hypersaline oil reservoir reduced sulfide levels by 40–60% in produced fluids from three production wells (McInerney *et al.*, 1991). A reduction of sulfide in produced fluids of two wells was detected 20 days after nitrate injection began and sulfide levels dropped to 25% of their pretreatment values 55 days after nitrate injection (Telang *et al.*, 1997). Reverse sample genome probing indicated that a sulfide-oxiding, nitrate-using bacterium became a dominant member of the microbial community implicating that sulfide oxidation to sulfate coupled to nitrate reduction was an important mechanism for the reduction in sulfide levels in production wells. The addition of 3 m^3 of calcium nitrate (0.3 mM nitrate) per day in the injection waters reduced the number of sulfate reducers by 50–90%, increased the numbers of nitrate-using bacteria, and decreased corrosion rates in biofilms that formed on metal coupons placed inside of the Gullfaks North Sea injection system (Sunde *et al.*, 2004; Thorstenson *et al.*, 2002). Hydrogen sulfide concentrations of eleven Gullfaks production wells decreased about 12–18 months after continuous nitrate injection began (Sunde and Torsvik, 2005). The decrease in sulfide levels coincided with time predicted based on the breakthrough of the nitrate-treated water. Fluids collected by backflowing nitrate-treated injection wells had barely detectable levels of sulfide (<1 mg/l) while those of biocide-treated wells had 39 mg/l of sulfide (Sunde and Torsvik, 2005). A reduction in sulfide concentration in production wells after continuous injection of 100–150 mg/l of nitrate was observed in the Halfdan North Sea field (Larsen *et al.*, 2004).

TABLE 6.2 Use of nitrate to control hydrogen sulfide production in reservoirs

Method	Results	References
Continuous NO_3^- injection till breakthrough in production wells	Decline in H_2S after breakthrough of treated water	Jenneman *et al.* (1999), Larsen *et al.* (2004), Sunde *et al.* (2004), Thorstenson *et al.* (2002)
Continuous injection of NO_3^- and PO_4^{2-} for 50 days	Reduction in sulfide levels; population of a sulfide-oxidizing, nitrate reducer increased	Telang *et al.* (1997)
Injection of NO_3^- and NO_2^-	Reduction in dissolved sulfide levels in production equipment and produced water	Hitzman *et al.* (2004)
Continuous injection of NH_4NO_3	40–60% decrease in sulfide levels decreased in 3 wells 45 days after injection began	McInerney *et al.* (1991)

The continuous injection of ammonium nitrate and sodium phosphate decreased sulfide concentrations and suppressed sulfate reducers in injection and production wells (Jenneman *et al.*, 1999). A mixture of nitrate, nitrite, and inorganic nutrients decreased sulfide levels in production facilities and hydrogen sulfide levels from a gas well (Hitzman and Sperl, 1994; Hitzman *et al.*, 2004). The use of nitrate to control souring shows that it is possible to manipulate the reservoir ecology in a predictable manner. In addition, the injection of nitrate can have a second benefit by slowing the natural decline in oil production (Brown, 2007; Brown *et al.*, 2002; Hitzman *et al.*, 2004).

While nitrate is effective in controlling souring, the effect of adding nitrate on corrosion is not as clear. The addition of nitrate to mixed cultures with SO-NRB (Nemati *et al.*, 2001c) or to up-flow packed-bed bioreactor studies (Hubert *et al.*, 2005) increased corrosion. The rate of corrosion was high during the time when sulfide was oxidized (Rempel *et al.*, 2006). The addition of nitrite, on the other hand, seems to have a less pronounced effect on corrosion when used in small concentrations and added gradually (Hubert *et al.*, 2005; Rempel *et al.*, 2006). The addition of a single high dose of nitrite completely eliminated microbial activity and hence corrosion (Hubert *et al.*, 2005). The inhibitory effect of high nitrite concentrations may be due to the production of nitric oxide (Kielemoes *et al.*, 2000).

V. MICROBIAL ACTIVITIES AND PRODUCTS USEFUL FOR OIL RECOVERY

Microorganisms produce a number of products (acids, solvents, gases, biosurfactants, biopolymers, and emulsifiers) and have activities (hydrocarbon metabolism and plugging) that are potentially useful for oil recovery (Table 6.3) (ZoBell, 1947a–c). Although the microbial mechanisms for oil recovery will be discussed independently, it is likely that microbial processes act synergistically. For example, significant mobilization of entrapped oil from sand-packed columns and sandstone cores required multiple products, an alcohol, a biosurfactant, and a polymer (Maudgalya *et al.*, 2004). The relative importance of each will depend on factors that limit oil production within a given reservoir, the strains of the microorganism involved, and the protocols used for injection of nutrients and inocula.

Microbial technologies are often grouped into three main categories, paraffin removal, microbial well stimulations, and microbially enhanced waterfloods (Knapp *et al.*, 1990). Hydrocarbon degradation is the mechanism for microbial paraffin removal (Table 6.3). Commercial companies use proprietary inocula and nutrients to degrade paraffins and other hydrocarbons that may have accumulated on production equipment, within the well, or within the reservoir (Fig. 6.1A and Table 6.4). The microbial treatment may be localized to the well-bore region or occur several meters to ten or more meters in the reservoir. For the sake of this discussion, we will call other microbial technologies that treat individual wells and do not involve *in situ* hydrocarbon metabolism well stimulations technologies. The objective of well stimulation technologies is to stimulate the production of large amounts of acids, gases, solvents, biosurfactants, and/or emulsifiers in the well and the near well region of the reservoir to improve oil production rates. The volumes of nutrients and cells injected are large enough to treat several to tens of meters of the resevoir. In addition to removing scale, paraffins, asphaltenes, and other debris, well stimulations may change wettability and flow patterns to allow more oil to flow to the well (Fig. 6.1A, Table 6.3). Microbially enhanced water flooding differs from the above in that nutrients with or without inocula are injected into one well in order to stimulate microbial activity in a large portion of the reservoir and the oil is recovered in wells different from that used for the injection (Fig. 6.1B and C, Table 6.3). The goal of microbially enhanced water flooding is to increase the ultimate oil

TABLE 6.3 Microbial products and activities useful for MEOR, their mechanism of action, the production problem they target, and the most suitable type of reservoir for their application

Microbial product/ activity	Microorganisms	Production problem	Mechanism of action	Type of formation/reservoir
Hydrocarbon metabolism	Aerobic hydrocarbon degraders	Paraffin deposition; poor microscopic displacement efficiency	Remove paraffin deposits; metabolites mobilize oil	Wells with paraffin deposition; mature waterflooded reservoirs
Gases (CO_2, CH_4)	Fermentative bacteria; Methanogens	Heavy oil	Reduce oil viscosity	Heavy oil-bearing formations (API <15)
Acids	Fermentative bacteria	Low porosity, poor drainage, formation damage	Dissolve carbonaceous minerals or deposits	Carbonate or carbonaceous reservoirs
Solvents	Fermentative bacteria	Heavy oil	Reduce oil viscosity	Heavy oil-bearing formations (API <15)
		Poor microscopic displacement efficiency	Alter wettability	Strongly oil-wet, waterflooded reservoirs

Emulsifiers	*Acinetobacter* sp., *Candida*, *Pseudomonas* sp., *Bacillus* sp.	Paraffin and oil sludge deposition, poor microscopic displacement efficiency	Emulsify oil to form o/w emulsions (or less commonly w/o emulsions)	Waxy oil (>C22 alkanes); paraffinic oil and asphaltene-bearing formations
Biosufactants	*Bacillus* sp., Pseudomonads, *Rhodococcus* sp.	Poor microscopic displacement efficiency	Lower interfacial tension	Sandstone or carbonate reservoirs with moderate temperatures (<50 °C) and relatively light oil (API >25)
Biomass/ polymer production	Many kinds	Poor volumetric sweep efficiency	Plug water channels and reduce permeability in water-swept regions by biofilm formation	Stratified reservoirs with variations in permeabilities

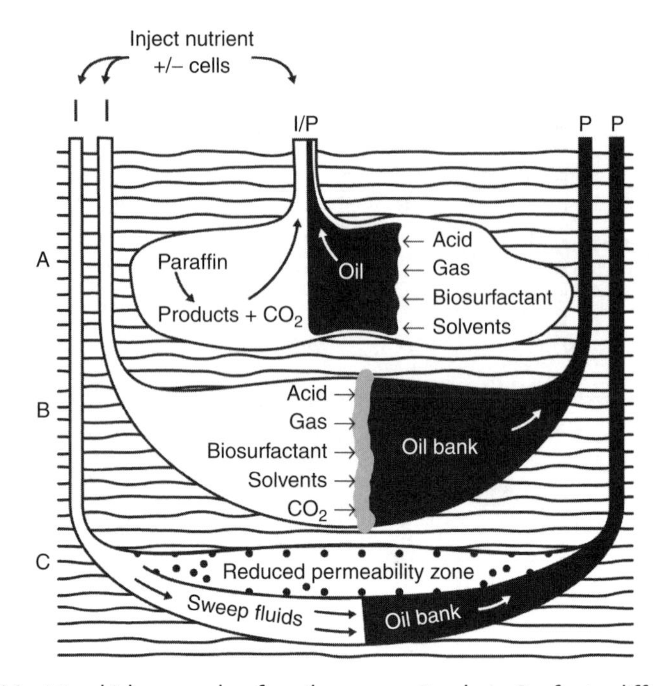

FIGURE 6.1 Microbial approaches for oil recovery. Panels A–C refer to different processes described in the text. (A) Microbial paraffin removal (left side) and microbial well stimulation (right side). Paraffin removal could be applied to either injection (I) or production (P) wells; well stimulations are done in production wells; most often, paraffin removal and microbial well stimulations involved the injection of nutrients and cells. (B) Microbially-enhanced water flooding where the stimulation of microbial metabolism creates useful products to mobilize oil; an inoculum maybe used. (C) Microbial selective plugging blocks high permeability zones (upper region) and redirects the recovery fluid into bypassed regions (lower) of the reservoir; nutrients and inoculum (if used) enter the high permeability zone (upper panel) and *in situ* microbial growth reduces permeability in this region.

recovery factor of the reservoir. This is done by improving the microscopic displacement efficiency through a reduction in the capillary forces that entrapped oil or by improving the volumetric sweep efficiency of the recovery fluid by blocking water channels and high permeability zones to push bypassed oil to production wells (Fig. 6.1B and C, Table 6.3).

Although well clean up and well stimulation technologies are not technically EOR processes, many times these processes extend the economic life of a field, either by reducing operating costs or increasing daily revenue (Brown *et al.*, 2002; Portwood, 1995b). By doing so, more oil is recovered from the reservoir than could have been recovered by conventional technology. Common usage in the microbial oil recovery discipline rarely distinguishes between microbial processes that improve oil production rate from those that improve the ultimate oil recovery factor and all microbial processes have been labeled MEOR. A recent analysis of numerous field trials concluded that MEOR is successful (Maudgalya *et al.*, 2007). Of the 403 MEOR projects that were analyzed, the overwhelming majority (96%) was considered to be successful by the investigators.

A. Paraffin control

The removal of paraffin and other deposits from the well and production equipment reduces operating costs and can improve the flow of oil into the well by altering drainage patterns and/or fluid saturations near the well. Expensive physical and chemical treatments are frequently needed to keep wells operative (Barker *et al.*, 2003; Becker, 2001; Etoumi, 2007; Ford *et al.*, 2000; Lazar *et al.*, 1999; Misra *et al.*, 1995). Chemical methods include the use of solvents, surfactants, dispersants, and wax crystal inhibitors. Thermal methods include the treatment of wells with hot fluids, usually hydrocarbons or water, to remove deposits. Formation damage may occur if the oil used as the solvent has paraffin content (Etoumi, 2007). Physical removal of paraffins by scraping is also used (Etoumi, 2007).

Stimulation of *in situ* hydrocarbon metabolism is the most common microbial approach to treat paraffin deposition problems (Fig. 6.1). Many hydrocarbon-degrading microorganisms have been isolated and there is a vast literature on the ability of microorganisms to degrade hydrocarbons aerobically (Van Hamme *et al.*, 2003) and anaerobically (Heider *et al.*, 1998; Spormann and Widdel, 2001). Usually, the procedure involves the injection of hydrocarbon degraders along with nutrients. Fluid production from the well is stopped (shut in) for several days to several weeks to allow *in situ* microbial growth and metabolism. The microbial treatments are repeated on regular schedule (every several weeks or monthly). Unfortunately, many of the published reports about microbial paraffin removal use proprietary mixtures of hydrocarbon-degrading bacteria and nutrients or "biocatalysts," making it difficult to provide a scientific assessment of the technology.

The stated mechanism for paraffin removal involves the conversion of long-chain hydrocarbons to short-chain hydrocarbons, resulting in oils with lower viscosities and improved mobilities (Brown *et al.*, 2005; Lazar *et al.*, 1999; Maure *et al.*, 2005; Nelson and Schneider, 1993; Smith and Trebbau, 1998). How long-chain alkanes are converted to short-chain alkanes is unclear. There are no microorganisms known to catalyze such a reaction. A number of studies report changes in the composition of the oil and its physical properties as a result of microbial activity. The proportion of low carbon number alkanes to high carbon number alkanes increases and viscosity decreases after microbial treatment (Brown, 1992; Deng *et al.*, 1999; He *et al.*, 2003; Nelson and Schneider, 1993; Partidas *et al.*, 1998; Smith and Trebbau, 1998; Trebbau *et al.*, 1999; Wankui *et al.*, 2006). Additional evidence for microbial metabolism such as the loss of electron acceptors and the production of metabolites are not provided. While the mechanism of action of commercial hydrocarbon-degrading microbial formulations is at least debatable, there is geochemical evidence that suggests that low molecular weight alkanes may be derived from fatty acids in certain oil reservoirs (Hinrichs *et al.*, 2006). There is also a number of reports of microbial paraffin degradation.

Etoumi (2007) found that *Pseudomonas* and *Actinomyces* spp. emulsified crude oil and hexadecane. Gas chromatographic (GC) analysis showed a decrease in the proportion of alkanes with carbon numbers greater than 22 and an increase in the proportion of alkanes between C13 and C21. Wax appearance temperature and crude oil viscosity decreased. Two *Bacillus* spp. and one *Pseudomonas* sp. isolated from fluids produced from the Liaohe field in China grew with wax as the sole carbon source, indicating their ability to degrade paraffins (He *et al.*, 2003). Bacteria isolated from hydrocarbon-polluted site degraded up to 84% of crude oil, 88% of semisolid or solid paraffin added to the cultures under aerobic conditions, and 47% of semisolid or solid paraffin under facultative anaerobic conditions (Lazar *et al.*, 1999). In laboratory flow experiments, the most active consortium did not significantly alter the total paraffinic content of crude oil but a decrease in viscosity was observed. Biosurfactants and biosolvents were detected during the test and may have contributed to the reduction in viscosity. A decrease in the apparent molecular weight of crude oil and its cloud point temperature indicated that microbial degration of heavy paraffinic hydrocarbons is possible (Sadeghazad and Ghaemi, 2003). Chemical and physical analyses showed that *P. aeruginosa* degraded normal alkanes (C16–C25) and *B. licheniformis* degraded cyclo and isoalkanes (C20–C30). Kotlar *et al.* isolated a strain identified as *Acinetobacter* sp. strain 6A2 from enrichments containing paraffin with melting temperature 52–54 °C (Kotlar *et al.*, 2007). The strain degraded alkanes with carbon numbers of C10–C40 and harbored enzymes involved in the degradation of high carbon number components.

Alterations of the physical properties of crude oil such as the viscosity, pour point (the lowest temperature where oil flows when cooled), and cloud point (the temperature where paraffins begin to precipitate from a liquid state) may indicate the production of emulsifiers or biosurfactants (Barkay *et al.*, 1999; Etoumi, 2007; Lazar *et al.*, 1999; Rosenberg *et al.*, 1983; Trebbau de Acevedo and McInerney, 1996). Another possible mechanism of action is the partial oxidation of hydrocarbons to alcohols, aldehydes, or fatty acids, which could act as solvents or surfactants (Pelger, 1991).

There is a large body of evidence that shows that hydrocarbons are incompletely metabolized with the production of alcohols, fatty acids, etc. (Abbott and Gledhill, 1971; Connan, 1984). The injection of oxygen or a chemical that can be converted to oxygen (hydrogen peroxide) is needed to stimulate *in situ* aerobic metabolism. It is not possible to know whether the commercial approaches provide the needed electron acceptor. Anaerobic metabolism is a possibility (Spormann and Widdel, 2001), but this is usually a very slow process and it is difficult to envision significant anaerobic hydrocarbon metabolism occurring in the 1- to 2-day shut in period. Another concern regarding commercial microbial paraffin-degrading formulations is that laboratory studies do not support the ability of the inocula to degrade hydrocarbons or to recover oil. Gieg *et al.* (2004) found some emulsification, but no evidence of hydrocarbon metabolism when several crude oils were incubated under a variety of conditions with a proprietary mixture of hydrocarbon-degrading bacteria according to manufacturer's guidelines. *In situ* growth of commercial formulations of bacteria and nutrients did not mobilize oil entrapped in sandstone (Lazar *et al.*, 1999; Rouse *et al.*, 1992).

Microbial paraffin removal has survived in the market place for many years and published information supports the effectiveness of the approach (Table 6.4). A large number of wells from many different reservoirs all over the world have been treated. The conclusions of these studies are that the use of the proprietary inocula reduces the frequency of physical and chemical paraffin control treatments (Brown, 1992; He *et al.*, 2000, 2003; Nelson and Schneider, 1993; Santamaria and George, 1991; Streeb and Brown, 1992), reduces other operating costs (pump current) (He *et al.*, 2003; Streeb and Brown, 1992), and increases oil production (Table 6.4). In many cases, the natural decline in oil production was slowed or stopped for periods of months to years (Table 6.4). In quantitative terms, the results can be impressive, for example, daily oil production rate improvements of 47–210 m^3/d (Abd Karim *et al.*, 2001) or increases in oil production of 1700% (Nelson and Schneider, 1993) (Table 6.4).

Two users of commercial microbial paraffin treatment products indicate that oil production did not change, but the microbial treatments were cheaper to use than chemical or physical remediation approaches (Ferguson *et al.*, 1996; Santamaria and George, 1991) (Table 6.4). The commercial formulations were tested in four production wells at the Department of Energy's Rocky Mountain oil storage facility (Giangiacomo, 1997). A slight increase in oil production and low costs of the microbial treatments were noted.

Several independent groups have developed their own hydrocarbon-degrading inocula (Table 6.4). Scientists in China used three hydrocarbon-degrading *Bacillus* spp. to treat a number of paraffin-laden wells in the Liaohe oilfield (He *et al.*, 2003; Wankui *et al.*, 2006). Forty-three of the sixty

TABLE 6.4 Field results of microbial treatments to control paraffin deposition

Method	Results	References
Use of proprietary inocula and nutrients	50 wells treated; 78% had an increase in oil production above 4 m^3/d; overall average increase was 10 m^3/d; better performance if well produced from one formation and if layer was <7 m thick; shift in alkane ratio and reduction in oil viscosity noted	Partidas *et al.* (1998), Smith and Trebbau, (1998), Trebbau *et al.* (1999)
	Less frequent chemical wells treatments; oil production increased by 0.2–0.6 m^3/d	Hitzman (1988), Lazar *et al.* (1993)
	Two fields treated; oil production increased from 0.2 to 0.6 m^3/d for about 1 year; less frequent servicing of wells needed	Bailey *et al.* (2001)
	6 fields treated; 20–1700% increase in reserves over 3 years; saved $348 per well in operating costs; reductions in cloud and pour point and viscosity noted	Nelson and Schneider (1993)
	3 fields treated; 160–600% increase in oil production for 7–9 months; less frequent hot oil treatments and reduced chemical costs	Pelger (1991)
	72 wells treated; reduced frequency of hot oil and chemical treatments; arrested the natural decline in oil production; oil viscosity decreased	Brown (1992)

TABLE 6.4 (*continued*)

Method	Results	References
	Numerous wells treated; reduced frequency of hot oil and chemical treatments; reduced fuel consumption; arrested the natural decline in oil production; 4760 m^3 incremental oil production in 1 year	Streeb and Brown (1992)
	Treated 27 wells; 0–48% increase in oil production; 2950 m^3 of incremental oil in 3–6 months. Treated another 20 wells; 18% increase in oil production for 15–30 days; wax content altered	Deng *et al.* (1999)
	Treated 2 wells in two different fields; 36% and 46% increase in oil production; 3080 and 2200 m^3 incremental oil	Maure *et al.* (2005)
	9 fields treated; oil production increased from 22% to 320% for 14–44 months; incremental oil recovery ranged from 340 to 4110 m^3	Portwood (1995a,b), Portwood and Hiebert (1992)
	3 of 4 wells had no response, one well had an 18% increase in production; less expensive than hot oil treatments	Ferguson *et al.* (1996)
	5 wells treated: no change in oil production, reduced frequency of hot oil treatments; savings of $8000 per month	Santamaria and George (1991)
	Unnamed supplier of microbial paraffin treatment; 4 wells treated; overall, 16% increase in production (8 months); 65 m^3 incremental oil; microbial treatment ($1031 per well) had lower cost than chemical treatment ($3414 per well)	Giangiacomo (1997)

(*continued*)

TABLE 6.4 (*continued*)

Method	Results	References
Hydrocarbon-degrading *Bacillus* strains	43 of 60 wells showed positive response; 10,630 m^3 incremental oil; oil viscosity and oil composition altered	Wankui *et al.* (2006)
	Stopped natural decline and increased oil production in 3 of 4 wells; 650 m^3 incremental oil; chemical treatments reduced from every 15 days to every 4 months; pump current reduced	He *et al.* (2003)
Anaerobic hydrocarbon degrader with kerosene	Five different fields treated; increases in oil production of 1–2 m^3/d noted for all fields	Nelson and Launt (1991)
Unspecified treatments	Inoculate and shut in 1–5 weeks; oil production in 3 wells increased by 47% on average for 5 months (47–210 m^3/d)	Abd Karim *et al.* (2001)
	Undescribed microbe reapplied every 28 days; 26 wells treated; 9 wells had less maintenance and increased oil production; inconclusive results or no change in other wells	Wilson *et al.* (1993)

treated wells showed a positive response (Wankui *et al.*, 2006). The need for frequent thermal and chemical treatments dropped markedly and oil production increased substantially. Another study showed that the injection of hydrocarbon degraders and kerosene into production wells from five different formations increased oil production (Nelson and Launt, 1991). The success of independently-derived hydrocarbon-degrading inocula and independently tested commercial formulations provides support that the stimulation of *in situ* hydrocarbon metabolism is an effective approach to improve operations and oil production. However, other investigators found that these formulations were ineffective or provided inconclusive results (Ferguson *et al.*, 1996; Wilson *et al.*, 1993). None of the

TABLE 6.5 Results of microbial processes not involving hydrocarbon metabolism applied to individual wells to improve performance

Method	Results	References
Early tests		
Injection of aerobic and anaerobic bacteria with peat biomass, silt extracts and aerated water	Oil production increased by 28–48 m^3/day	Hitzman (1983), Senyukov *et al.* (1970)
Injection of a mixed culture with 4% molasses and a 6-month shut in period	Oil production increased 3.5 m^3/month for 3 months	Hitzman (1983), Senyukov *et al.* (1970)
Stimulation of fermentative metabolism: acid, gas, and solvent production		
Injection of a mixed culture of anaerobic microorganisms and molasses	Oil production increased 230% for 7 months	Lazar (1991)
Injection of an adapted mixed microbial culture with 2–4 % molasses	Oil production increased 300% from 0.1 to 0.35 m^3/day and 500% from 0.2 to 1.1 m^3/day	Lazar *et al.* (1993)
Injection of *Bacillus* sp. and *Clostridium* sp. with 4% molasses	350% increase in oil production if done correctly; most effective in carbonate wells with 15–39 API gravity crude oil and less than 10% salt	Hitzman (1983)
Injection of clostridial strains and 20 ton of molasses into carbonate formation	480% increase in oil production from 0.6 to 2.9 m^3/day	Wagner (1991)
Injection of clostridial spores and 9% molasses with a 30-day shut in period	Oil production increased from 0.16 to 0.32 m^3/day in one well	Grula *et al.* (1985)

(*continued*)

TABLE 6.5 *(continued)*

Method	Results	References
Injection of clostridial spores with sucrose, molasses, and NH_4NO_3 and a 3-month shut in period	Slight increase in oil production for 12 weeks	Hitzman (1983), Lazar (1991)
Injection of a mixed culture with 3.2% sugar and a 40-day shut in period	2086 m^3 of incremental oil; microbial numbers increased and metabolites detected	Wang *et al.* (1993)
Injection of a mixed culture with 4–5% molasses and a 15–21-day shut in period	42 of 44 wells tested had positive response with 33–733% increase in daily oil production; 261 m^3 of incremental oil per well; microbial numbers increased; CO_2 gas increased; microbial metabolites detected	Wang *et al.* (1995)
Injection of facultative anaerobic and anaerobic polymer producers and molasses	Cell numbers increased in the produced water	Lazar (1991)
Injection of molasses with 7-day shut in period	Oil production increased 216% (10 m^3/day)	Yusuf and Kadarwati (1999)
Stimulation of fermentative metabolism and biosurfactant production		
Injection of surfactant, alcohol, and polymer producers with sugar, molasses, yeast extract, $PO_4^{=}$, and NO_3^- and a 3-week shut in period	Oil production increased from 2.4 to 6.3 m^3/ day	Zaijic (1987)
Injection of a mixed culture of acid, gas, solvent and biosurfactant	Oil production increased by 79%	Bryant *et al.* (1988), Lazar (1991)

TABLE 6.5 (*continued*)

Method	Results	References
producers with 4% molasses		
Injection of *Pseudomonas aeruginosa*, *Xanthomonas campestris*, and *Bacillus licheniformis* with unspecified nutrients and a 40–64-day shut in period	Oil production increased by 2.3 and 3.4 m^3/day for 8 and 18 months, respectively, in two treated wells; 1140 m^3 of incremental oil	Zhang and Zhang (1993)
Injection of two *Bacillus* strains and one pseudomonad with waste fluids from a fermentation industry and a 7-day shut in period	Oil production increased 60% (1.9 m^3/day) in one well; no change in other well	He *et al.* (2000)
Injection of biosurfactant-producing *Bacillus* sp. and clostridia with unspecified nutrients and a 7-dayshut in period	Oil production increased in 2 wells (percent increase, incremental oil (m^3), duration in days): 50%, 56 m^3, 70 days and 40%, 137 m^3, 137 days; no change in 3 other wells	Buciak *et al.* (1994)
Injection of *B. licheniformis* and *Bacillus subtilis* subsp. *subtilis spizizenii* with glucose-nitrate-trace metals and a 4-day shut in period	Oil production increased 30–100% in two wells for 100 days; 38 m^3 of incremental oil; increase in microbial numbers, microbial metabolites including the lipopeptide biosurfactants detected	Simpson *et al.* (2007)
Stimulation of fermentative metabolism, polymer production, or growth to plug flow channels		
Injection of polymer-producing *Enterobacter cloacae* and	7 of 12 wells had increased oil production; 1730 m^3 of	Maezumi *et al.* (1998)

(*continued*)

TABLE 6.5 (*continued*)

Method	Results	References
biosurfactant-producing *Bacillus licheniformis* with 10–20% molasses and an 11–21-day shut in period	incremental oil recovered	
Injection of *Enterobacter cloacae* followed by 1% molasses in 2 wells or 5% molasses in 2 wells and *E. cloacae* and 5% molasses coinjected into 2 wells, each followed by a 10-day shut in period	4 wells showed increased oil production (1.7 m^3/day) for about 1 year; no change in two wells	Nagase *et al.* (2001)
Injection of *Enterobacter cloacae* or a polymer-producing *Bacillus* sp. and 5 or 10% molasses with and without $PO_4^{=}$ and a 14–21-day shut in period	7 of 14 wells that received molasses and an inoculum had increased oil production and decreased water production; pH decreased, water viscosity increased, and microbial metabolites detected; only 1 of 4 wells that received only molasses had an increase in oil production	Ohno *et al.* (1999)
Injection of unspecified inorganic nutrients	Oil production in one well increased by 30% (0.5 m^3/day)	Sheehy (1990)
Injector clean-out: 6 tons of molasses and adapted mixed culture of microbes	Injection pressure decreased, oil viscosity decreased, and cell numbers increased 1–4 logs	Lazar *et al.* (1991)
Microbial fracturing of carbonate formations: treat with unspecified	Increased oil production by 20% for about 30	Moses *et al.* (1993)

TABLE 6.5 (*continued*)

Method	Results	References
nutrients and microbes; shut in 7 days	days then returned to pretreatment levels	
Fracture damage repair: treat with microbes that degrade guar gum-based gels	Oil production restored in two wells that had lost production due to previous fracturing treatment	Bailey *et al.* (2001)
Use of unspecified microbial strains and nutrients		
Unspecified	24 wells treated with a 42% average increase in oil production; 75% showed an increase in pressure	Hitzman (1983)
Unspecified strains and nutrients	Increased oil production in 20 wells for about 15–30 days	Deng *et al.* (1999)
Unspecified strains and nutrients with a 40-day shut in period	Oil production of well increased from 4 to 6.4 m^3/day for 8 months; 556 m^3 of incremental oil	Hitzman (1988)

studies mentioned so far address the issue of whether an inoculum is required or if indigenous microorganisms caused the changes observed in the field.

B. Biogenic acid, solvent, and gas production

End products of anaerobic sugar fermentation include gases (CO_2 and H_2), acids (acetate, propionate, butyrate, valerate, and lactate), and solvents (ethanol, propanol, butanol, acetone, and 2, 3 butanediol) (Nakano *et al.*, 1997). H_2 is produced in large quantities in anaerobic ecosystems, but H_2 is quickly used by many different kinds of microorganisms. Acid, gas, and solvent production are used to improve oil production from individual wells (Fig. 6.1A, Table 6.5) or to mobilize entrapped oil during waterfloods (Fig. 6.1B, Table 6.6). In both approaches, readily fermentable carbohydrates with or without an inoculum are injected into the formation. If sufficient CO_2 and CH_4 are made, these gases will result in

TABLE 6.6 Field results of microbial processes to enhance oil production in oil reservoirs

Process	Method	Results	References
Multiple products (acids, gases, solvents)	Injection of *Clostrdium acetobutylicum* and molasses into poorly consolidated sand with carbonate minerals	Oil production increased 250% and water production decreased for 7 months; 22,000 m^3 of CO_2 and 11, 340 kg of fermentation acids accounting for 78% of the sugar carbon were produced	Yarbrough and Coty (1983)
	Injection of *Clostrdium tyrobutyricum* and molasses into carbonate reservoir	Water production decreased by ~20% and oil production increased by 50–65%; 2550–4900 m^3 of incremental oil recovered after 3 years; calcium and bicarbonate increased indicating dissolution of the rock; 350,000 m^3 of CO_2; 6 g/l organic acid concentration; isotopically light methane detected	Nazina *et al.* (1999b), Wagner *et al.* (1995)
	Injection of an adapted mixed culture of facultative and anaerobic microorganisms and 2–4% molasses followed by 2–4% molasses until molasses detected in production wells	Increased oil production by 400–600% (1–2 m^3/day/well) in two calcareous sandstone formations for 3 years; cell numbers increased and microbial metabolites detected in production wells. For carbonate or calcareous sandstone formations	Lazar (1992, 1993), Lazar *et al.* (1991, 1993)
	Injection of an adapted mixture of facultative and anaerobic microorganisms with molasses, four times at 4-month intervals	Around 15–50% increase in oil production rate in 2 of 8 reservoirs treated; large pH decease and microbial metabolites detected. Effective in carbonaceous reservoirs	Lazar (1987), Lazar and Constantinescu (1985), Lazar *et al.* (1988)

Injection of anaerobes derived from the field with molasses, NH_4Cl, polyphosphate and soda into carbonate reservoir; shut in until pressure increases at production wells	Oil production increased from 300% to 600% for nine months; 7000 m^3 incremental oil recovered; water production decreased 30–40%; 2 MPa increase in wellhead pressure; microbial metabolites and increase in cell numbers detected in production wells	Wagner (1991)
Indigenous microorganisms stimulated by two injections of local food manufacturing waste (sugars, proteins, amino acids) with N, P and K; shut in for 40 days	Oil production rate increased for 1.5 years; one well doubled oil production; 360 m^3 of incremental oil recovered; microbial metabolites detected; aerobic hydrocarbon degraders and heterotrophs increased; decrease of the C12 to C22 alkane fraction in oil.	Murygina, *et al.* (1995)
Injection of molasses with mixed microbial population containing sulfate reducers and pseudomonads	Significant increase in cell numbers in wells 500 m distant from injection wells; oil viscosity decreased; oil production increased 5–35% in some wells	Dostalek and Spurny (1957a,b), Hitzman (1983)
Injection of mixed microbial population (facultative and strict anaerobes, sulfate reducers, and pseudomonads), molasses, sucrose, NO_3^-, and $PO_4^=$	Oil production increased from 10% to 120% for several weeks to 18 months in some production wells; large decreases in oil viscosity and pH observed; CO_2 detected	Dienes and Jaranyi, (1973), Hitzman (1983), Jaranyi (1968)

(continued)

TABLE 6.6 *(continued)*

Process	Method	Results	References
	Injection of mixed microbial population from soils or sewage with molasses; shut in wells for 4–6 months	Oil production increased by 28–340% in some wells for 2–8 years	Hitzman (1983), Karaskiewicz (1975)
Multiple products with biosurfactant production	Injection of a biosurfactant-producing *Bacillus* sp., a *Clostridium* sp. and molasses; shut in for 2 weeks; periodic injection of molasses for 2 years	Oil production increased 13% and water to oil ratio decreased 35%; 88 m^3 of incremental oil recovered at $20/$m^3$; surface tension of produced water decreased	Bryant *et al.* (1990)
	Injection of a biosurfactant-producing *Bacillus* sp. a *Clostridium* sp. and molasses; continuous molasses injection for 1 year	Oil production increased 19% for 3 years; 400 m^3 of incremental oil recovered at $15/$m^3$	
Stimulation of indigenous hydrocarbon metabolism	Injection of aerated mineral salts solution for 3–6 months	Oil production rate increased 18–45% in 3 of 4 reservoirs; 5660, 12,440, 83,400, and 105,800 m^3 of incremental oil produced	Ibatullin (1995), Matz *et al.* (1992)
	Cyclic injection of aerated water with N and P sources	Oil production rate increased 10–46%; 2670–48,160 m^3 of incremental oil produced at $6/$m^3$; metabolites, δ-C^{13} of bicarbonate increased and δ-C^{13} of CH_4 decreased; increased methanogenesis; acetate detected in production wells. Multiple formations treated	Belyaev *et al.* (1998, 2004)

	Cyclic injection of aerated water with N and P sources, but treated immediately before wells scheduled for shut in	376,000 m^3 of incremental oil produced from 118 wells (59 injection wells treated) over 11-year period	Belyaev *et al.* (2004)
	Cyclic injection of aerated water with N and P sources with addition of crude oil	4830, 35,100, and 67,000 m^3 of incremental oil produced from 3 different formations in 4, 12, and 7 years, respectively	Belyaev *et al.* (2004)
	Inject air–water mixture with mineral salts 2–3 days every month for 3 years	16200 m^3 of incremental oil produced in 3.5 years from Dagang oil field; inter-well permeability profile modified; methanogens, sulfate reducers and aerobes increase 1–2 logs; metabolites detected and surface and interfacial tension decreased	Nazina *et al.* (2007b)
Stimulation of hydrocarbon metabolism using an inoculum	Cyclic injection of aerated water with N and P sources with halophilic hydrocarbon degraders	2760, 6180, and 44380 m^3 of incremental oil recovered from 3 formations over 4, 3, and 8 years, respectively	Belyaev *et al.* (2004)
	Two cycles of injections of *Bacillus cereus* and *Brevibacillus brevis* hydrocarbon-degrading strains	Oil production in Daqing oil field increased 165% for about 200 days; 6700 m^3 of incremental oil from 10 producers; oil viscosity decreased, alkane profile changed, microbial counts increased 2 logs	Wankui *et al.* (2006)

(*continued*)

TABLE 6.6 *(continued)*

Process	Method	Results	References
	Periodic injection of a proprietary mixture of microorganisms, inorganic nutrients (N, P, and trace metals), and a "biocatalyst"	Multiple reports: 78% of the projects either increased oil production or arrested natural decline in oil production; oil production increased by an average of 36%; 1740–3086 m^3 of incremental oil; oil viscosity decreased; cost of $1.6–12.50 per m^3 of incremental oil	Dietrich *et al.* (1996), Maure *et al.* (1999, 2001, 2005) Portwood (1995a, b), Portwood and Hiebert (1992), Strappa *et al.* (2004), Yu and Schneider (1998)
Permeability profile modification	Indigenous microorganisms stimulated by weekly injections of molasses, NO_3^-, and $PO_4^=$	Increased oil production in 13 of 19 production wells in the treated area; oil from 9 wells had altered oil composition; extended economic lifetime of field by 5–11 years; 63,600 m^3 of incremental oil recovered at a cost of $8.30/$m^3$	Brown (2007), Brown *et al.* (2002)
	Indigenous denitrifiers stimulated by injection of nitrate, nitrite, and other unspecified nutrients	Slowed natural decline of oil production in two fields; 1315 and 1225 m^3 of incremental oil recovered in 7–10 months; sulfide levels in brine decreased	Hitzman *et al.* (2004)
	Indigenous microorganisms stimulated with 3 batch injections of molasses and NH_4NO_3, each followed by cessation of injection fluid for 2–4 weeks	Major flow channel blocked; inter-well permeability variation reduced; oil production reinitiated (0.16 m^3/day); microbial metabolites detected	Knapp *et al.* (1992)

	Indigenous microorganisms stimulated with maltodextrins and ethyl acid phosphate followed by cessation of injection fluid for 2 weeks	Fluid injection into upper portion of reservoir, which initially took 28% of the injected fluid, stopped; microbial metabolites detected	Jenneman *et al.* (1996)
	Injection of *Bacillus* spores, sucrose, yeast extract, polyphosphate, and NO_3^- followed by cessation of injection fluid for 3 weeks	One injection zone blocked, another zone reduced by 50%, and 7 new injection zones detected; biofilm stable for 8 months	Lee Gullapalli *et al.* (2000)
	Injection of polymer-producing *Enterobacter* strain for one week followed by 10% molasses for 2 months	Flow diversion occurred based on tracer analysis; water production decreased and oil production increased in seven of eight wells in the patterns; PCR analysis detected injected strains in produced fluids	Nagase *et al.* (2002)

swelling of crude oil and reduce its viscosity (Bryant and Burchfield, 1989). *In situ* gas production may also lead to repressurization of oil reservoirs and hence improve oil recovery especially in mature reservoirs, but very large volumes would have to be made (Bryant and Lockhart, 2002). Organic acid production can lead to the dissolution of carbonates in source rocks, increasing porosity and permeability, and enhancing oil migration (Adkins *et al.*, 1992a; Udegbunam *et al.*, 1991). Given that large amounts of acids can be made by *in situ* microbial fermentation, it is possible that microbial processes could replace conventional acid treatments (Coleman *et al.*, 1992). Solvents alter the rock wettability at the oil–rock interface, releasing the oil from the porous matrix. Solvents could also dissolve in oil and lower its viscosity (McInerney *et al.*, 2005a).

1. Microorganisms

The most common microorganisms used for acid, gas, and/or solvent production for MEOR processes include members of the genera *Bacillus* and *Clostridium* (Bryant, 1988; Bryant and Douglas, 1988; Chang, 1987; Donaldson and Clark, 1982; Tanner *et al.*, 1993; Udegbunam *et al.*, 1991; Wagner, 1985; Wagner *et al.*, 1995). Spore production by these species is an advantage because spores survive harsh conditions and penetrate deep into petroleum reservoirs. *Clostridium* spp. produce gases (CO_2 and H_2), alcohols (ethanol and butanol), solvents (acetone), and acids (acetate and butyrate). *Bacillus* spp. produce acids (acetate, formate, lactate, etc.), gas (CO_2), alcohols (ethanol and 2,3-butanediol), and biosurfactants. Lactic acid bacteria (LAB) have also been used in oil recovery operations (Coleman *et al.*, 1992). Homofermentative LAB produce only lactate from sugars, while heterofermentative LAB produce ethanol and CO_2 beside lactate. Methane production by methanogens could potentially aid in oil release (Belyaev *et al.*, 2004).

2. Laboratory flow studies

A number of studies show that *in situ* production of acid, solvent, and gas increases oil recovery from laboratory models (Almeida *et al.*, 2004; Behlulgil and Mehmetoglu, 2002; Bryant, 1988; Bryant and Burchfield, 1989; Bryant and Douglas, 1988; Chang, 1987; Desouky *et al.*, 1996; Jinfeng *et al.*, 2005; Marsh *et al.*, 1995; Rauf *et al.*, 2003; Wagner, 1985; Wagner *et al.*, 1995). Most of these studies employed allochthonous microorganisms and molasses or some type of readily-fermentable carbohydrate.

Gas production has often been mentioned as an important mechanism for oil recovery (Jack *et al.*, 1983). The authors pointed out that what works effectively in the laboratory might not be as effective in the field due to low gas transfer. Isolates that showed promise for field applications include *Enterobacter* sp. with production of 1.6 moles of gas per mole of sucrose utilized (Jack *et al.*, 1983) and *Clostridium* strains that metabolize at

5–7.5% salt concentrations (Grula *et al.*, 1983). *Vibrio* sp. and *Bacillus polymyxa* were found to be the most proficient gas-producing strains under conditions that simulated actual oil reservoirs conditions (Almeida *et al.*, 2004). *In situ* growth of a consortium containing *B. polymyxa* in sand-packed column recovered 18% of the residual oil (which could not be recovered by extensive water flooding). The consortium containing *Vibrio* sp. recovered 16% of the residual oil. *In situ* growth of *Streptococcus* sp., *Staphylococcus* sp., or *Bacillus* sp. with molasses or glucose in sand-packed columns produced more gas (CO_2, H_2), and higher pressures than that observed by the stimulation of indigenous microorganisms with the same substrates (Desouky *et al.*, 1996). Large recoveries of residual oil (>50%) occurred when an inoculum was used. Although the cultures were selected based on gas production, the large decrease in interfacial tension to ≤ 1 mN/m suggests that the main mechanism for oil recovery may have been biosurfactant production rather than gas production.

Clostridium acetobutylicum and related species have been used in a number of studies due to the production of copious amounts of gases, acids, and solvents produced from carbohydrate substrates (Wagner, 1985; Wagner *et al.*, 1995). Large pressure increases and oil viscosity decreases were observed (Behlulgil and Mehmetoglu, 2002). Flow experiments with crushed limestone showed that *in situ* growth and metabolism recovered 49% of the residual oil. The core pressure increased, the pH decreased by 3 units, and the weight of crushed limestone decreased after microbial growth occurred. Residual oil was recovered when cores were treated with cell-free culture fluids (containing acids and alcohols) probably due to changes in wettability and oil viscosity. Larger oil recoveries (up to 30%) were observed due to the *in situ* growth and metabolism of *Clostridium tyrobutyricum* (Wagner, 1985; Wagner *et al.*, 1995) or *C. acetobutylicum* in sandstone cores (Marsh *et al.*, 1995). In the latter study, the cores were incubated at an initial pore pressure of about 7000 kPa to mimic actual reservoir conditions where free gas phases are often not present. Large amounts of acetate, butyrate, butanol, ethanol, and CO_2 were made and their production coincided with oil recovery. The injection of cell-free culture fluids that contained the acids and solvents did not recover residual oil, consistent with *in situ* CO_2 production being the main mechanism for oil recovery. However, large reductions in permeability occurred, making it likely that multiple mechanisms were involved in oil recovery. *In situ* growth of a *Clostridium* sp. (acid and gas producer) or a *Bacillus* sp. (biosurfactant and gas producer) in a medium containing sodium pyrophosphate mobilized entrapped oil from sand-packed columns and sandstone or limestone cores (Chang, 1987).

Acid production by microorganisms may be an important mechanism for oil recovery from carbonates. The *in situ* growth of a halophilic, acid-producing bacterium in columns packed with crushed limestone lead to the dissolution of the carbonate minerals as evidenced by the presence of Ca^{2+} in the core effluents and significant amounts of carbonate particulates in the dissected cores (Adkins *et al.*, 1992b). Significant amounts of residual oil were recovered. Dissolution of the carbonate matrix was confirmed by studying the effect of *in situ* microbial growth and metabolism on the pore entrance size distribution of carbonate and sandstone cores (Udegbunam *et al.*, 1991). *Clostridium acetobutylicum* and a polymer-producing *Bacillus* strain were unable to penetrate the carbonate core. However, an unidentified halophilic, acid-producing anaerobe was able to grow through the carbonate core and electrical conductivity, permeability, porosity, and capillary pressure measurements showed pore enlargement and porosity increase due to acid dissolution of carbonate mineral. Acid production by *Lactobacillus* sp. and *Pediococcus* sp. dissolved $CaCO_3$ and iron scales in media containing goethite and magnetite (Coleman *et al.*, 1992).

3. Well stimulations

The first report of the use of acid, solvent, and gas production for oil recovery was the patent granted to Updegraff (Davis and Updegraff, 1954; Updegraff, 1956) for the use of *Clostridium* sp. and molasses for oil recovery. Later, Russian scientists tested whether gas production would decrease oil viscosity and improve oil recovery (Kuznetsov, 1962; Kuznetsov *et al.*, 1963; Senyukov *et al.*, 1970; Updegraff, 1990) (Table 6.5). They injected 54 m^3 of a mixed bacterial culture in 4% molasses into a well in the Sernovodsk oil field, which was then shut in for 6 months. A large increase in well pressure (1. 5 atm) indicated that *in situ* metabolism occurred. However, oil viscosity increased rather than decreased and only a slight increase in oil production was observed (3.5 m^3 for 3 months) (Table 6.5). Another early approach used a geobioreagent (190 m^3) consisting of a mixture of aerobic and anaerobic bacteria and a nutrient solution containing peat and silt biomass (Hitzman, 1983; Senyukov *et al.*, 1970). This was followed by the injection of fresh water (650 m^3) (Table 6.5). Analyses of the produced fluid from the well showed that microbial populations in the produced fluids had changed and the pH increased from 5–5.6 to 6.5–8.3 and the gas/oil ratio increased from 17 to 70 m^3/ton. All of these studies showed that it was possible to stimulate *in situ* microbial growth and metabolism. The geobioreagent treatment increased oil production from 28 to 48 m^3 per day (Table 6.5).

Improvements to well treatment technologies continued for several decades and lead to increases in the size of the inoculum and nutrients and the use of mixed cultures adapted to the nutrients and the environmental conditions of the reservoir (Grula *et al.*, 1985; Hitzman, 1983; Lazar, 1991; Petzet and Williams, 1986; Wang *et al.*, 1993, 1995) (Table 6.5).

These improvements decreased shut-in periods. *In situ* acid, gas, and solvent formation appears to be most effective in carbonate wells with an oil gravity of 875–965 kg/m^3, salinity less than 100 g/l, and a temperature around 35–40 °C (Hitzman, 1983). The data indicate that the injection of *Clostridium* and *Bacillus* spp. with molasses can increase oil recovery reproducibly, if done correctly (Hitzman, 1983). Seventy-five percent of the wells (24 wells total) treated in one study showed an increase in well-head pressure and an increase in oil production for 3–6 months (Hitzman, 1983). Another study states that 64 of 80 wells that were treated showed an increase in well-head pressure indicating that *in situ* microbial growth and metabolism occurred and more than 40 of these wells showed some increase in oil production (Petzet and Williams, 1986). Forty-two of forty-four wells had an increase in oil production of 33–733% (261 m^3 of incremental oil per well) (Wang *et al.*, 1993, 1995). The use of an adapted, mixed culture of microorganisms or clostridial strains with molasses increased oil production by 300–500% in low productivity wells (Lazar, 1987, 1991, 1992, 1993, 1998; Lazar and Constantinescu, 1985; Lazar *et al.*, 1988, 1991 1993; Wang *et al.*, 1993, 1995). One study indicates that an inoculum may not be needed as oil production increased by 216% with the addition of molasses only (Yusuf and Kadarwati, 1999), but details of this test are limited and control wells that received an inoculum were not used.

In situ microbial production of acids, gases, and solvents can also remove scale and debris in injection wells and increase injectivity (Table 6.5). Microorganisms have also been used to degrade guar gum and other injected chemicals to correct the damage caused by nonmicrobial treatments (Table 6.5).

4. Fermentatively-enhanced water flooding

Analogous to well stimulation approaches, fermentative bacteria and carbohydrate-based nutrient (usually molasses) are injected deep into the reservoir and fluid injection is stopped to allow for *in situ* growth and metabolism (Fig. 6.1B). The large number of cells in the inoculum along with the large amount of readily degradable carbohydrate provides a competitive advantage for the inoculum and selects for its growth and metabolism (Table 6.6).

One of the earliest and best-documented tests of MEOR was done by Mobil Oil Company in 1954 in the Upper Cretaceous Nacatoch Formation in Union County, Arkansas (Yarbrough and Coty, 1983) (Table 6.6). Laboratory studies showed that *C. acetobutylicum* produced 8–30 volumes of gas (CO_2 and H_2) per volume of 2% molasses medium and large amounts of acids (formic, acetic, and butyric) and solvents (acetone and butanol) in oil-saturated sand packed columns or cores (Updegraff, 1990). Based on the encouraging laboratory results, a field test of the technology was conducted. The Nacatoch sand formation is a loosely consolidated sand of high permeability and porosity about 700 m deep with bottom hole temperatures of 35–39 °C and a salinity of about 4.2%. The temperature and salinity were ideal for the growth of the *C. acetobutylicum*, but the high permeability and porosity and low residual oil saturation (4.5–8.5%) made the field undesirable in terms of oil release studies (Yarbrough and Coty, 1983). Two percent beet molasses was injected into a well at a rate of 16 m^3 per day for 5.5 months. A heavy inoculum of *C. acetobutylicum* was injected in 18 separate batches of 0.8 m^3 each (15.1 m^3 total) over a 4-month period. Breakthrough of the injection fluids in the production well (well 31), 88 m from the injection well, occurred after 70 days; fermentation products and sucrose were detected between 80 and 90 days after injection began. Fermentation products were detected 286 days after injection in another well that was 220 m from the injection well. Short chain organic acids in the produced fluids accounted for about 59% of the sugar injected. About 22,000 m^3 of CO_2 gas was produced from the well, which accounted for 19% of the sugar added. Some of the CO_2 was probably made by neutralization of fermentation acids by the carbonate minerals. Oil production in well 31 increased about the same time that microbial products were detected from about 0.1 m^3 per day prior to treatment to about 0.34 m^3 per day after treatment (250% increase) and lasted for at least 7 months. The formation of a free gas phase by CO_2 production was not likely given the reservoir pressure, but fermentation acids and CO_2 production at the surface of the sands may have lead to oil release.

Although the Nacatoch sand trial clearly showed that it was possible to generate large amounts of microbial metabolites throughout an oil reservoir and more oil was recovered, the results did not generate much excitement in the U. S for MEOR. Oil prices were low and the amount of oil recovered from the Nacatoch sand trial was low because of the low residual oil recovery and high porosity. A series of tests were conducted in Eastern Europe throughout the 1960s where the amounts and type of nutrients and inocula were optimized (Table 6.6). In general, residual oil recovery in many of these trials increased by 10–340% for 2–8 years (Dienes and Jaranyi, 1973; Dostalek and Spurny, 1957a,b; Hitzman, 1983; Jaranyi, 1968; Karaskiewicz, 1975; Lazar, 1987, 1991). One approach involved the use of sulfate reducers and pseudomonads and improved oil production slightly (Table 6.6). Concern over the detrimental effects of sulfate reducers limited the application of this approach elsewhere. These studies did provide evidence that it is possible to propagate the growth of microorganisms throughout the reservoir as the numbers of sulfate reducers in production wells 500 m from the injection well increased significantly (Dostalek and Spurny, 1957a,b; Hitzman, 1983).

Improvements to the technology included the use of mixture cultures adapted to the nutrients and reservoir salinities and temperatures and much larger volumes of nutrients (Table 6.6). One test involved the use of a mixed culture of anaerobes obtained from the reservoir, molasses, NH_4^+, and pyrophosphate (Wagner *et al.*, 1995). Oil production increased by 300–600% for 7 months (7000 m^3 incremental oil recovered) and large amounts of CO_2 and organic acids were detected. The injection of a mixed culture of anaerobes adapted to molasses and reservoir conditions and large amounts of molasses increased oil production by 400–600% (1–2 m^3/d/well) in two calcareous sandstone formations for 3 years (Lazar, 1987, 1991, 1992, 1993, 1998; Lazar and Constantinescu, 1985; Lazar *et al.*, 1988, 1991, 1993). A large decrease in pH (1–2 units) and microbial metabolites were detected in production wells. Lazar concludes that this approach works best in carbonate or carbonaceous reservoirs. Based on the amount of molasses used, the amount of incremental oil recovered and the price of sucrose, we estimate that the cost was about $19 per m^3 of incremental oil recovered.

Two studies used large amounts of molasses and *Clostridium tyrobutyricum* to treat carbonate reservoirs (Nazina *et al.*, 1999b; Wagner *et al.*, 1995). Both studies report large incremental oil recoveries (2550 and 4900 m^3 in 3 years). Each study provides very strong evidence that links microbial activity to oil recovery. Large amounts of CO_2 were detected in the gas phase (3,50,000 m^3) (Wagner *et al.*, 1995). The concentration of calcium and bicarbonate in production fluids increased consistently with a partial dissolution of the rock matrix. Organic acids (up to 6 g/l) were also detected in produced fluids. Highest increases in oil production were detected in wells where the greatest isotopic fractionation of methane was detected, that is, where the strongest evidence for biological activity was present (Nazina *et al.*, 1999b; Wagner *et al.*, 1995).

The feasibility of *in situ* carbon dioxide production for oil recovery has been questioned because it is doubtful that sufficient amounts of CO_2 can be made to create free gas phase and that the amount of biomass needed to generate large quantities of CO_2 would lead to serious plugging problems (Bryant and Lockhart, 2002; Sarkar *et al.*, 1989). Nonetheless, the approach appears to be effective in carbonate and carbonaceous sandstone formations (Hitzman, 1983, 1988; Lazar and Constantinescu, 1985; Tanner *et al.*, 1993; Wagner *et al.*, 1995). In these types of formations, the production of organic acids could dissolve carbonate minerals and alter the pore structure (Udegbunam *et al.*, 1991, 1993), which would release oil from surfaces of the rock (Yarbrough and Coty, 1983). Consistent with this mechanism, many field trials report a large reduction in the pH (1–2 units) of produced fluids, large concentrations of organic acids in produced fluids, and increased calcium concentrations in produced fluids (Lazar, 1987; Lazar *et al.*, 1988; Yarbrough and Coty, 1983). The production of solvents (butanol, acetone, and isopropanol) may also be beneficial because solvents are known to decrease oil viscosity and alter rock wettability, both of which improve oil mobility. One study reported that large amounts of butanol were generated after the injection of fermentative anaerobes and carbohydrate-based nutrients (Davidson and Russell, 1988). However, no improvement in oil production was observed suggesting that solvent production alone may not be a viable approach for MEOR.

C. Biosurfactant production

Biosurfactants are surface-active agents produced by a wide variety of microorganisms. Biosurfactants are low molecular weight amphiphilic compounds that form micelles. Due to the presence of hydrophilic and lipophilic moieties in their structure, biosurfactants are able to partition at the oil–air or the oil–water interfaces and to lower surface or interfacial tension, respectively. This property makes them good candidates for MEOR (Youssef *et al.*, 2004). Towards the end of the secondary stage of oil recovery, the high capillary pressure traps crude oil in small pores within the rock matrix. To recover this entrapped or residual oil, a large decrease in interfacial tension between the oil and aqueous phases is needed (see Section III) (McInerney *et al.*, 2005a). Biosurfactants are ideal agents because they partition at the oil–rock interface and promote the mobilization of oil from the rock by the displacing fluid.

The most common biosurfactants used in MEOR are lipopeptides produced by *Bacillus* and some *Pseudomonas* spp., glycolipids (rhamnolipids) produced by *Pseudomonas* sp., and trehalose lipids produced by *Rhodococcus* sp. (Banat, 1995a,b; Bodour and Miller-Maier, 2002; Youssef *et al.*, 2004) (Fig. 6.2). Lipopeptides and rhamnolipid biosurfactants lower interfacial tension between the hydrocarbon (crude oil or pure hydrocarbons) and aqueous phases to values of 0.1 mN/m or lower (Lin *et al.*, 1994; Maier and Soberon-Chavez, 2000; McInerney *et al.*, 1990; Nguyen *et al.*, 2008; Wang *et al.*, 2007). These low IFT values are sufficient to significantly lower the capillary number and mobilize significant amounts of oil (see Section III). The critical micelle concentrations of biosurfactants are orders of magnitude lower than synthetic surfactants, indicating that biosurfactants are effective at much lower concentrations (Georgiou *et al.*, 1992; McInerney *et al.*, 2005b; Youssef *et al.*, 2004).

Several laboratory studies investigated the improvement of biosurfactant yield and/or activity via cultivation conditions-related or genetic manipulation methods (for detailed review see Abu-Ruwaida *et al.*, 1991; Bordoloi and Konwar, 2007; Das and Mukherjee, 2005; Davis *et al.*, 1999; Joshi *et al.*, 2008a; Makkar and Cameotra, 1998; Mukherjee *et al.*, 2006; Schaller *et al.*, 2004).

1. *Ex situ* biosurfactant flooding

Even though the low interfacial tensions and critical micelle concentrations exhibited by various biosurfactants strongly argue for their effectiveness in oil recovery, there was considerable skepticism about their use in EOR (Bryant and Lockhart, 2002; McInerney *et al.*, 2002).

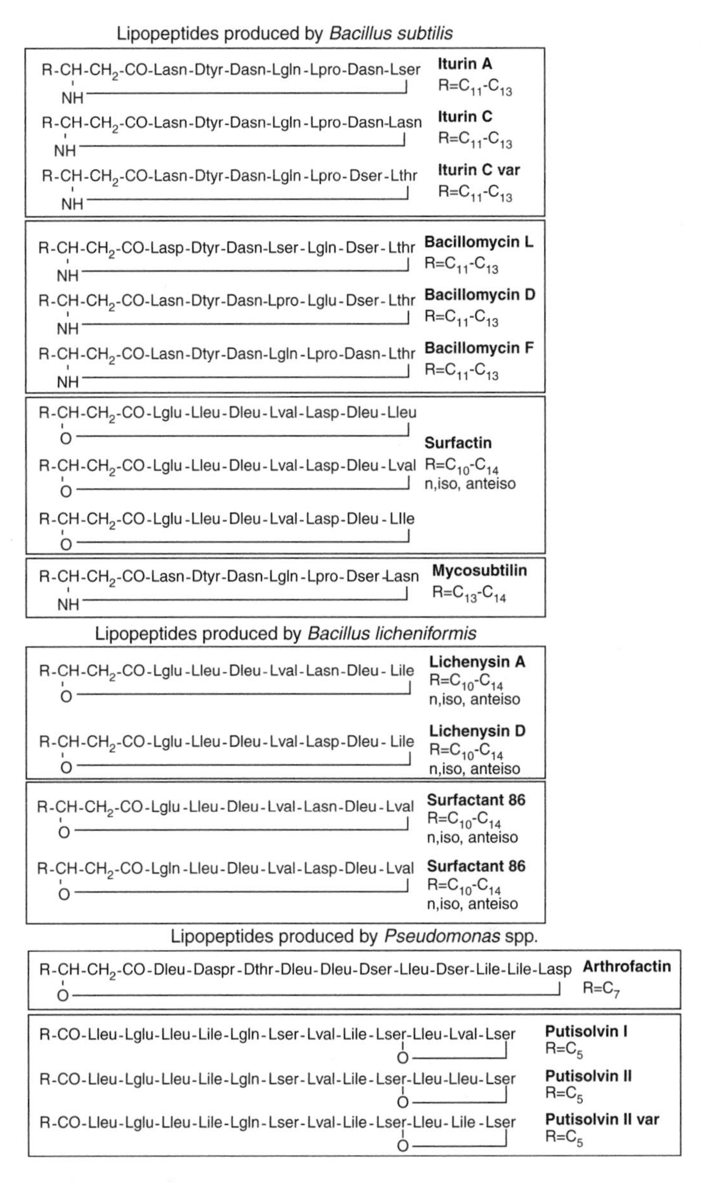

FIGURE 6.2 (Continued)

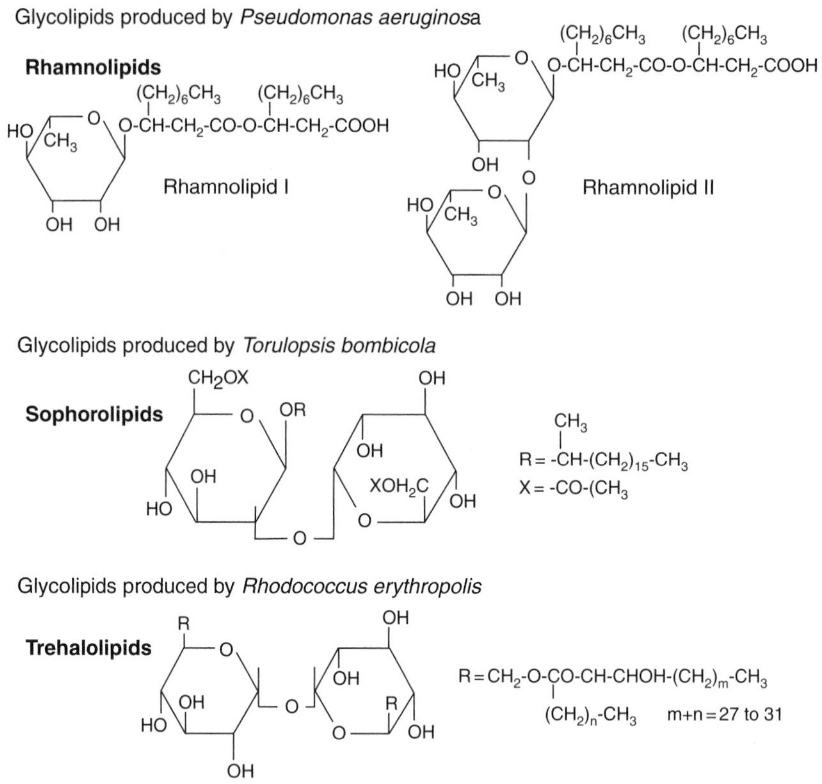

Glycolipids produced by *Pseudomonas aeruginosa*

Rhamnolipids

$O\text{-}CH(-(CH_2)_6CH_3)\text{-}CH_2\text{-}CO\text{-}O\text{-}CH(-(CH_2)_6CH_3)\text{-}CH_2\text{-}COOH$

Rhamnolipid I

Rhamnolipid II

Glycolipids produced by *Torulopsis bombicola*

Sophorolipids

$R = \text{-}CH(CH_3)\text{-}(CH_2)_{15}\text{-}CH_3$

$X = \text{-}CO\text{-}(CH_3$

Glycolipids produced by *Rhodococcus erythropolis*

Trehalolipids

$R = CH_2\text{-}O\text{-}CO\text{-}CH((CH_2)_n\text{-}CH_3)\text{-}CHOH\text{-}(CH_2)_m\text{-}CH_3$ $m+n = 27$ to 31

FIGURE 6.2 Structures of microbial biosurfactants. (A) Lipopeptide biosurfactants made by *Bacillus* and *Pseudomonas* sp... (B) Glycolipids including rhamnolipids, sophorolipids, and trehalolipids.

The lipopeptide-producing *Bacillus* strain JF-2 (Javaheri *et al.*, 1985) has been used in core flood experiments and in field trials (Bryant and Douglas, 1988; Bryant *et al.*, 1990), but oil recoveries were low and inconsistent. It is important to note that strain JF-2 has been reclassified as a strain of *Bacillus mojavensis* (Folmsbee *et al.*, 2006). The JF-2 lipopeptide was partially purified and tested in flow experiments with sand-packed columns and Berea sandstone cores (Knapp *et al.*, 2002; Maudgalya *et al.*, 2004, 2005; McInerney *et al.*, 2005b) (Table 6.7). Even very high concentrations of the JF-2 biosurfactant alone (12.3 g/l) were not effective in oil mobilization. The formation and subsequent disappearance of an oil bank in sand-packed columns during flooding suggested that mobility control was needed. Pre- and post-flushing of the column with a viscous solutions of 1% partially hydrolyzed polyacrylamide (PHPA) and the addtion of a cosurfactant (10 mM of 2,3 butanediol) to the biosurfactant solution

TABLE 6.7 The effect of biosurfactant addition on interfacial tension and residual oil recovery in model porous systems

Microorganism	Biosurfactant	Type of experiment	Effect on IFT, wettability, and/or residual oil recovery	References
Aerobic mesophilic hydrocarbon-degrading bacteria	Unidentified	Core flood	IFT lowered; Wettability alteration	Kowalewski *et al.* (2006)
Isolates from Egyptian and Saudi oil fields	Unidentified	Berea sandstone core and sand-packed columns	IFT lowered; Wettability alteration; Increased oil recovery	Sayyouh (2002)
Thermophilic bacterial mixtures obtained from UAE water tanks	Unidentified	Core flood under reservoir conditions	IFT of 0.07 mN/m against four crude oils; Average residual oil recovery of 15–20%	Abdulrazag *et al.* (1999)
Five microorganisms from Persian reservoirs	Unidentified	Glass micromodels and carbonate rock with or without fracture	IFT reduction; Wettability alteration	Nourani *et al.* (2007)
Indigenous microorganisms from Persian reservoirs (45 °C)	Unidentified, Lipopeptides	Core flood	Residual oil recovery of 14.3%	Abhati *et al.* (2003)

(continued)

TABLE 6.7 *(continued)*

Microorganism	Biosurfactant	Type of experiment	Effect on IFT, wettability, and/or residual oil recovery	References
Bacillus subtilis and *Pseudomonas* strain	Unidentified	Crushed limestone-packed column	IFT of 0.052 mN/m, Injection pressure decreased 5–40%, Residual oil recovery of 5–10%	Li *et al.* (2002)
Facultative anaerobes from Daqing oil field	Unidentified	Anaerobic core flood	IFT lowered; pH decreased; oil viscosity decreased; light alkane proportion increased; residual oil recovery of 10%	Peihui *et al.* (2001)
Anaerobic enrichments from high-temperature oil reservoir	Unidentified	Sand-packed column at reservoir conditions	Residual oil recovery of 22%	Banwari *et al.* (2005)
Biosurfactant-producing microorganisms from Indonesian oil fields	Unidentified	Native and model core floods	Residual oil recovery of 10–60%	Sugihardjo and Pratomo (1999)
Bacillus mojavensis strain JF-2	Lipopeptide	Sand-packed columns	Residual oil recovery increased	McInerney *et al.* (1985b)

		Sand-packed columns and Berea sandstone cores	Residual oil recovery proportional to biosurfactant concentration; residual oil recovery of 83% with 920 mg/l biosurfactant + 10mM butanediol + 1% PHPA	Maudgalya *et al.* (2004, 2005), McInerney *et al.* (2005b),
		Crushed limestone columns	Residual oil recovery of 27%; calcium carbonate minerals dissolved	Adkins *et al.* (1992a,b)
Bacillus subtilis	Lipopeptide	Sand-packed columns flooded with sodium pyrophosphate	Residual oil recovery of 35%	Chang (1987)
Bacillus subtilis strain MTCC1427	Lipopeptide	Sand-packed column with kerosene	Residual kerosene recovery of 56% with 100 ml of 1 mg/ml crude biosurfactant	Makkar and Cameotra (1998)
	Lipopeptide	Sand-packed column with crude oil	Residual oil recovery of 34–39%	Makkar and Cameotra (1997)
Bacillus subtilis strains DM03, DM04 (thermophilic)	Lipopeptide	Sand-packed column	Residual oil recovery of 56–60%	Das and Mukherjee (2007)
Bacillus subtilis 20B, *B. licheniformis* K51, *B. subtilis* R1, *Bacillus* strain HS3	Lipopeptide	Sand-packed columns	Residual oil recovery of 25–33%	Joshi *et al.* (2008a,b)

(continued)

TABLE 6.7 (*continued*)

Microorganism	Biosurfactant	Type of experiment	Effect on IFT, wettability, and/or residual oil recovery	References
B. subtilis	Surfactin	Adsorption to carbonates	Wettability alteration; surfactin adsorbed	Johnson *et al.* (2007)
Acinetobacter calcoaceticus	Unidentified	Sand-packed column at 73 °C	IFT lowered; residual oil recovery of 36.4%	Sheehy (1992)
Engineered strains of *Pseudomonas aeruginosa* and *Escherichia coli*	Rhamnolipids	Sand-packed column	Residual oil recovery of 50% with 4 pore volumes of 250 mg/l rhamnolipid solution	Wang *et al.* (2007)
Pseudomonas strain	Glycolipid and phospholipids	Sand-packed column	Residual oil recovery of 52%	Okpokwasili and Ibiene (2006)
Pseudomonas strains	Glycolipid	Sand-packed column	Residual oil recovery of 64%	Das and Mukherjee (2005)
Pseudomonas aeruginosa strains	Glycolipid	Sand-packed column	Residual oil recovery of 50–60%	Bordoloi and Konwar (2007)
Rhodococcus strain	Glycolipid	Sand-packed column	Residual oil recovery of 86% with 5 pore volumes of broth	Abu-Ruwaida *et al.* (1991)

resulted in very large oil recoveries (up to 83%) that were proportional to the biosurfactant concentration. 2,3 Butanediol was chosen as the cosurfactant because it is an end product of glucose fermentation by *Bacillus* sp. (Nakano *et al.*, 1997), including JF-2 (Folmsbee *et al.*, 2006). The lipopeptide is effective at very low concentrations so long as 2,3 butanediol and a PHPA were present (Maudgalya *et al.*, 2005). Culture fluids with as little as 16 mg/l recovered 22% of the residual oil recovery and concentrations above 40 mg/l recovered >40% of the residual oil from sandstone cores. Residual oil recovered was proportional to biosurfactant concentration in what appears to be a curvilinear relationship, which provides a relationship to predict oil recovery; on average, 2.2 ml of crude oil can be recovered per mg of biosurfactant used (Youssef *et al.*, 2007b).

Others have also shown that low concentrations of the biosurfactants can recover residual oil from model porous systems at elevated temperatures and salinities (Table 6.7). Very low concentrations (1 pore volume of 1 mg/l solution) of lipopeptides purified from *B. subtilis* MTCC 1427 cultures recovered 56% of the residual kerosene from sand-packed columns (Makkar and Cameotra, 1998). The effective concentration of the lipopeptide is much lower than that of rhamnolipid biosurfactants (Wang *et al.*, 2007). The lipopeptide lowered the surface tension to 28 mN/m and was stable at pH ranges of 3–11 and temperatures up to 100 °C. The lipopeptide produced from molasses-grown cells lower surface tension to 29 mN/m and recovered 34–39% of residual oil from sand-packed columns (Makkar and Cameotra, 1997). Two thermophilic *Bacillus subtilis* strains DM-03 and DM-04 produced a lipopeptide biosurfactant when grown with cheap nutrients (potato peel) that lowered surface tension to 32–34 mN/m and recovered 56–60% of residual oil from sand-packed columns (Das and Mukherjee, 2007). Joshi *et al.* (2008a,b) used *Bacillus subtilis* strain 20B, *B. licheniformis* K51, *B. subtilis* R1, and *B.* strain HS3 to produce lipopeptide biosurfactants from different carbon sources including molasses and whey. The biosurfactant-containing culture broth lowered the surface tension to 29.5 mN/m and recovered 25–33% of residual oil from sand-packed columns. Lipopeptides are active over a wide range of environmental conditions often present in oil reservoirs, temperatures up to 100 °C, pH from 6 to 10, and salt concentration up to 8% (Cameotra and Makkar, 1998; Jenneman *et al.*, 1983; Joshi *et al.*, 2008b; Makkar and Cameotra, 1997, 1998; McInerney *et al.*, 1990). Surfactin produced by *B. subtilis* was more effective than sodium lauryl sulfate (an anionic chemical surfactant) in changing the rock wettability from oil–wet to water–wet system (Johnson *et al.*, 2007). The use of environmental scanning electron microscope showed that only small pores in flooded areas duing MEOR have altered wettability (Kowalewski *et al.*, 2005).

Glycolipids, in particular, rhamnolipids produced by *Pseudomonas aeruginosa*, are also effective in lowering IFT and recovering residual oil (Table 6.7). *Escherichia coli* and a *Pseudomonas* strain were genetically engineered to produce rhamnolipids from cheap renewable substrates (Wang *et al.*, 2007). The engineered *Pseudomonas* strain produced two kinds of rhamnolipids (with one or two rhamnose sugars) while the *E. coli* strain produced only rhamnolipids with one rhamnose. Crude preparations of biosurfactants from both strains reduced IFT at different pH and NaCl concentrations and recovered about 50% of the residual oil entrapped in sand-packed columns. A *Pseudomonas* strain that produced a mixture of glycolipid and phospholipids biosurfactants recovered 52% of the residual oil from sand-packed columns (Okpokwasili and Ibiene, 2006). A *Rhodococcus* strain isolated from oily soil produced a glycolipid biosurfactant that reduced surface tension to <30 mN/m. The biosurfactant was stable at high temperatures (120 °C), at high salt concentrations (up to 10%), and over a wide range of pH (2–12) (Abu-Ruwaida *et al.*, 1991). In sand-packed columns, 5 pore volumes of the biosurfactant-containing broth recovered 86% of residual oil. In a similar study, two *Pseudomonas* strains produced biosurfactants stable at pH 3–12 and temperatures up to 100 °C and recovered 64% of residual oil from sand-packed columns (Das and Mukherjee, 2005). The injection of cells of four biosurfactant-producing *Pseudomonas aeruginosa* strains into sand-packed columns resulted in clogging (Bordoloi and Konwar, 2007). Cell-free biosurfactant-containing culture fluids did not clog and recovered 50–60% of the residual oil. The authors suggested the use of *ex situ* produced biosurfactant to treat high-temperature reservoirs as most biosurfactant producers are mesophilic and would not be able to grow in thermophilic reservoirs.

A number of studies with glycolipids were directed towards bioremediation. Rhamnolipids removed 79% of the crude oil from oil-contaminated soil (Urum *et al.*, 2003). Removal efficiencies of 80–95% were observed with biosurfactant-containing broths from *Rhodococcus* strain ST-5, a thermophilic *Bacillus* AB-2, and a proprietary strain Pet1006 compared to 58% with the synthetic surfactant, petroleum sulfonate (Banat, 1995a). Very low concentrations of biosurfactant produced by *Bacillus* strain C-14 released oil from contaminated sand. The biosurfactant produced by *Rhodococcus ruber* removed 80% crude oil from contaminated soil despite the adsorption of the biosurfactant to clay present in the soil (Kuyukina *et al.*, 2005).

The addition of rhamnolipids to synthetic surfactant used for alkaline surfactant polymer (ASP) flooding reduced the amount of the synthetic surfactant required for the efficiency of oil recovery (Daoshan *et al.*, 2004) and adding biosurfactant-containing culture fluids decreased the amount of synthetic surfactant and alkali required to generate low IFT against crude oil (Feng *et al.*, 2007).

2. Efficacy of *in situ* biosurfactant production

Biosurfactant producers isolated from a number of oil reservoirs are effective in mobilizing residual oil from a variety of laboratory test systems (Table 6.7). *In situ* growth of *B. mojavensis* strain JF-2 recovered residual oil from sand-packed columns (McInerney *et al.*, 1985b) and columns packed with crushed unconsolidated viola limestone (Adkins *et al.*, 1992b) (Table 6.7). In the latter study, oil recovery increased by 27% and increase in dissolved calcium suggested that some of the rock matrix had been dissolved. A number of other studies report effectiveness of lipopeptide-producing *Bacillus* strains for oil recovery (Table 6.7). The growth of *B. subtilis* in a sand-packed columns recovered 35% of the residual oil compared to 21% when only nutrients were added (Chang, 1987). *In situ* growth of biosurfactant-producing bacteria also recovered residual oil from sandstone cores (Thomas *et al.*, 1993; Yakimov *et al.*, 1997) (Table 6.7). In general, oil recoveries were low (<20%) and multiple pore volumes were needed. One study reported a residual oil recovery of 39% (Zekri *et al.*, 1999). Isolates from Saudi Arabian and Egyptian oilfields reduced IFT and altered rock wettability (Sayyouh, 2002). A thermophilic bacterial mixture reduced IFT to 0.07 mN/m at temperatures up to 100 °C and salinities up to 10% and recovered 15–20% of the residual oil from cores (Abdulrazag *et al.*, 1999). Five microorganisms isolated from a Persian reservoir reduced IFT, changed wettability, and recovered oil from glass micromodel systems and carbonated cores (Abhati *et al.*, 2003; Nourani *et al.*, 2007). Hydrocabon degraders from the Daqing oil field in China decreased the IFT between crude oil and water to ultra low values (0.052 mN/m) and recovered 5–10% of the residual oil from columns packed with crushed limestone (Li *et al.*, 2002; Peihui *et al.*, 2001). Anaerobic enrichments of biosurfactant producers from fluids produced from a reservoir with temperature between 70 and 90 °C recovered about 22% of the residual oil recovery from sand-packed columns (Banwari *et al.*, 2005). Alternating cycles of nutrient starvation may be an approach to stimulate *in situ* biosurfactant production (Sheehy, 1992). This approach increased residual oil recovery from sand-packed columns by 36.4%.

Combining multiple microbial mechanisms by the use of a consortium of microbes with different properties is clearly an effective strategy for oil recovery (Bryant, 1988; Bryant and Burchfield, 1989; Bryant and Douglas, 1988). The combination of a biosurfactant producer (*B. licheniformis*), an acid, gas, and solvent-producer (*Clostridium* sp.) and a facultatively anaerobic, gram-negative rod increased oil recovery by 60% from etched glass micromodels and 28% from Berea sandstone. Residual oil recoveries of 39% and 10–60% from sand-packed columns were observed with two *Bacillus* strains grown with molasses (Rauf *et al.*, 2003). A consortium of three strains, an *Arthrobacter* and a *Pseudomonas* spp. both known to degrade oil and a biosurfactant-producing *Bacillus* sp. (Jinfeng *et al.*, 2005), when grown at 73 °C in a medium with molasses and crude oil, produced a biosurfactant that lowered the IFT between the oil and aqueous phases and decreased oil viscosity and cloud point. However, residual oil recovery was low, only 5%. Stimulation of *in situ* hydrocarbon metabolism by injection of oxygenated water into model core systems lowered the interfacial tension between oil and aqueous phases and changed the matrix from strongly water-wet to less water-wet which led to residual oil recovery (Kowalewski *et al.*, 2006).

3. Field trials

While stimulation of acid, gas, and solvent production appears to be very effective in recovering residual oil in carbonate reservoirs, fermentative metabolism alone may not be as effective in sandstone formations. One would predict that biosurfactant production would be needed based on laboratory studies (Table 6.7). Surprisingly, very little field information is available on the efficacy of biosurfactant use. Several studies have used biosurfactant-producing bacteria with those that make acid, gas, and solvent and report oil production increases of 30–100% for up to 18 months (Table 6.5).

Recently, it has been conclusively shown that large amounts of a lipopeptide biosurfactant can be made *in situ* and that that inoculation of the wells with lipopeptide-producing strains is required (Simpson *et al.*, 2007; Youssef *et al.*, 2007b). Although the test was conducted as a proof of principle for *in situ* biosurfactant production, incremental oil was recovered (Simpson *et al.*, 2007; Youssef *et al.*, 2007b).

Two tests made use of a biosurfactant producer during water flooding (Bryant and Burchfield, 1991; Bryant *et al.*, 1990, 1993) (Table 6.6). In both tests, a mixed culture of bacteria was used and the biosurfactant producer was *Bacillus* strain JF-2 (Javaheri *et al.*, 1985). The surface tension of produced fluids from production wells decreased 6 weeks after injection suggesting that a biosurfactant was made. However, the identity of the surface-active agent in the produced fluids was not determined. When molasses was periodically added to four injection wells, oil production from the field increased by 13% after the microbial treatment and improvements in the water to oil ratio were noted in the treated area (Bryant and Burchfield, 1991; Bryant *et al.*, 1990). In a different field where continuous injection of molasses was used, oil production of the field increased by about 19% for 3 years (Bryant *et al.*, 1993). Total incremental oil recoveries in each case were low, 88 and 400 m^3 over a 2-year span, which means that the daily oil productivities (m^3 of oil per day) were low and probably not that much different from pretreatment values. The low incremental oil recoveries and small increases in daily productivity probably made the economics of the process unattractive.

D. Emulsifiers

Bioemulsifiers are high molecular weight amphiphilic compounds that are produced by a wide variety of microorganisms (Banat *et al.*, 2000; Bognolo, 1999; Dastgheib *et al.*, 2008; McInerney *et al.*, 2005a). Bioemulsifiers form stable emulsions with hydrocarbons (usually oil-in-water and less commonly water-in-oil) (Dastgheib *et al.*, 2008). Compared to biosurfactants, bioemulsifiers may not lower surface or interfacial tension (Dastgheib *et al.*, 2008). Members of the genus *Acinetobacter* produce the most commonly used bioemulsifier called emulsan (Rosenberg and Ron, 1999). Emulsan is an anionic heteropolysaccharide and protein complex. Other bioemulsifiers are heteropolysaccharides such as those produced by *Halomonas eurihalina* and *Pseudomonas tralucida*, protein complexes such as those produced by *Methanobacterium thermoautotrophicus*, protein–polysaccharide–lipid complex such as those produced by *Bacillus stearothermophilus*, carbohydrate–protein complex as liposan produced by *Candida lipolytica*, mannan protein as that produced by *Saccharomyces cerevisiea*, and others (Rosenberg and Ron, 1999).

Emulsan emulsifies hydrocarbon mixtures, but not pure hydrocarbons (Rosenberg *et al.*, 1979a,b). Emulsan did not adsorb to sand or limestone saturated with oil, emulsified a wide range of hydrocarbons and crude oils, and recovered 90% of the crude oil present in oil-contamined sand and 98% of crude oil saturating crushed limestone ($CaCO_3$) in shake flask experiments (Gutnick *et al.*, 1986). A bioemulsifer obtained from a *Bacillus licheniformis* strain emulsified different hydrocarbons in shake flask experiments, but did not recover residual oil from sand-packed columns (Dastgheib *et al.*, 2008). However, *in situ* growth of the organism resulted in residual oil recoveries of 22%. The mechanism of oil recovery may have been profile modification rather than emulsification. Another possible application for bioemulsifier in oil recovery involves its use in preventing paraffin deposition (McInerney *et al.*, 2005a; Wentzel *et al.*, 2007). While bioemulsifiers alone may not be effective for oil recovery, their use in oil storage tanks cleanup has been well documented (Banat, 1995a; Banat *et al.*, 2000). A proprietary bioemulsifying strain removed hydrocarbons from the sludge floating on top of an oil storage tank (Banat, 1995a; Banat *et al.*, 2000) and emulsan reduced the viscosity of Venezuelan heavy oil and increased its mobility in the pipeline (Bognolo, 1999).

E. Exopolymer production and selective plugging

Microbial polymers have been used as mobility control agents to reduce viscous fingering of waterfloods and EOR processes for many years (Craig, 1974; Dabbous, 1977; Dabbous and Elkins, 1976; Harrah *et al.*, 1997;

Malachosky and Herd, 1986; Trushenski *et al.*, 1974). The use of biopolymer for mobility control will not be discussed here. Instead, we will focus on the role of biopolymers and/or microbial biomass as plugging agents to improve volumetric sweep efficiency. Gelled polymer systems are widely used in petroleum reservoirs to reduce flow in high permeability zones and thereby redirect the displacement fluid into previously by-passed portions of the reservoir (Abdul and Farouq Ali, 2003; Ali and Barrufet, 1994; Kantzas *et al.*, 1995; Vossoughi, 2000). Curdlan is a β-1,3-D-glucan polymer that is soluble at alkaline pH and forms an insoluble gel as the pH decreases (Bailey *et al.*, 2000; Buller and Vossoughi, 1990; Harrah *et al.*, 1997). The gelation of the polymer can be induced by acid production by alkaliphilic bacteria. This approach was tested in sandstone cores and was shown to reduce permeability by two to four orders of magnitude and to divert flow from a high permeability core to a low permeability core that were connected in parallel (Bailey *et al.*, 2000). A number of biopolymers (xanthan gum, poly-β-hydroxybutyrate (PHB), guar gum, polyglutamic acid, and chitosan) were tested in a laboratory-pressurized flow system (Khachatoorian *et al.*, 2003). All of the polymers reduced the permeability of the sand-pack with PHB being the most effective.

Another mechanism to rectify permeability variation and to improve volumetric sweep efficiency is to stimulate the *in situ* growth of microorganisms in high permeability zones (Brown, 1984; Crawford, 1962, 1983; Jenneman *et al.*, 1984; McInerney *et al.*, 1985b) (Fig. 6.1C). Selectivity of a microbial plugging process is controlled by water movement. Most of the fluid injected into a reservoir flows through regions of high permeability (Crawford, 1962, 1983). Because of this, most of the nutrients injected into the formation will enter the high permeability regions and little will enter the low permeability regions. Thus, microbial growth will preferentially occur in the high permeability regions because these regions receive most of the nutrients. The growth of microorganisms in high permeability zones of the reservoir or dominant flow channels will reduce the movement of water in these regions and divert the water into regions of the reservoir with higher oil saturations (Brown, 1984; McInerney *et al.*, 1985a). The microbial selective plugging is a generic one that can be applied to almost any reservoir because all that is required is to stimulate the growth of indigenous microorganisms by nutrient injection. It does not depend on the production of a specific chemical or the growth of a specific bacterium.

1. Laboratory studies on plugging and oil recovery

The growth of microorganisms in high permeability zones of the reservoir or dominant flow channels will reduce the movement of water in these regions and divert the water into regions of the reservoir with higher oil saturations (Brown, 1984; McInerney *et al.*, 1985a). A number of studies have shown that the *in situ* growth of bacteria in sandstone cores or other reservoir model systems results in significant reductions in permeability (Bae *et al.*, 1996; Cusack *et al.*, 1990; Jack and DiBlasio, 1985; Jack and Steheier, 1988; MacLeod *et al.*, 1988; Raiders *et al.*, 1986b, 1989). The injection of sucrose-nitrate medium stimulated the growth of indigenous microorganisms and resulted in large permeability reductions (>90% of the initial permeability) of Berea sandstone cores (Jenneman *et al.*, 1984; McInerney *et al.*, 1985a; Raiders *et al.*, 1985, 1986b, 1989). Large amounts of gas were produced and it is likely that a free gas phase formed and blocked water movement in some of these experiments. Later experiments with cores inoculated at reservoir pressure where a free gas phase would not form confirmed that *in situ* biomass production results in large permeability reductions (Knapp *et al.*, 1991). Two cores of differing permeability connected in parallel and two slabs of sandstone with different permeabilities layered on top of each other to allow crossflow between the slabs severed as model systems to test the selectivity of the microbial plugging process (Raiders *et al.*, 1986b, 1989). The injection of nutrients followed by an incubation period stimulated *in situ* microbial growth and permeabitiy reductions preferentially occurred in the high permeability core or slab. After *in situ* growth, most of the fluid was diverted to the low permeability core or slab. *In situ* growth of *Klebsiella pneumoniae* preferentially occurred in high permeability regions of a model porous system (Cusack *et al.*, 1990, 1992). The use of less readily degradable carbohydrates (maltodexrins vs. glucose) and sodium trimetaphosphate, which adsorbs less to rock surfaces than phosphate allowed more uniform microbial growth and permeability reductions throughout the core (Davey *et al.*, 1998).

Residual oil recoveries ranging 8–35% were observed with sandstone cores in the absence of biosurfactant production (Raiders *et al.*, 1986b, 1989). The mechanism for oil recovery was postulated to be microscopic sweep efficiency where microbial growth blocks large pores and diverts fluid flow into smaller pores. In this mechanism, *in situ* growth of microorganisms would occur preferentially in large pores because these pores receive most of the nutrients and thus support most of the biomass production. Pore size distribution analysis of sandstone cores (Torbati *et al.*, 1986) and fused-glass columns (Stewart and Fogler, 2002) showed that microbial growth preferentially plugged the large pores or flow channels. The distribution of microbial activity in a shallow aquifer is strongly correlated with regions of high porosity where nutrient levels would be high (Musslewhite *et al.*, 2007).

Inocula have also been used for microbial selective plugging. Spores of a halotolerant mesophilic biopolymer-producing penetrated deeper into cores and, upon germination, permeability reduction and biofilm formation were observed uniformly throughout the core rather than only at the core inlet when vegetative cells were used (Bae *et al.*, 1996). Starved ultramicrocells of *K. pneumoniae* transported uniformly throughout the core and, when resuscitated, uniform reductions in permeability and biofilm formation throughout the porous model systems were observed rather than at the inlet end when vegetative cells were used (Lappin-Scott *et al.*, 1988; MacLeod *et al.*, 1988). Attachment and transport of *Lactobacillus casei* in a vertical, two-dimensional, packed-bed flow system depended on the concentration of injected cells concentration and the flow rate (Yang *et al.*, 2005).

Significant reduction in permeability in high permeability sand-packed and fused glass columns requires biopolymer production in addition to cell biomass (Geesey *et al.*, 1987; Jack and Steheier, 1988; Jenneman *et al.*, 2000; Lappan and Fogler, 1992; Robertson, 1998; Shaw *et al.*, 1985). Bacteria capable of secreting extracellular polymers and forming biofilms, for example, *Cytophaga*, *Arcobacter*, and *Rhizobium* were able to plug fractured limestone cores (Ross *et al.*, 2001). *Leuconostoc mesenteroides* produces the polysaccharide, dextran, when grown with sucrose (Lappan and Fogler, 1994). Batch experiments showed a correlation between the concentration of sucrose and other growth conditions and the amount of dextran produced (Lappan and Fogler, 1994, 1996; Wolf and Fogler, 2001). Visualization of the plugging process by use of glass bead-micromodel systems showed that plugging occurred in three phases: (1) an induction phase characterized by initiation of dextran production, (2) a plugging phase characterized by biofilm and dextran production in large flow channels, and (3) plug propagation phase characterized by sequential development and breakthrough of plugs (Stewart and Fogler, 2001). *In situ* growth and polymer prodution of *L. mesenteroides* selectively plugged the high permeability core of a parallel core system (Lappan and Fogler, 1996). Injection of *L. mesenteroides* into fractured etched-glass micromodels and subsequent dextran production led to the plugging of matrix–fracture interface (Soudmand-asli *et al.*, 2007). Extracellular polymer production by four *B. licheniformis* strains was observed when sandstone cores were analyzed by scanning electron microscopy (Yakimov *et al.*, 1997). Biogenic acid production and microscopic selective plugging resulted in 9–22% residual oil recoveries. The permeability of sand-packd columns and sandstone cores was reduced by 65–95% reduction by the *in situ* growth and polymer production of two strains of *B. licheniformis* (Silver *et al.*, 1989). The degree of permeability reduction depended on the amount of polymer produced. The authors suggested the use of spores pretreated with lysozyme for deeper penetration and showed the importance of adding sodium triphosphate, citric acid, and aluminum for polymer formation.

2. Field trials

Surprisingly, the use of polymer-producing bacteria is effective in stimulating oil recovery from individual wells. The injection of the polymer producer, *Enterobacter cloacea*, and the biosurfactant-producer, *B. licheniformis*, with 10–20% molasses increased oil production in 7 out of 12 treated wells with the total incremental oil recovery of 1730 m^3 (Maezumi *et al.*, 1998) (Table 6.5). The effect of an inoculum (*E. cloacae*) and molasses concentration was studied (Nagase *et al.*, 2001; Ohno *et al.*, 1999). Seven of the 14 wells that received an inoculum and molasses had an increase in oil production and a reduction in the amount of water produced while only 1 of 4 wells treated with molasses had an increase in oil production (Table 6.5). Molasses concentrations greater than 5% were more effective than when 1% molasses was used. The produced fluids from the wells after treatments had a decreased pH and an increase in CO_2 concentration and H_2O viscosity compared to the fluids prior to treatment. Fermentations acids (acetic, propionic, and butyric) and 2,3-butanediol were detected. The presence of 2,3-butanediol and the biopolymer, known products of the metabolism of *E. cloacae*, indicated that the inoculum was active in the reservoir. The injection of inorganic nutrients to stimulate *in situ* microbial growth increased oil production by 30% for 6 months (Sheehy, 1990, 1991). Economic analysis indicated that additional oil was recovered at less than two Australian dollars per barrel (12.5 Australian dollars per m^3). The suggested mechanism of action is the blockage of water channels. Microbial plugging in the vicinity of the well probably alters the relative permeability of the rock to oil and water and improves the migration of oil to the well (Donaldson, 1985).

The *in situ* growth of microorganisms in a hypersaline sandstone formation blocked a major channel and reduced interwell permeability variation in the reservoir (Fig. 6.3) (Knapp *et al.*, 1992). Large amounts of molasses and nitrate were injected until sucrose was detected in production well fluids. The further injection of brine was stopped for about 1 month to allow time for the microorganisms to grow. Evidence for *in situ* microbial activity included an increase in alkalinity and an increase in microbial numbers in the produced fluids (Knapp *et al.*, 1992). Prior to nutrient injection, the inter-well permeabilities between the injection well 7-2 and production wells 5-1 and 5-2 were 2.5 and 3 times that between well 7-2 and well 7-1 (Fig. 6.3). After nutrient treatment, the inter-well permeability between 7-2 and 5-1 and 5-2 decreased resulting in more uniform permeability throughout this region of the reservoir (Fig. 6.3). In addition, trace studies showed that a major water channel between the injection well 7-2 and the production well 1A-9 was blocked. The recovery of residual oil occurred. The reservoir brine contained large concentrations of divalent cations (Mg^{2+} and Ca^{2+}) and the interactions of these ions with the CO_2 made microbially probably resulted in the formation of

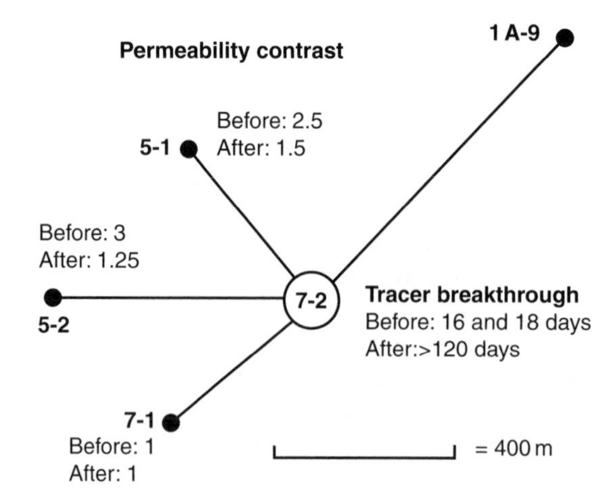

FIGURE 6.3 Efficacy of a microbial selective plugging process. Well 7-2 is an injection well that received molasses and nitrate; wells 1A-9, 5-1, 5-2, and 7-1 are production wells. Permeability contrast was calculated by dividing the inter-well permeabilities between wells 7-2 and 5-1 and 7-2 and 5-2 (before treatment permeabilities of 0.15 and 0.18 μm^2, respectively) by the interwell permeability between wells 7-2 and 7-1 (before treatment permeability of 0.06 μm^2). The time for tracers to travel between well 7-2 and well 1A-9 is given in days (Knapp *et al.*, 1992).

calcium/magnesium carbonate minerals that could have blocked flow channels and reduced permeability. Such a mechanism has been proposed to explain the reduction in groundwater flow of a drinking water aquifer (Chapelle and Bradley, 1997) and biogenic sealing of carbonate reservoirs (Ashirov and Sazonova, 1962).

The injection of a spore preparation of a *B. licheniformis* strain with molasses and nitrate blocked one highly transmissive zone and drastically reduced fluid intake by another zone (Gullapalli *et al.*, 2000) (Table 6.6). Well logs showed that fluid entered the formation by seven new zones after microbial treatment and that the biofilm was stable for 8 months. The stimulation of indigenous microorganisms with the injection of maltodextrins and ethyl acid phosphate followed by cessation of injection fluid for 2 weeks reduced fluid injection into one transmissive zone, but the biofilm was not stable for long periods of time (Jenneman *et al.*, 1996).

The injection of a polymer-producing *Enterobacter* CJF-002 for 1 week followed by the injection of 10% molasses for 2 months altered the inter-well flow pattern of a reservoir as indicated by changes in tracer breakthrough times (Nagase *et al.*, 2002) (Table 6.6). The change in flow patterns lead to a large increase in oil production (38%) for 6 months and the amount of incremental oil recovered was large, 2144 m^3. Produced fluids contained large amounts of polymer and microbial metabolites. The numbers of CJF-002, estimated by most probable number analysis and restriction fragment length polymorphism, showed that CJF-002 propagated throughout the reservoir.

Several studies have used LAB as inocula to improve sweep efficiency of water floods (Jack and Steheier, 1988; Jenneman *et al.*, 1995; von Heiningen *et al.*, 1958; Yulbarisov, 1990). In one case, a reduction in the water to oil ratio of the produced fluids and an increase in water viscosity occurred (von Heiningen *et al.*, 1958). In the other studies, the permeability reductions were not large enough for significant flow diversion or oil recovery, probably because the permeabilies were very large (>1 darcy, 1 μm^2) (Jack and DiBlasio, 1985; Jack and Steheier, 1988).

Several studies show that stimulating indigenous denitrifiers will result in improved oil recovery (Brown *et al.*, 2002; Hitzman *et al.*, 2004) (Table 6.6); the mechanism for one of these studies is believed to be selective plugging (Vadie *et al.*, 1996). The injection of nitrate, phosphate, and molasses into portions of an oil field undergoing water flooding to stimulate *in situ* growth and metabolism and block water channels slowed the natural decline in oil production (Brown, 2007; Brown *et al.*, 2002). Thirteen of the nineteen production wells in the treated area had an increase in oil production while only two production wells from the control pattern that did not receive nutrients had an increase in oil production. Most of the production wells in the control patterns exhibited their normal decline in oil production or were shut-in due to low oil production during the test. The production of oil with a different composition in the treated wells compared to the control wells suggested that oil previously by-passed by the waterflood was recovered. The economic lifetime of the field was extended by 5–11 years and resulted in the recovery of 63,600 m^3 of incremental oil at a cost of \$8.30/m^3 (Brown, 2007; Brown *et al.*, 2002). The stimulation of indigenous denitrifiers by the addition of nitrate, nitrite, and unspecified inorganic nutrients slowed the natural decline in oil production in two fields and recovered 1315 and 1225 m^3 of incremental oil in 7–10 months (Hitzman *et al.*, 2004).

F. *In situ* hydrocarbon metabolism

The *in situ* stimulation of hydrocarbon metabolism by injection of oxygen and inorganic nutrients was also an early approach to recover additional oil (Andreevskii, 1959, 1961). This idea has been studied intensively in Russia and very strong evidence links microbial activity with oil recovery (Belyaev *et al.*, 1982, 1998, 2004; Ivanov and Belyaev, 1983; Ivanov *et al.*, 1993; Nazina *et al.*, 1995a, 1999a, 2000a,b; Rozanova and Nazina, 1980; Rozanova *et al.*, 2001a; Yulbarisov, 1976, 1981, 1990; Yulbarisov and Zhdanova, 1984). In this approach, the stimulation of aerobic hydrocarbon metabolism in the vicinity of the injection well results in the production of acetate, other organic acids, and alcohols. These metabolites are converted to methane by methanogenic consortia deeper in the reservoir. Methane production would swell the oil and make it more mobile. Several push–pull tests were conducted to test this mechanism (Belyaev *et al.*, 1982; Ivanov and Belyaev, 1983). In a push–pull test, fluid is injected into a well for a period of time and then produced back from the same well. High concentrations of aerobic hydrocarbon degraders and organic acids were detected in fluids close to the injection well. Fluids recovered further away from the injection well had much lower numbers of aerobic hydrocarbon degraders and high rates of methanogenesis. The carbon in the carbonate pool was substantially heavier (average $\delta\ ^{13}C$ of −10.1) than that from untreated areas (average $\delta\ ^{13}C$ of −24.9) and the carbon of methane was substantially lighter (average $\delta\ ^{13}C$ of −64.1) than that from untreated areas (average $\delta\ ^{13}C$ of −55.7) (Belyaev *et al.*, 1998; Ivanov and Belyaev, 1983; Nazina *et al.*, 2007a). These data are consistent with the recent biological origin of some of the methane.

Laboratories studies showed that aerobic hydrocarbon degraders obtained from the produced fluids from the reservoir produced a mixture of organic acids and alcohols when incubated in minimal medium with crude oil and limiting amounts of oxygen and that methanogenic consortia converted these products of aerobic hydrocarbon metabolism to methane (Belyaev *et al.*, 1982; Groudeva *et al.*, 1993; Nazina *et al.*, 1985, 1995a; Rozanova and Nazina, 1980). Laboratory core studies showed that the continued injection of oxygenated brine and inorganic nutrients decreased the residual oil saturation in sandstone cores (Kulik *et al.*, 1985; Nazina *et al.*, 1985, 1995a; Sunde *et al.*, 1992). When sufficient oxygen was supplied, aerobic hydrocarbon-degrading cultures formed metabolites that decreased oil/aqueous phase interfacial tension by four orders of magnitude (Kowalewski *et al.*, 2005, 2006).

A number of field tests of this technology have been conducted and are summarized in Table 6.6. The results of one field trial are given in Fig. 6.4. Analysis of the microbial populations showed that aerobic hydrocarbon degraders were present near the injection well and their concentration increased with the injection of aerated water. The numbers of anaerobes and methanogens increased in production well fluids. Rates of methanogenesis increased and high concentrations of acetate and bicarbonate were detected in production well fluids (Belyaev *et al.*, 1998; Nazina *et al.*, 2007a) (Fig. 6.4). Isotopically heavy carbon in carbonates and isotopically light carbon in methane was detected in the production well fluids consistent with the idea that some of the methane was of recent biological origin. Oil production increased by 10–45% in many different formations (Ibatullin, 1995; Ivanov *et al.*, 1993; Matz *et al.*, 1992; Murygina *et al.*, 1995; Nazina *et al.*, 2007b). The amounts of incremental oil recovered were very large (Table 6.6). The process is very economic with treatment costs near \$6 per m^3 of incremental oil recovered (Belyaev *et al.*, 2004).

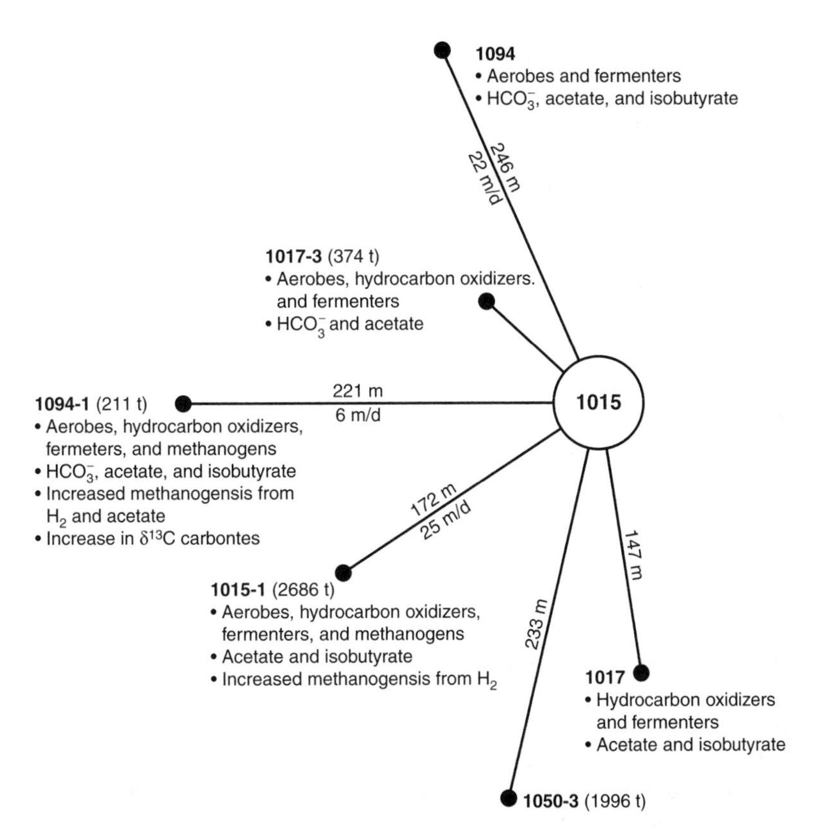

FIGURE 6.4 Efficacy of an *in situ* hydrocarbon metabolism process. Well 1015 is an injection well; wells 1094-1, 1017-3, 1094, 1015-1, 1050-3, and 1017 are production wells. Aerated water with nitrogen and phosphorous sources was injected two to three times a month from March to September over a 3-year period (15 treatments total) into well 1015. Inter-well distances and rates of fluid flow are given above and below the lines connecting the wells, respectively. The amount of additional oil recovered (metric tons) is given in parentheses next to the well number. Evidence for microbial activity is given under the well number. Only those parameters that had a large change from pretreatment values are given. Data are from Nazina *et al.* (2007a).

As with microbial paraffin removal, there are many reports on the effectiveness of proprietary inocula and nutrients to increase oil recovery (Table 6.6). The use of commercial formulations slowed or arrested the natural decline in oil production and large incremental recoveries of oil have been reported (Bailey *et al.*, 2001; Brown *et al.*, 2005; Dietrich *et al.*, 1996; Maure *et al.*, 1999, 2001, 2005; Portwood, 1995b; Strappa *et al.*, 2004; Yu and Schneider, 1998). Supporting evidence for the involvement of microorganisms is limited to changes in the physical properties of the oil, reductions in oil viscosity, cloud point, or pour point, and an alteration in the alkane composition of the oil (Table 6.6) (Dietrich *et al.*, 1996; Strappa *et al.*, 2004; Wankui *et al.*, 2006). The cost of the treatment was usually less than \$20 per m^3 of incremental oil recovered. A computer simulation approach showed that the reductions in oil viscosity and relative permeability could explain the observed oil recoveries by the commercial formulations (Dietrich *et al.*, 1996). Independently, Wankui *et al.* (2006) used two hydrocarbon-degrading strains, *Bacillus cereus* and *Brevibacillus brevis*, to treat part of the Daqing oil field. Oil production increased by 165% for about 200 days and about 6700 m^3 of incremental oil was produced from 10 production wells (Table 6.6). A decrease in oil viscosity, alteration in the alkane profile, and an increase in microbial cell concentration by two orders of magnitude support the role of microorganisms in oil recovery. Halophilic hydrocarbon-degrading inocula have also been used to treat formation with high salinities (Belyaev *et al.*, 2004). Coinjection of these inocula with aerated water containing nitrogen and phosphorous nutrients resulted in large incremental oil recoveries (Table 6.6).

VI. IMPLEMENTATION OF MEOR

As shown in Table 6.3, there are a number of mechanisms that could be used to increase oil production (Bryant, 1991; McInerney *et al.*, 2005a). Surveys of the environmental conditions indicated that about 27% of the reservoirs in the U. S. could be candidates for MEOR (Bryant, 1991; Clark *et al.*, 1981). The initial screen criteria to determine the suitability of a reservoir for application of MEOR (Bryant, 1991; Clark *et al.*, 1981; Sayyouh *et al.*, 1993) include: temperatures less than 80 °C; rock permeability >75 mDarcies (0.075 μm^2); salinities <10%; depth <2500 m; crude oil viscosity >15 °C API; As, Hg, Ni, and Se <10–15 ppm; and residual oil saturation >25%. The problem is to match the MEOR process with the production problem. This is often done heuristically based on sound petroleum engineering analysis of historical production data. Once the the problem is defined, then the appropriate microbial approach can be selected and a treatment strategy developed (Bryant, 1991).

A. Treatment strategies

Regardless of the MEOR process, three general strategies exist for the implementation of MEOR: (1) injection of nutrients to stimulate indigenous microorganisms, (2) in absence of the suitable indigenous population, the injection of exogenous microorganisms(s) and nutrients, or (3) in cases where the reservoir conditions are too harsh to allow growth of exogenously added microorganisms and the absence of suitable indigenous microorganisms, the injection of *ex situ* produced products (Banat *et al.*, 2000).

1. Injection of nutrients to stimulate indigenous microorganisms

This strategy requires the presence of indigenous microorganisms that perform the desired function (e.g., plugging, gas, solvent, acid, or biosurfactant production). To choose this approach, one must first determine if the appropriate microorganism or activity is present and then decide on how to stimulate the microbe or activity. Often, this decision is based on the analysis of produced fluids. However, core material should be considered if available. A number of procedures are available for sampling core material that can minimize the problems associated with contamination (Kieft *et al.*, 2007). Both molecular and microbiological techniques can be used to verify the presence of the appropriate microorganism or activity. Molecular techniques require the removal of oil from the samples and concentration of the cells usually by filtration to provide sufficient biomass for DNA extraction (Youssef *et al.*, 2007b). Once the DNA is available, the presence of specific genes can be detected by PCR or quantitative PCR (qPCR). For some approaches such as biosurfactant production, identification of suitable target genes such as *srfA* for surfactin, *licA* for lichenysin, *rhlR* for rhamnolipid production is straightforward. For others such as acid, solvent, and gas production, the choice of the appropriate gene target is less clear. The analysis of the 16S rRNA content using universal bacterial primers could be valuable in identifying useful members of the community in cases where activity is linked to phylogeny (e.g., methanogenesis).

Once the appropriate indigenous microorganisms are identified, further tests are needed to confirm the production of the desired metabolite or activity. This requires a suitable medium to support growth and metabolism of the organism in question. The use of undefined media to grow indigenous microorganisms might not be the best approach as high concentrations of nutrients can be inhibitory to indigenous microorganisms. Also, the presence of multiple carbon compounds in these media may make it hard to control the process *in situ*. Systematic amendment of C, N, and P sources and other nutrients (trace metals, vitamins, etc.) to produced fluids incubated under conditions that stimulate reservoir conditions as closely as possible is an effective approach to detect the appropriate microbe of activity (Harvey and Harms, 2001). This approach also defines the nutrients that need to be injected into the reservoir. Subjecting indigenous microorganisms to nutrient-limiting conditions may increase the production of surface-active metabolites (Sheehy, 1992). Periodic cycles of nutrient-excess and nutrient-limitation are suggested because microorganisms become hydrophobic and cell wall components act as surface-active agents when starvation is induced.

2. Injection of exogenous microorganisms and nutrients

If the appropriate microorganisms or activity is absent, then inoculation of the reservoir with exogenous microorganism is needed. The use of exogenous microorganisms may also be an effective way to establish the appropriate activity quickly in the reservoir. Long incubations times may be needed before indigenous microorganisms grow sufficiently to produce the amounts required. The problem here is that long incubations time mean economic loss as the operator is not receiving revenue and significantly add to the cost of the treatment. Foremost, the exogenous microorganism must be able to grow under the environmental conditions present in the reservoir and in presence of competing indigenous population (Bryant, 1991). Some exogenous microorganisms may be adapted to reservoir conditions by stepwise challenging the microorganism to differing temperature and salt regimes (Banwari *et al.*, 2005). One advantage of the use of exogenous microorganisms is that a nutrient package can be designed specifically to stimuate their growth and metabolism in the reservoir (Youssef *et al.*, 2007b).

A critical factor is the transport abilities of the exogenous microorganism. Ideally, the injected microorganisms should have minimal adsorption to reservoir rock material. Some studies provide conflicting recommendations on the use of starved versus nonstarved cells (Camper *et al.*, 1993; Cunningham *et al.*, 2007). However, starved cells are smaller and smaller cells have less retention and higher transport efficiency than larger cells (Fontes *et al.*, 1991). A number of laboratory studies show that starved cells penetrate porous material more effectively than vegetative cells (Lappin-Scott *et al.*, 1988; MacLeod *et al.*, 1988). Spores are also desirable in this respect (Bae *et al.*, 1996; Jang *et al.*, 1983; Gullapalli *et al.*, 2000; McInerney *et al.*, 2005b). Alternatively, addition of nutrients may allow microorganisms to grow throughout the reservoir (Jenneman *et al.*, 1985; Sharma and McInerney, 1994; Sharma *et al.*, 1993).

3. Injection of *ex situ*-produced metabolites (Biosurfactant or polymer)

If indigenous microorganisms are not suitable for the desired outcome and conditions in the reservoir are too harsh for survival of exogenous microorganisms, the last resort is to add *ex situ*-produced products. Lipopeptides and rhamnolipids are stable at high temperatures (100 °C), a wide range of pH (2–12), and at high salt concentrations (up to 10%) (Abu-Ruwaida *et al.*, 1991; Bordoloi and Konwar, 2007; Das and Mukherjee, 2005; Joshi *et al.*, 2008a,b; Makkar and Cameotra, 1998). Some biopolymers show some deterioration at reservoir conditions, but not as extensively as synthetic polymer (Buller and Vossoughi, 1990).

Loss of the injected chemical is a major concern and a problem that plagues chemical EOR approaches (Green and Willhite, 1998; Strand *et al.*, 2003; Weihong *et al.*, 2003). Adsorption of surfactin and rhamnolipids to rock was higher than synthetic surfactants (Daoshan *et al.*, 2004; Johnson *et al.*, 2007). Studies on biopolymer loss due to adsorption have been conducted (Huh *et al.*, 1990). Synergistic effects of biosurfactants and chemical surfactants have been reported (Daoshan *et al.*, 2004; Feng *et al.*, 2007; Singh *et al.*, 2007), which would decrease the amount of chemical surfactant required and account for loss due to adsorption of the biosurfactant. Biosurfactants and polymers can be produced from cheap renewable sources in amounts sufficient for injection without extensive purification (Maneerat, 2005; Mukherjee *et al.*, 2006). There is also the possibility of designing recombinant strains that over-produce the biosurfactant or produce biosurfactant with specific structure or quality (Mukherjee *et al.*, 2006; Wang *et al.*, 2007).

B. Nutrients selection

The choice of nutrients largely depends on the desired outcome and the organism(s) involved. When exogenous microorganisms are used, the extensive knowledge base generated by studying the organism can be used to develop nutrient packages to maximize growth and/or product formation. For other MEOR approaches, there are a few guidelines that can be used. Selective plugging requires biomass production in the reservoir. Biomass formation can be maximized by supplying nitrate as the electron acceptor and its presence will also help control detrimental activities such as souring (Section V). Many LAB such a *Leuconostoc* sp. produced biopolymers (dextran) only when sucrose is added (Jenneman *et al.*, 2000; Lappan and Fogler, 1994). Nutrient manipulation showed that *B. licheniformis* could produce a biosurfactant or a biopolymer (Gabitto and Barrufet, 2005). Acid, solvent, and gas formation occurs by anaerobic fermentation of carbohydrates so the injection of large amounts of a readily degraded carbohydrate will create the conditions needed. Subjecting indigenous microorganisms to cycles of nutrient-excess and nutrient-limiting conditions may be a mechanism to promote the production of bioemulsifiers or other surface-active metabolites (Sheehy, 1992).

Biosurfactant production, on the other hand, requires fine balance between carbon and nitrogen. Davis *et al.* (1999) showed that surfactin production by *Bacillus subtilis* was enhanced when nitrate was limiting ($Y_{p/x}$ of 0.075 g surfactin per gram of biomass compared to 0.012 g surfactin per gram of biomass when nitrate vs. ammonium was limiting in the growth medium, respectively). Nitrogen limitation causes overproduction of the biosurfactant made by *Candida tropicalis*, and C/N ratio of >11 maximized rhamnolipid production by *Pseudomonas* sp.

(Cameotra and Makkar, 1998; Gautam and Tyagi, 2006). Extensive laboratory study is worth the effort as the systematic testing of different media components to maximize lipopeptide production from two *Bacillus* sp. under anaerobic conditions (McInerney *et al.*, 2005b) gave success in the field (Youssef *et al.*, 2007b). One important aspect to consider is the partial loss of nutrients due to adsorption to rock material. Concentration of the nutrient or a different form (organophosphates instead of inorganic phosphates) (Jenneman and Clark, 1994a,b) can minimize this problem.

C. Monitoring the success of MEOR field trials

The ultimate goal of MEOR is to increase in the amount of oil recovered or the rate of oil production, or to alleviate a production problem. It is important to monitor the microbiology to ensure that the process worked. In cases where both microorganisms and nutrients are injected, the process can be monitored by assessing the strain survivability, usually through a cell counting approaches, or by quantifying the loss in substrates injected and the appearance of metabolites (Youssef *et al.*, 2007b). When nutrients are injected to stimulate indigenous microorganisms, the process can be monitored by the increase in the number of indigenous population following treatment, the loss of substrate, and appearance of specific products of metabolism (Nazina *et al.*, 2007a,b).

VII. CURRENT AND FUTURE DIRECTIONS

A. Biosurfactant formulations

Significant improvements in microscopic displacement efficiency will require the use of biosurfactants that generate very low interfacial tensions against the crude oil present in the reservoir. Biosurfactants vary in their hydrophobicity/hydrophilicity such that a hydrophilic biosurfactant partitions in the aqueous phase and a hydrophobic biosurfactant partitions in the oil phase rather than at the interface between the two phases. The success of biosurfactant-mediated oil recovery depends on formulating biosurfactant mixtures with the appropriate hydrophobicity/hydrophilicity so that they partition at the interface. This can be done by measuring the activity of the biosurfactants against hydrocarbons with different equivalent alkane carbon number (Nguyen *et al.*, 2008; Youssef *et al.*, 2007a). The biosurfactant structure and the properties of the displacing fluid can then be manipulated to maximize IFT reduction. Table 6.8 shows the different methods that could be used to change the structure of lipopeptide biosurfactants. Most of these methods are simple manipulations of the growth medium to obtain a more hydrophobic or

TABLE 6.8 Methods to manipulate lipopeptide biosurfactant structure and its activity

Manipulation method	Effect on structure	Effect on surface activity	Reference
Growth in Landy's medium with valine as nitrogen source	Replacement of leucine in position 7 with the less hydrophobic valine, [Val7]-surfactin variant	Decreases the hydrophobicity and increases the CMC of surfactin	Peypoux and Michel (1992)
Growth in Landy's medium with leucine or isoleucine as nitrogen sources	Replacement of the valine in position 4 with more hydrophobic leucine or isoleucine makes the molecule more hydrophobic, [Leu4]- or [Ile4]-surfactin variant	Improves surfactant activity	Bonmatin *et al.* (1995)
Growth in Landy's medium with isoleucine as nitrogen source	[Ile4, 7]-surfactin variant; [Ile2, 4, 7]-surfactin variant	Increased hydrophobicity, CMC decreased from 220 to 90 μM, increased affinity for calcium ions	Grangemard *et al.* (1997)
Genetic manipulation by module swapping between fungal and bacterial genes	[Val7]-, [Phe7], [Orn7], and [Cys7]-surfactin variants	Hemolytic activity lowered; surface activity not tested	Stachelhaus *et al.* (1995)

(continued)

TABLE 6.8 *(continued)*

Manipulation method	Effect on structure	Effect on surface activity	Reference
Addition of precursors of branched chain fatty acids (valine, leucine, isoleucine) to the growth medium of *Bacillus licheniformis*	Lichenysin with variable fatty acid tail composition	The increased proportion of branched chain fatty acid of lichenysin a lowered the surface activity. Normal chain C14 fatty acid is important for surface activity	Yakimov *et al.* (1996)
Addition of precursors of branched chain fatty acids (valine, leucine, isoleucine) to the growth medium of *Bacillus subtilis* subsp. *subtilis*	Surfactin with variable fatty acid tail composition	Oil diplacement activity against crude oil increased with the increase in the ratio of iso to normal even numbered fatty acid in surfactin	Youssef *et al.* (2005)

more hydrophilic lipopepetide. If altering the structure does not yield a biosurfactant with the appropriate properties, for example, very low IFT against the crude oil in question, mixtures of two or more biosurfactants or synthetic surfactants can be used to obtain very low IFT against crude oil. Table 6.9 shows the effect of different biosurfactant/synthetic surfactant mixtures on IFT (Youssef *et al.*, 2007a). The salt concentration and the pH of the displacement fluid can be manipulated to enhance interfacial activity (Nguyen *et al.*, 2008).

B. Understanding the microbial ecology of oil reservoirs

Our understanding of the phylogenetic diversity, metabolic capabilities, ecological roles, and community dynamics of oil reservoir microbial communities is far from complete, and several fundamental ecological

TABLE 6.9 Formulating effective mixtures of biosurfactants and synthetic surfactants to minimize interfacial tension

Mixture components	Effect on IFT	References
Lipopeptides with different fatty acid tail composition	IFT against a hydrophilic hydrocarbon (toluene) was lowered when the mixture has less than 70% of the FA as C14 and C15, and the ratio of C16 to C18 FA is ≥ 8	Youssef *et al.* (2007a)
Lipopeptides and rhamnolipids	IFT against toluene lowered when the hydophilicity of the mixture was increased. Rhamnolipids are more hydrophilic than lipopeptide and hydrophilic lipopeptides were required	Youssef *et al.* (2007a)
Lipopeptides with hydrophobic synthetic surfactant	IFT lowered against more hydrophobic hydrocarbons (e.g., hexane and decane) when the hydrophobicity of the mixture increased	Youssef *et al.* (2007a)
Rhamnolipid with hydrophobic synthetic surfactants	IFT against more hydrophobic hydrocarbons (e.g., hexadecane) was lowered as the hydrophobicity of the mixture increased	Nguyen *et al.* (2008)

questions remain partly or completely unanswered. There are three general areas of research in which we believe significant advances could be achieved by using traditional and molecular ecological approaches. It is important to note that at the heart of all current and future opportunities is the issue of sampling.

1. Detailed understanding and comparative analysis of microbial diversity in oil reservoirs

As outlined in Section IV, none of the published studies has satisfactorily documented a complete or near complete census of the bacterial or archaeal communities in a reservoir and the relatively small number of clones sequenced per study prevents statistical estimation of species richness, evenness, or diversity patterns. Detailed high throughput investigations of the bacterial and archaeal communities (e.g., studies generating >1000 clones) in multiple oil reservoirs are needed to obtain the needed information. Such effort will also result in the identification of less abundant members of the community, but ones that could potentially be useful for MEOR. In addition, coupling high throughput investigations with thorough documentation of the geochemical conditions in oil reservoirs will allow for detailed statistical analysis correlating presence and/or abundance of specific groups of microorganisms to specific environmental conditions, and provide a means to manipulate the appropriate microbial activity.

The utilization of group specific primers to provide an in-depth view of the microbial community is a commonly used approach but rarely applied to oil reservoirs. Lineage specific 16S rRNA gene-based primers are either currently available, or could readily be designed (e.g., using Greengenes database or ARB software package, (DeSantis *et al.*, 2006; Ludwig *et al.*, 2004)). Also, primers targeting genes other than the 16S rRNA could be used to study the diversity within specific metabolic groups, for example, *dsr* gene for the sulfate reducers (Wagner *et al.*, 1998), *mcrA* gene for methanogens (Luton *et al.*, 2002), *soxB* genes for sulfide oxidizers (Petri *et al.*, 2001). The higher level of resolution obtained by using group specific primers usually results in identifying microorganisms that escaped detection by bacterial and archaeal 16S rRNA gene clone libraries.

A novel, powerful tool for microbial community analysis is the high-density oligonucleotide 16S rRNA gene-based microarray (phylochip) (DeSantis *et al.*, 2007). This microchip was developed by scientists in Lawrence Berkley National Laboratory and uses the constantly updated global 16S rRNA gene database Greeengenes (DeSantis *et al.*, 2006; Wilson *et al.*, 2002) to design high-density microarrays that could theoretically detect all phylogenetic lineages available in Greengenes. The most recent phylochip, to be introduced in 2008 will contain ~30,000 probes derived from 16S rRNA genes of pure cultures as well as uncultured sequences currently available (Todd DeSantis, personal communication). The phylochip achieves higher sensitivity than small-sized clone libraries (e.g., less than 1000 clones) as recent validation studies have clearly demonstrated (Brodie *et al.*, 2006, 2007).

2. Monitoring changes in oil reservoir community composition in response to manipulations

Another promising area of research is to identify changes in community structure that occur during various stages of production or in response to specific manipulations, for example, water flooding, injection of nutrients, and/or exogenous microorganisms. Various molecular typing approaches, for example, differential or temperature gradient gel electrophoresis (DGGE or TGGE), internal transcribed spacer (ITS) amplification, and restriction fragment length polymorphism (RFLP) that have routinely been used in microbial ecology, but rarely in oil reservoirs, could be used to monitor changes in microbial community composition. In addition, comparing microbial communities using the previously described phylochip could be a powerful tool for monitoring changes in communities. More specific source tracking goals, for example, monitoring the survivability of an injected exogenous microorganism, or monitoring genes expression *in situ* could be achieved with quantitative or real time PCR.

3. Metagenomic analysis of oil-reservoir communities

Metagenomics (direct random sequencing of environmental DNA fragments from the environment) represents another tool that could potentially be extremely powerful in oil-reservoir community studies. Metagenomic analysis of a specific oil reservoir, or comparative genomics between reservoirs with different geochemical properties could provide important insights into the reservoir's ecology. For example, metagenomics could identify and compare relative abundance and diversity of genes involved in specific processes in an ecosystem (Venter *et al.*, 2004), document the occurrence or the potential of occurrence of processes not yet encountered in the ecosystem (Hallam *et al.*, 2004), identify novel functions in microorganisms not previously known to mediate these processes (Beja *et al.*, 2000), or provide diagnostic genes associated with an ecosystem in general, or with specific processes occurring within this system in particular (Tringe *et al.*, 2005).

VIII. CONCLUSIONS

Oil reservoirs are home to phylogenetically and metabolically diverse microbial communities. Our understanding of the phylogenetic diversity, metabolic capabilities, ecological roles, and community dynamics of oil reservoir microbial communities is far from complete. The lack of a complete consensus of the number of species or phylotypes prevents the statistical estimation of species richness, evenness, or diversity patterns needed to understand how the community changes over time and responds to different exploitation practices (water flooding, etc.). Even simple questions such as whether an organism is autochthonous or allochthonous are difficult to answer without a more complete description of the microbial ecology of the reservoir. The lack of appreciation of the microbiology of oil reservoirs often leads to detrimental consequences such as souring or plugging. However, an understanding of the microbiology can be used to enhance operations. It is clear that biotechnology can also be used to mobilize entrapped oil in reservoirs. Laboratory and field studies clearly show that (1) nitrate and/or nitrite addition control H_2S production, (2) oxygen injection stimulates hydrocarbon metabolizers that feed methanogenic communities to make methane and help mobilize crude oil, (3) injection of fermentative bacteria and carbohydrates generates large amounts of acids, gases, and solvents that increase oil recovery, particularly in carbonate formations, and oil production rates of individual wells, and (4) nutrient injection stimulates microbial growth preferentially in high permeability zones and improves volumetric sweep efficiency and oil recovery. Other work shows that biosurfactants significanly lower the interfacial tension between oil and water and that large amounts of biosurfactant can be made *in situ*. However, there have not been enough field trials of biosurfactant-mediated oil recovery to determine the efficacy of the process. There are still many questions that need to be resolved particularly whether an inoculum is needed or not. Metagemonic and high-throughput community analysis should be able to provide an answer to this and other questions concerning the metabolic capabilities of the microbial community.

Many of the commerical microbial technologies have been shown to slow the rate of decline in oil production and extend the operational life of marginal oil fields. With marginal oil fields, the goal is to keep the well producing rather than maximizing the ultimate amount of oil recovered from the reservoir. The risk for implementing MEOR is low in marginal fields as these fields are near the end of their economic lives. The data needed to assess performance, for example, oil production rates and operating costs, are relatively easy to obtain. With larger, more productive oil fields, increasing the ultimate recovery factor is the goal, but this requires more extensive analysis of the reservoir and mathematical models to predict the outcomes of treatments. Many companies simply do not have the microbiogical expertise to obtain the information needed to make these assessments. Microbial oil recovery processes will only gain more widespread acceptance and application when quantitative measures of performance can be reliably obtained. Given the future demand for energy and the likely dependence on petroleum resources to meet this demand, petroleum engineering and microbiology disciplines must come together to develop the needed technologies.

ACKNOWLEDGMENTS

We thank Neil Q. Wofford for preparing the figures and Ryan T. Vaughan for research and editorial assistance. Our work on microbial oil recovery was supported by U. S. Department of Energy contracts DE-FC26-02NT15321 and DE-FC26-04NT15522.

REFERENCES

Abbott, B. J., and Gledhill, W. E. (1971). The extracellular accumulation of metabolic products by hydrocarbon-degrading microorganisms. *Adv. Appl. Microbiol.* **14,** 249–388.

Abd Karim, M. G., Salim, M. A. H., Zain, Z. M., and Talib, N. N. (2001). "Microbial Enhanced Oil Recovery (MEOR) Technology in Bokor Field, Sarawak." SPE 72125. *Proceedings of SPE Asia Pacific Improved Oil Recovery Conference,* Society of Petroleum Engineers, Richardson, Texas.

Abdul, H. J., and Farouq Ali, S. M. (2003). Combined polymer and emulsion flooding methods for oil reservoirs with a water leg. *J. Can. Pet. Technol.* **42,** 35–40.

Abdulrazag, Y., Almehaideb, R. A., and Chaalal, O. (1999). "Project of Increasing Oil Recovery from UAE Reservoirs Using Bacteria Flooding." SPE 56827. *Proceedings of the SPE Annual Technical Conference,* Society of Petroleum Engineers, Richardson, Texas.

Abhati, N., Roostaazad, R., and Ghadiri, F. (2003). "Biosurfactant Production in MEOR for Improvement of Iran's Oil Reservoirs Production Experimental Approach." SPE 84907. *Proceedings of SPE International Improved Oil Recovery Conference in Asia Pacific.* Society of Petroleum Engineers, Richardson, Texas.

Abu-Ruwaida, A. S., Banat, I. M., Haditiro, S., Salem, A., and Kadri, M. (1991). Isolation of biosurfactant-producing bacteria: Product characterization and evaluation. *Acta Biotechnol.* **11,** 315–324.

Adkins, J. P., Cornell, L. A., and Tanner, R. S. (1992a). Microbial composition of carbonate petroleum reservoir fluids. *Geomicrobiol. J.* **10,** 87–97.

Adkins, J. P., Tanner, R. S., Udegbunam, E. O., McInerney, M. J., and Knapp, R. M. (1992b). Microbially enhanced oil recovery from unconsolidated limestone cores. *Geomicrobiol. J.* **10,** 77–86.

Ali, L., and Barrufet, M. A. (1994). Profile modification due to polymer adsorption in reservoir rocks. *Energy Fuels* **8,** 1217–1222.

Almeida, P. F., Moreira, R. S., Almeida, R. C. C., Guimaraes, A. K., Carvalho, A. S., Quintella, C., Esperidia, M. C. A., and Taft, C. A. (2004). Selection and application of microorganisms to improve oil recovery. *Eng. Life Sci.* **4,** 319–325.

Andreevskii, I. L. (1959). Application of oil microbiology to the oil-extracting industry. *Trudy Vses. Nauch.-Issled. Geol.-Razved. Inst.* **131,** 403–413.

Andreevskii, I. L. (1961). The influence of the microflora of the third stratum of the Yaregskoe oil field on the composition and properties of oil. *Trudy Inst. Mikrobiol.* **9,** 75–81.

Anonymous (2006). 2006 worldwide EOR survey. *Oil Gas J.* **104,** 45–58.

Ashirov, K. B., and Sazonova, K. B. (1962). Biogenic sealing of oil deposits in carbonate reservoirs. *Microbiology* **31,** 555–557.

Atiken, C. M., Jones, D. M., and Larter, S. R. (2004). Anaerobic hydrocarbon biodegradation in deep subsurface oil reservoirs. *Nature* **431,** 291–294.

Bae, J. H., Chambers, K. T., and Lee, H. O. (1996). Microbial profile modification with spores. *Soc. Pet. Eng. Reservoir Eng.* **11,** 163–167.

Bailey, S. A., Bryant, R. S., and Duncan, K. E. (2000). Design of a novel alkalophilic bacterial system for triggering biopolymer gels. *J. Ind. Microbiol. Biotechnol.* **24,** 389–395.

Bailey, S. A., Kenney, T. M., and Schneider, D. R. (2001). "Microbial Enhanced Oil Recovery: Diverse Successful Applications of Biotechnology in the Oil Field." SPE 72129. *Proceedings of SPE Asia Pacific Improved Oil Recovery Conference.* Society of Petroleum Engineers, Richardson, Texas.

Banat, I. M. (1995a). Biosurfactants production and possible uses in microbial enhanced oil recovery and oil pollution remediation: A review. *Biores. Technol.* **51,** 1–12.

Banat, I. M. (1995b). Characterization of biosurfactants and their use in pollution removal-State of the art (review). *Acta Biotechnol.* **15,** 251–267.

Banat, I. M., Makkar, R. S., and Cameotra, S. S. (2000). Potential commercial applications of microbial surfactants. *Appl. Microbiol. Biotechnol.* **53,** 495–508.

Banwari, L. A. L., Reedy, M. R. V., Agnihotri, A., Kumar, A., Sarbhai, M. P., Singh, M., Khurana, R. K., Khazanchi, S. K., and Misra, T. R. (2005). A process for enhanced recovery of crude oil from oil wells using novel microbial consortium. World Intellectual Property Organization. Patent No. WO/2005/005773.

Barkay, T., Navon-Venezia, S., Ron, E. Z., and Rosenberg, E. (1999). Enhancement of solubilization and biodegradation of polyaromatic hydrocarbons by bioemulsifier alasan. *Appl. Environ. Microbiol.* **65,** 2697–2702.

Barker, K. M., Waugh, K. L., and Newberry, M. E. (2003). "Cost-Effective Treatment Programs For Paraffin Control." SPE 80903. Proceedings of the SPE Productions & Operations Symposium. Society of Petroleum Engineers, Richardson, Texas.

Bass, C. J., and Lappin-Scott, H. M. (1997). The bad guys and the good guys in petroleum microbiology. *Oilfield Rev.* **9,** 17–25.

Bastin, E. S., Greer, F. E., Merritt, C. A., and Moulton, G. (1926). The presence of sulphate-reducing bacteria in oil field waters. *Science* **63,** 21–24.

Becker, J. R. (2001). Paraffin treatments: Hot oil/hot water vs. crystal modifiers. *J. Pet. Technol.* **53,** 56–57.

Beeder, J., Nilsen, R. K., Rosnes, J. T., Torsvik, T., and Lien, T. (1994). *Archaeoglobus fulgidus* isolated from hot North sea oil field water. *Appl. Environ. Microbiol.* **60,** 1227–1231.

Beeder, J., Torsvik, T., and Lien, T. (1995). *Thermodesulforhabdus norvegicus* gen. nov., sp. nov., a novel thermophilic sulfate-reducing bacterium from oil field water. *Arch. Microbiol.* **164,** 331–336.

Behlulgil, K., and Mehmetoglu, M. T. (2002). Bacteria for improvement of oil recovery: A laboratory study. *Energy Sources* **24,** 413–421.

Beja, O., Aravind, L., Koonin, E. V., Suzuki, M. T., Hadd, A., Nguyen, L. P., Jovanovich, S. B., Gates, C. M., Feldman, R. A., Spudich, J. L., Spudich, E. N., and DeLong, E. F. (2000). Bacterial rhodopsin: Evidence for a new type of phototrophy in the sea. *Science* **289,** 1902–1906.

Beliakova, E. V., Rozanova, E. P., Borzenkov, I. A., Turova, T. P., Pusheva, M. A., Lysenko, A. M., and Kolganov, T. V. (2006). The new facultatively chemolithoautotrophic, moderately halophilic, sulfate-reducing bacterium *Desulfovermiculus halophilus* gen. nov., sp. nov., isolated from an oil field. *Microbiology (Eng. Tr.)* **75,** 161–171.

Belyaev, S. S., Borzenkov, I. A., Glumov, I. F., Ibatullin, R. R., Milekhina, E. I., and Ivanov, M. V. (1998). Activation of the geochemical activity of stratal microflora as basis of a biotechnology for enhancement of oil recovery. *Microbiology (Eng. Tr.)* **67,** 708–714.

Belyaev, S. S., Borzenkov, I. A., Nazina, T. N., Rozanova, E. P., Glumov, I. F., Ibatullin, R. R., and Ivanov, M. V. (2004). Use of microorganisms in the biotechnology for the enhancement of oil recovery. *Microbiology* **73,** 590–598.

Belyaev, S. S., and Ivanov, V. (1983). Bacterial methanogenesis in underground waters. *Ecol. Bull.* **35,** 273–280.

Belyaev, S. S., Laurinavichus, K. S., Obraztsova, A. Y., Gorlatov, S. N., and Ivanov, M. V. (1982). Microbiological processes in the critical zone of injection wells. *Mikrobiologiya* **51,** 997–1001.

Belyaev, S. S., Obraztsova, A. Y., Laurinavichus, K. S., and Bezrukova, L. V. (1986). Characteristics of rod-shaped methane-producing bacteria from oil pool and description of *Methanobacterium ivanovii*. *Microbiology (Eng. Tr.)* **55,** 821–826.

Bhupathiraju, V. K., McInerney, M. J., Woese, C. R., and Tanner, R. S. (1999). *Haloanaerobium kushneri* sp. nov., an obligately halophilic, anaerobic bacterium from an oil brine. *Int. J. Syst. Bacteriol.* **49,** 953–960.

Bhupathiraju, V. K., Oren, A., Sharma, P. K., Tanner, R. S., Woese, C. R., and McInerney, M. J. (1994). *Haloanaerobium salsugo* sp. nov., a moderately halophilic, anaerobic bacterium from a subterranean brine. *Int. J. Syst. Bacteriol.* **44,** 565–572.

Birkeland, N.-K. (2005). Sulfate-reducing bacteria and archaea. *In* "Petroleum Microbiology" (B. Ollivier and M. Magot, Eds.), pp. 35–54. ASM, Washington, DC.

Bodour, A. A., and Miller-Maier, R. M. (2002). Biosurfactants: Types, screening methods, and applications. *In* "Encyclopoedia of Environmental Microbiology" (G. Bitton, Ed.), pp. 750–770. Wiley-Interscience, New York, NY.

Bognolo, G. (1999). Biosurfactants as emulsifying agents for hydrocarbons. *Colloids Surf. A. Physicochem. Eng. Aspects* **152,** 41–52.

Bonch-Osmolovskaya, E. A., Miroshnichenko, M. L., Lebedinsky, A. V., Chernyh, N. A., Nazina, T. N., Ivoilov, V. S., Belyaev, S. S., Boulygina, E. S., Lysov, Y. P., Perov, A. N., Mirzabekov, A. D., Hippe, H., *et al.* (2003). Radioisotopic, culture-based, and oligonucleotide microchip analyses of thermophilic microbial communities in a continental high-temperature petroleum reservoir. *Appl. Environ. Microbiol.* **69,** 6143–6151.

Bonilla Salinas, M., Fardeau, M.-L., Thomas, P., Cayol, J.-L., Patel, B. K. C., and Ollivier, B. (2004). *Mahella australiensis* gen. nov., sp. nov., a moderately thermophilic anaerobic bacterium isolated from an Australian oil well. *Int. J. Syst. Evol. Microbiol.* **54,** 2169–2173.

Bonmatin, J. M., Labbé, H., Grangemard, I., Peypoux, F., Maget-Dana, R., Ptak, M., and Michel, G. (1995). Production, isolation, and characterization of [Leu4]- and [Ile4]-surfactins from *Bacillus subtilis. Lett. Peptide Sci.* **2,** 41–47.

Bordoloi, N. K., and Konwar, B. K. (2007). Microbial surfactant-enhanced mineral oil recovery under laboratory conditions. *Colloids Surf. B. Biointerfaces* **63**(1), 73–82, doi:10.1016/j.colsurfb.2007.1011.1006.

Brodie, E. L., DeSantis, T. Z., Joyner, D. C., Baek, S. M., Larsen, J. T., Andersen, G. L., Hazen, T. C., Richardson, P. M., Herman, D. J., Tokunaga, T. K., Wan, J. M., and Firestone, M. K. (2006). Application of a high-density oligonucleotide microarray approach to study bacterial population dynamics during uranium reduction and reoxidation. *Appl. Environ. Microbiol.* **72,** 6288–6298.

Brodie, E. L., DeSantis, T. Z., Parker, J. P. M., Zubietta, I. X., Piceno, Y. M., and Andersen, G. L. (2007). Urban aerosols harbor diverse and dynamic bacterial populations. *Proc. Natl. Acad. Sci. USA* **104,** 299–304.

Brown, L. R. (1984). Method for increasing oil recovery. U. S. Patent Office. Patent No. 4,475,590.

Brown, F. G. (1992). "Microbes: The Practical and Environmental Safe Solution to Production Problems, Enhanced Production, and Enhanced Oil Recovery." SPE 23955. *Proceedings of the 1992 Permian Basin Oil and Gas Recovery Conference.* Society of Petroleum Engineers, Richardson, Texas.

Brown, L. R. (2007). The relationship of microbiology to the petroleum industry. *Soc. Ind. Microbiol. News* **57,** 180–190.

Brown, F., Maure, A., and Warren, A. (2005). Microbial-induced controllable cracking of normal and branched alkanes in oils. U. S. Patent Office. Patent No. 6905870.

Brown, L. R., Vadie, A. A., and Stephens, J. O. (2002). Slowing production decline and extending the economic life of an oil field: New MEOR technology. *Soc. Pet. Eng. Reservoir Eval. Eng.* **5,** 33–41.

Bryant, R. S. (1988). Microbial enhanced oil recovery and compositions thereof. US Patent office. Patent No. 4,905,761.

Bryant, R. S. (1991). MEOR screening criteria fit 27% of U.S. oil reservoirs. *Oil Gas J.* **15,** 56–59.

Bryant, R. S., and Burchfield, T. E. (1989). Review of microbial technology for improving oil recovery. *Soc. Pet. Eng. Reservoir Eng.* **4,** 151–154.

Bryant, R. S., and Burchfield, T. E. (1991). Microbial enhanced waterflooding: A pilot study. *Dev. Pet. Sci.* **31,** 399–420.

Bryant, R. S., Burchfield, T. E., Dennis, M. D., and Hitzman, D. O. (1988). "Microbial Enhanced Waterflooding Mink Unit Project." SPE 17341. *Proceedings of SPE/DOE Enhanced oil Recovery Symposium*, Society of Petroleum Engineers, Richardson, Texas.

Bryant, R. S., Burchfield, T. E., Dennis, D. M., and Hitzman, D. O. (1990). Microbial-Enhanced waterflooding: Mink Unit project. *Soc. Pet. Eng. Reservoir Eng.* **5,** 9–13.

Bryant, R. S., and Douglas, J. (1988). Evaluation of microbial systems in porous media for EOR. *Soc. Pet. Eng. Reservoir Eng.* **3,** 489–495.

Bryant, S. L., and Lockhart, T. P. (2002). Reservoir engineering analysis of microbial enhanced oil recovery. *Soc. Pet. Eng. Reservoir Eval. Eng.* **5,** 365–374.

Bryant, R. S., Stepp, A. K., Bertus, K. M., Burchfield, T. E., and Dennis, M. (1993). Microbial-enhanced waterflooding field pilots. *Dev. Pet. Sci.* **39,** 289–306.

Buciak, J. M., Vazquez, A. E., Frydman, R. J., Mediavilla, J. A., and Bryant, R. S. (1994). "Enhanced Oil Recovery by Means of Microorganisms: Pilot Test." SPE 27031. *Proceedings of the 3rd Latin American and Caribbean Petroleum Engineering Conference*, Society of Petroleum Engineers, Richardson, Texas.

Buller, C. S., and Vossoughi, S. (1990). Subterranean preliminary modification by using microbial polysaccharide polymers. US patent office. Patent No. 4,941533.

Burger, E. D. (1998). Method for inhibiting microbially influenced corrosion. US Patent office. Patent No. 5,753,180.

Cameotra, S. S., and Makkar, R. S. (1998). Synthesis of biosurfactants in extreme conditions. *Appl. Microbiol. Biotechnol.* **50,** 520–529.

Camper, A. K., Hayes, J. T., Strurman, P. J., Jones, W. L., and Cunningham, A. B. (1993). Effects of motility and adsorption rate coefficient on transport of bacteria through saturated porous media. *Appl. Environ. Microbiol.* **59,** 3455–3462.

Carpenter, W. T. (1932). Sodium nitrate used to control nuisances. *Water Sew. Works* **79,** 175–176.

Cayol, J. L., Fardeau, M. L., Garcia, J. L., and Ollivier, B. (2002). Evidence of interspecies hydrogen transfer from glycerol in saline environments. *Extremophiles* **6,** 131–134.

Cayol, J. L., Ollivier, B., Patel, B. K., Ravot, G., Magot, M., Ageron, E., Grimont, P. A., and Garcia, J. L. (1995). Description of *Thermoanaerobacter brockii* subsp. *lactiethylicus* subsp. nov., isolated from a deep subsurface French oil well, a proposal to reclassify *Thermoanaerobacter finnii* as *Thermoanaerobacter brockii* subsp. *finnii* comb. nov., and an emended description of *Thermoanaerobacter brockii*. *Int. J. Syst. Bacteriol.* **45,** 783–789.

Chang, Y. (1987). Preliminary studies assessing sodium pyrophosphate effects on microbially mediated oil recovery. *Ann. N. Y. Acad. Sci.* **506,** 296–307.

Chapelle, F. H., and Bradley, P. M. (1997). Alteration of aquifer geochemistry by microorganisms. *In* "Manual of Environmental Microbiology" (C. J. Hurst, G. R. Kunudsen, M. J. McInerney, L. D. Stetzenbach and M. V. Wlater, Eds.), pp. 558–564. ASM, Washington, DC.

Chapelle, F. H., O'Neill, K., Bradley, P. M., Methe, B. A., Ciufo, S. A., Knobel, L. L., and Loveley, D. R. (2002). A hydrogen-based subsurface community dominated by methanogens. *Nature* **415,** 312–325.

Chen, C.-I., Reinsel, M. A., and Mueller, R. F. (1994). Kinetic investigation of microbial souring in porous media using microbial consortia from oil reservoirs. *Biotechnol. Bioeng.* **44,** 263–269.

Cheng, L., Qiu, T.-L., Yin, X.-B., Wu, X.-L., Hu, G.-Q., Deng, Y., and Zhang, H. (2007). *Methermicoccus shengliensis* gen. nov., sp. nov., a thermophilic, methylotrophic methanogen isolated from oil-production water, and proposal of *Methermicoccaceae* fam. nov. *Int. J. Syst. Evol. Microbiol.* **57,** 2964–2969.

Christensen, B., Torsvik, T., and Lien, T. (1992). Immunomagnetically captured thermophilic sulfate-reducing bacteria from North Sea oil field waters. *Appl. Environ. Microbiol.* **58,** 1244–1248.

Clark, J. B., Munnecke, D. M., and Jenneman, G. E. (1981). *In situ* microbial enhancement of oil production. *Dev. Ind. Microbiol.* **22,** 695–701.

Coleman, J. K., Brown, M. J., Moses, V., and Burton, C. C. (1992). "Enhanced Oil Recovery." World Intellectual Property Organization. Patent No. WO 92/15771.

Connan, J. (1984). Biodegradation of crude oils in reservoirs. *In* "Advances in Petroleum Geochemistry," pp. 299–333. Academic Press, London, England.

Craig, F. F., Jr. (1974). Waterflooding. *In* "Secondary and Tertiary Oil Recovery Processes," pp. 1–30. Interstate Oil Compact Commission, Oklahoma City, Oklahoma.

Craig, F. F., Jr. (1980). "The Reservoir Engineering Aspects of Waterflooding." Society of Petroleum Engineers, Richardson, Texas.

Cravo-Laureau, C., Matheron, R., Cayol, J.-L., Joulian, C., and Hirschler-Rea, A. (2004). *Desulfatibacillum aliphaticivorans* gen. nov., sp. nov., an *n*-alkane- and *n*-alkene-degrading, sulfate-reducing bacterium. *Int. J. Syst. Evol. Microbiol.* **54,** 77–83.

Crawford, P. B. (1962). Continual changes observed in bacterial stratification rectification. *Water Technol.* **26,** 12.

Crawford, P. B. (1983). Possible reservoir damage from microbial enhanced oil recovery. *In* "Proceedings of the 1982 International Conference on Microbial Enhancement of Oil Recovery" (E. C. Donaldson and J. B. Clark, Eds.), pp. 76–79. U. S. Dept. of Energy, Bartellesville, Okla.(CONF-8205140).

Cunningham, A. B., Sharp, R. R., Caccavo, F., Jr., and Gerlach, R. (2007). Effects of starvation on bacterial transport through porous media. *Adv. Water Res.* **30,** 1583–1592.

Cusack, F., Lappin-Scott, H., Singh, S., De Rocco, M., and Costerton, J. W. (1990). Advances in microbiology to enhance oil recovery. *Appl. Biochem. Biotechnol.* **24–25,** 885–898.

Cusack, F., Singh, S., McCarthy, C., Grieco, J., De Rocco, M., Nguyen, D., Lappin-Scott, H. M., and Costerton, J. W. (1992). Enhanced oil recovery-three dimensional sandpack simulation of ultramicrobacteria resuscitation in reservoir formation. *J. Gen. Microbiol.* **138,** 647–655.

Dabbous, M. K. (1977). Displacement of polymers in waterflooded porous media and its effects on a subsequent micellar flood. *Soc. Pet. Eng. J.* **17,** 358–368.

Dabbous, M. K., and Elkins, L. E. (1976). "Preinjection of Polymers to Increase Reservoir Flooding Efficiency." SPE 5836. *Proceedings of Improved Oil Recovery Symposium of the SPE of AIME,* Society of Petroleum Engineers, Richardson, Texas.

Dahle, H., and Birkeland, N. -K. (2006). *Thermovirga lienii* gen. nov., sp. nov., a novel moderately thermophilic, anaerobic, amino-acid-degrading bacterium isolated from a North Sea oil well. *Int. J. Syst. Evol. Microbiol.* **56,** 1539–1545.

Dahle, H., Garshol, F., Madsen, M., and Birkeland, N.-K. (2008). Microbial community structure analysis of produced water from a high-temperature North Sea oil-field. *Antonie van Leeuwenhoek* **93,** 37–49.

Daoshan, L., Shouliang, L., Yi, L., and Demin, W. (2004). The effect of biosurfactant on the interfacial tension and adsorption loss of surfactant in ASP flooding. *Colloids Surf. A. Physicochem. Eng. Aspects* **244,** 53–60.

Das, K., and Mukherjee, A. K. (2005). Characterization of biochemical properties and biological activities of biosurfactants produced by *Pseudomonas aeruginosa* mucoid and non-mucoid strains isolated from hydrocarbon-contaminated soil samples. *Appl. Microbiol. Biotechnol.* **69,** 192–199.

Das, K., and Mukherjee, A. K. (2007). Comparison of lipopeptide biosurfactants production by *Bacillus subtilis* strains in submerged and solid state fermentation systems using a cheap carbon source: Some industrial applications of biosurfactants. *Process Biochem.* **42,** 1191–1199.

Dastgheib, S. M. M., Amoozegar, M. A., Elahi, E., Asad, S., and Banat, I. M. (2008). Bioemulsifier production by a halothermophilic *Bacillus* strain with potential applications in microbially enhanced oil recovery. *Biotechnol. Lett.* **30,** 263–270.

Davey, M. E., Gevertz, D., Wood, W. A., Clark, J. B., and Jenneman, G. E. (1998). Microbial selective plugging of sandstone through stimulation of indigenous bacteria in a hypersaline oil reservoir. *Geomicrobiology* **15,** 335–352.

Davey, M. E., Wood, W. A., Key, R., and Nakamura, K. (1993). Isolation of three species of *Geotoga* and *Petrotoga*: Two new genera, representing a new lineage in the bacterial line of descent distantly related to the "*Thermotogales*." *Syst. Appl. Microbiol.* **16,** 191–200.

Davidova, I. A., Duncan, K. E., Choi, O. K., and Suflita, J. M. (2006). *Desulfoglaeba alkanexedens* gen. nov., sp. nov., an *n*-alkane-degrading, sulfate-reducing bacterium. *Int. J. Syst. Evol. Microbiol.* **56,** 2737–2742.

Davidova, I., Hicks, M. S., Fedorak, P. M., and Suflita, J. M. (2001). The influence of nitrate on microbial processes in oil industry production waters. *J. Ind. Microbiol. Biotechnol.* **27,** 80–86.

Davidson, S. W., and Russell, H. H. (1988). A MEOR pilot test in the Loco field. *In* "Proceedings of the Symposium on the Application of Microorganisms to Petroleum Technology" (T. E. Burchfield and R. S. Byrant, Eds.), pp. VII 1–VII 12. National Technical Information Service, Springfield, VA.

Davis, D. A., Lynch, H. C., and Varley, J. (1999). The production of surfactin in batch culture by *Bacillus subtilis* ATCC 21332 is strongly influenced by the conditions of nitrogen metabolism. *Enzyme Microb. Technol.* **25,** 322–329.

Davis, J. B., and Updegraff, D. M. (1954). Microbiology in the petroleum industry. *Bacteriol. Rev.* **18,** 215–238.

Davydova-Charakhch'yan, I. A., Kuznetsova, V. G., Mityushina, L. L., and Belyaev, S. S. (1992a). Methane-forming bacilli from oil-fields of Tataria and western Siberia. *Microbiology (Eng. Tr.)* **61,** 299–305.

Davydova-Charakhch'yan, I. A., Mileeva, A. N., Mityushina, L. L., and Belyaev, S. S. (1992b). Acetogenic bacteria from oil fields of Tataria and western Siberia. *Microbiology (Eng. Tr.)* **61,** 306–315.

DeLong, E. F. (2005). Microbial community genomics in the ocean. *Nat. Rev. Microbiol.* **3,** 459–469.

Deng, D., Li, C., Ju, Q., Wu, P., Dietrich, F. L., and Zhou, Z. H. (1999). "Systematic Extensive Laboratory Studies of Microbial EOR Mechanisms and Microbial EOR Application Results in Changqing Oil Field." SPE 54380. *Proceedings of SPE Asia Pacific Oil and Gas Conference and Exhibition*. Society of Petroleum Engineers, Richardson, Texas.

DeSantis, T. Z., Brodie, E. L., Moberg, J. P., Zobieta, I. X., Piceno, Y. M., and Anderson, G. L. (2007). High-Density universal 16S rRNA microarray analysis reveals broader diversity than typical clone libraries when sampling the environment. *Microb. Ecol.* **53,** 371–383.

DeSantis, T. Z., Hugenholtz, P., Larsen, N., Rojas, M., Brodie, E. L., Keller, K., Huber, T., Dalevi, D., and Andersen, G. L. (2006). Greengenes, a chimera-checked 16S rRNA gene database and workbench compatible with ARB. *Appl. Environ. Microbiol.* **72,** 5069–5072.

Desouky, S. M., Abdel-Daim, M. M., Sayyouh, M. H., and Dahab, A. S. (1996). Modelling and laboratory investigation of microbial enhanced oil recovery. *J. Pet. Sci. Eng.* **15,** 309–320.

Dienes, M., and Jaranyi, I. (1973). Increase of oil recovery by introducing anaerobic bacteria into the formation, Demjen field, Hungary. *Koolaj es Foldgaz* **106,** 205–208.

Dietrich, F. L., Brown, F. G., Zhou, Z. H., and Maure, M. A. (1996). "Microbial EOR Technology Advancement: Case Studies of Successful Projects." SPE 36746. *Proceedings of the SPE Annual Technical Conference and Exhibition*. Society of Petroleum Engineers, Richardson, Texas.

Donaldson, E. C. (1985). "Use of Capillary Pressure Curves for Analysis of Production Formation Damage." SPE 13809. *Proceedings of the SPE Production Operations Symposium*. Society of Petroleum Engineers, Richardson, Texas.

Donaldson, E. C., and Clark, J. B. (1982). Conference focuses on microbial enhancement of oil recovery. *Oil Gas J.* **80,** 47–52.

Dostalek, M., and Spurny, M. (1957a). Release of Oil by microorganisms. I. Pilot experiment in an oil deposit. *Cechoslovenska Mikrobiologie* **2,** 300–306.

Dostalek, M., and Spurny, M. (1957b). Release of oil through the action of microorganisms. II: Effect of physical and physical-chemical conditions in oil-bearing rock. *Cechoslovenska Mikrobiologie* **2,** 307–317.

Eckford, R. E., and Fedorak, P. M. (2002). Chemical and microbiological changes in laboratory incubations of nitrate amendment ''sour'' produced waters from three western Canadian oil fields. *J. Ind. Microbiol. Biotechnol.* **29,** 243–254.

Energy Information Administration (2006). International Energy Outlook DOE/EIA-0484 (2006). Report to the U.S. Department of Energy Washington, DC.

Energy Information Administration (2007). Annual Energy Outlook DOE/EIA-0383 (2007). Report to the U.S. Department of Energy Washington, DC.

Etoumi, A. (2007). Microbial treatment of waxy crude oils for mitigation of wax precipitation. *J. Pet. Sci. Eng.* **55,** 111–121.

Euzeby, J. P. (2008). List of prokaryotic names with standing in nomenclature; Available from http://www.bacterio.cict.fr/.

Fardeau, M.-L., Salinas, M. B., L'Haridon, S., Jeanthon, C., Verhe, F., Cayol, J.-L., Patel, B. K. C., Garcia, J.-L., and Ollivier, B. (2004). Isolation from oil reservoirs of novel thermophilic anaerobes phylogenetically related to *Thermoanaerobacter subterraneus*: Reassignment of *T. subterraneus, Thermoanaerobacter yonseiensis, Thermoanaerobacter tengcongensis* and *Carboxydibrachium pacificum* to *Caldanaerobacter subterraneus* gen. nov., sp. nov., comb. nov. as four novel subspecies. *Int. J. Syst. Evol. Microbiol.* **54,** 467–474.

Feng, W., Guocheng, D., Qun, Y., and Zhenggang, C. (2007). Activities of surface-active broths produced by isolated microbes for application in enhanced oil recovery. *Sciencepaper online* http://www.paper.edu.cn/en/paper.php?serial_number=200708-186.

Ferguson, K. R., Lloyd, C. T., Spencer, D., and Hoeltgen, J. (1996). ''Microbial Pilot Test for the Control of Paraffin and Asphaltenes at Prudhoe Bay.'' SPE 36630 *Proceedings of the SPE Annual Technical Conference.* Society of Petroleum Engineers, Richardson, Texas.

Finnerty, W. R., and Singer, M. E. (1983). Microbial enhancement of oil recovery. *Bio/Technology* **1,** 47–54.

Fisher, J. B. (1987). Distribution and occurrence of aliphatic acid anions in deep subsurface waters. *Geochim. Cosmochim. Acta* **51,** 2459–2468.

Folmsbee, M., Duncan, K. E., Han, S.-O., Nagle, D. P., Jennings, E., and McInerney, M. J. (2006). Re-identification of the halotolerant, biosurfactant-producing *Bacillus licheniformis* strain JF-2 as *Bacillus mojavensis* strain JF-2. *Syst. Appl. Microbiol.* **29,** 645–649.

Fontes, D. E., Mills, A. L., Hornberger, G. M., and Herman, J. S. (1991). Physical and chemical factors influencing transport of microorganisms through porous media. *Appl. Environ. Microbiol.* **57,** 2473–2481.

Ford, W. G., Glenn, R. V., Herbert, J., and Buller, J. (2000). Dispersant solves paraffin problems. *Am. Oil Gas Rep.* **43,** 91–94.

Gabitto, J. F., and Barrufet, M. (2005). Combined microbial surfactant–polymer system for improved oil mobility and conformance control. Report to the Department of Energy. OSTI ID: 881858, http://www.osti.gov/energycitations/servlets/purl/881858-t9HxQJ/10.2172/881858.

Galushko, A. S., and Rozanova, E. P. (1991). *Desulfobacterium cetonicum* sp. nov., a sulfate-reducing bacterium which oxidizes fatty acids and ketones. *Microbiology (Eng. Tr.)* **60,** 742–746.

Gautam, K. K., and Tyagi, V. K. (2006). Microbial surfactants: A review. *J. Oleo Sci.* **55,** 155–166.

Geesey, G. G., Mittelman, M. W., and Lieu, V. T. (1987). Evaluation of slime-producing bacteria in oil field core flood experiments. *Appl. Environ. Microbiol.* **53,** 278–283.

Georgiou, G., Lin, S.-C., and Sharma, M. M. (1992). Surface-active compounds from microorganisms. *Bio/Technology* **10,** 60–65.
Gevertz, D., Telang, A. J., Voordouw, G., and Jenneman, G. E. (2000). Isolation and characterization of strains CVO and FWKO B, two novel nitrate-reducing, sulfide-oxidizing bacteria isolated from oil field brine. *Appl. Environ. Microbiol.* **66,** 2491–2501.
Giangiacomo, L. (1997). Chemical and microbial paraffin control project (FC9544/96PT12). Report to the Rocky Mountain Oil Field Testing Center, US Department of Energy, Gemantown, VA.
Gieg, L., McInerney, M. J., Jenneman, G. E., and Suflita, J. M. (2004). "Evaluation of Commercial, Microbial-based Products to Treat Paraffin Deposition in Oil Production Equipment." *Proceedings of 11th Annual International Petroleum Environmental Conference.* University of Tulsa, Tulsa, Okla.
Grabowski, A., Blanchet, D., and Jeanthon, C. (2005a). Characterization of long-chain fatty-acid-degrading syntrophic associations from a biodegraded oil reservoir. *Res. Microbiol.* **156,** 814–821.
Grabowski, A., Nercessian, O., Fayolle, F., Blanchet, D., and Jeanthon, C. (2005b). Microbial diversity in production waters of a low-temperature biodegraded oil reservoir. *FEMS Microbiol. Ecol.* **54,** 427–443.
Grabowski, A., Tindall, B. J., Bardin, V., Blanchet, D., and Jeanthon, C. (2005c). *Petrimonas sulfuriphila* gen. nov., sp. nov., a mesophilic fermentative bacterium isolated from a biodegraded oil reservoir. *Int. J. Syst. Evol. Microbiol.* **55,** 1113–1121.
Grangemard, I., Peypoux, F., Wallach, J., Das, B. C., Labbé, H., Caille, A., Genest, M., Maget-Dana, R., Ptak, M., and Bonmatin, J. M. (1997). Lipopeptides with improved properties: Structure by NMR, purification by HPLC, and structure-activity relationships of new isoleucyl-rich surfactins. *J. Peptide Sci.* **3,** 145–154.
Grassia, G. S., McLean, K. M., Glenat, P., Bauld, J., and Sheehy, A. J. (1996). A systematic survey for thermophilic fermentative bacteria and archaea in high temperature petroleum reservoir. *FEMS Microbiol. Ecol.* **21,** 47–58.
Green, D. W., and Willhite, G. P. (1998). "Enhanced Oil Recovery." Society of Petroleum Engineers, Richardson, TX.
Greene, A. C., Patel, B. K., and Sheehy, A. J. (1997). *Deferribacter thermophilus* gen. nov., sp. nov., a novel thermophilic manganese- and iron-reducing bacterium isolated from a petroleum reservoir. *Int. J. Syst. Bacteriol.* **47,** 505–509.
Griffin, W. T., Phelps, T. J., Colwell, F. S., and Fredrickson, J. K. (1997). Methods for obtaining deep subsurface microbiological samples by drilling. *In* "The microbiology of the Terrestrial Deep Subsurface" (P. S. Amy and D. A. Haldeman, Eds.), pp. 23–44. CRC press LLC, New York.
Groudeva, V. I., Ivanova, I. A., Groudev, S. N., and Uzunov, G. G. (1993). Enhanced oil recovery by stimulating the activity of the indigenous microflora of oil reservoirs. *Biohydrometall. Technol. Proc. Int. Biohydrometall. Symp.* **2,** 349–356.
Grula, E. A., Russel, H. H., Bryant, D., Kenaga, M., and Hart, M. (1983). Isolation and screening of clostridia for possible use in microbially enhanced oil recovery. *In* "Proceedings of the 1982 International Conference on Microbial Enhancement of Oil Recovery" (E. C. Donaldson and J. B. Clark, Eds.), pp. 43–47. U. S. Dept. of Energy, Bartellesville, Okla. (CONF-8205140).
Grula, E. A., Russell, H. H., and Grula, M. M. (1985). Field trials in central Oklahoma using clostridial strains for microbial enhanced oil recovery. *In* "Microbes and Oil Recovery" (J. E. Zajic and E. C. Donaldson, Eds.), pp. 144–150. Bioresources Publications, El Paso, Texas.
Gullapalli, I. L., Bae, J. H., Heji, K., and Edwards, A. (2000). Laboratory design and field implementation of microbial profile modification process. *Soc. Pet. Eng. Reservoir Eval. Eng.* **3,** 42–49.

Gutnick, D. L., Rosenberg, E., Belsky, I., Zosim, Z., and Shabtai, Y. (1986). Extracellular microbial lipoheteropolysaccharides and derivatives, their preparation and compositions containing them, and their uses. European patent Office. Patent No. 0016546.

Hall, C., Tharakan, P., Hallock, J., Cleveland, C., and Jefferson, M. (2003). Hydrocarbons and the evolution of human culture. *Nature* **426,** 318–322.

Hallam, S. J., Putnam, N., Preston, C. M., Detter, J. C., Rokhsar, D., Richardson, P. M., and DeLong, E. F. (2004). Reverse methanogenesis: Testing the hypothesis with environmental genomics. *Science* **305,** 1457–1462.

Harms, G., Zengler, K., Rabus, R., Aeckersberg, F., Minz, D., Rossello-Mora, R., and Widdel, F. (1999). Anaerobic oxidation of o-xylene, m-xylene, and homologous alkylbenzenes by new types of sulfate-reducing bacteria. *Appl. Environ. Microbiol.* **65,** 999–1004.

Harrah, T., Panilaitis, B., and Kaplan, D. (1997). ''Microbial Exopolysaccharides.'' Springer-Verlag, NewYork. http://www.springerlink.com/content/t260334242t7g674/.

Harvey, R. W., and Harms, H. (2001). Transport of microorganisms in the terrestial subsurface: *In situ* and laboratory methods. *In* ''Manual of Environmental Microbiology'' (C. J. Hurst, G. R. Knudsen, M. J. McInerney, L. D. Stetzenback and R. L. Crawford, Eds.), pp. 753–776. ASM, Washington, DC.

Haveman, S. A., Greene, E. A., Stilwell, C. P., Voordouw, J. K., and Voordouw, G. (2004). Physiological and gene expression analysis of inhibition of *Desulfovibrio vulgaris* Hildenborough by nitrite. *J. Bacteriol.* **186,** 7944–7950.

He, Z., Mei, B., Wang, W., Sheng, J., Zhu, S., Wang, L., and Yen, T. F. (2003). A pilot test using microbial paraffin-removal technology in Liaohe oilfield. *Pet. Sci. Technol.* **21,** 201–210.

He, Z., She, Y., Xiang, T., Xue, F., Mei, B., Li, Y., Ju, B., Mei, H., and Yen, T. F. (2000). MEOR pilot sees encouraging results in Chinese oil field. *Oil Gas J.* **98,** 46–51.

Head, I. M., Jones, D. M., and Larter, S. R. (2003). Biological activity in the deep subsurface and the origin of heavy oil. *Nature* **426,** 344–352.

Heider, J., Spormann, A. M., Beller, H. R., and Widdel, F. (1998). Anaerobic bacterial metabolism of hydrocarbons. *FEMS Microbiol. Rev.* **22,** 459–473.

Heukelelekian, H. (1943). Effect of the addition of sodium nitrate to sewage on hydrogen sulfide production and B. O. D. reduction. *Sew. Works J.* **15,** 225–261.

Hinrichs, K.-U., Hayes, J. M., Bach, W., Spivack, A. J., Hmelo, L. R., Holm, N. G., Johnson, C. G., and Sylva, S. P. (2006). Biological formation of ethane and propane in the deep marine subsurface. *Proc. Natl. Acad. Sci. USA* **103,** 14684–14689.

Hitzman, D. O. (1983). Petroleum microbiology and the history of its role in enhanced oil recovery. *In* ''Proceedings of the 1982 International Conference on Microbial Enhancement of Oil Recovery'' (E. C. Donaldson and J. B. Clark, Eds.), pp. 162–218. U. S. Dept. of Energy, Bartellesville, Okla. (CONF-8205140).

Hitzman, D. O. (1988). Review of microbial enhanced oil recovery field tests. *In* ''Proceedings of the Symposium on Applications of Microorganisms to Petroleum Technology'' (T. E. Burchfield and R. S. Bryant, Eds.), pp. VI 1–VI41. National Technical Information Service, Springfield, VA.

Hitzman, D. O., Dennis, M., and Hitzman, D. C. (2004). ''Recent Successes: MEOR Using Synergistic H_2S Prevention and Increased Oil Recovery Systems.'' SPE 89453. *Proceedings of Fourteenth symposium on Improved Oil Recovery*. Society of Petroleum Engineers, Richardson, Texas.

Hitzman, D. O., and Sperl, G. T. (1994). "New Microbial Technology for Enhanced Oil Recovery and Sulfide Prevention and Reduction." SPE/DOE 27752. *Proceedings of the 9th Symposium on Improved Oil Recovery*. Society of Petroleum Engineers, Richardson, Texas.

Hubert, C., Nemati, M., Jenneman, G. E., and Voordouw, G. (2003). Containment of biogenic sulfide production in continuous up-flow packed-bed bioreactors with nitrate or nitrite. *Biotechnol. Prog.* **19,** 338–345.

Hubert, C., Nemati, M., Jenneman, G. E., and Voordouw, G. (2005). Corrosion risk associated with microbial souring control using nitrate or nitrite. *Appl. Microbiol. Biotechnol.* **68,** 272–282.

Hubert, C., and Voordouw, G. (2007). Oil field souring control by nitrate-reducing *Sulfurospirillum* spp. that outcompete sulfate-reducing bacteria for organic electron donors. *Appl. Environ. Microbiol.* **73,** 2644–2652.

Hubert, C., Voordouw, G., Nemati, M., and Jenneman, G. E. (2004). Is souring and corrosion by sulfate-reducing bacteria in oil fields reduced more efficiently by nitrate or nitrite? Corrosion 04762. *Proceedings of Corrosion 2004.* NACE, Houston, Texas.

Huh, C., Lange, E. A., and Cannella, W. J. (1990). "Polymer Retention in Porous Media." SPE 20235. *Proceedings of SPE/DOE Enhanced Oil Recovery Symposium.* Society of Petroleum Engineers, Richardson, Texas.

Hutchinson, C. A. (1959). Reservoir inhomogeneity assessment and control. *Pet. Eng.* B19–B26.

Huu, N. B., Denner, E. B. M., Ha Dang, T. C., Wanner, G., and Stan-Lotter, H. (1999). *Marinobacter aquaeolei* sp. nov., a halophilic bacterium isolated from a Vietnamese oil-producing well. *Int. J. Syst. Bacteriol.* **49,** 367–375.

Ibatullin, R. R. (1995). A microbial enhanced oil recovery method modified for water-flooded strata. *Microbiology* **64,** 240–241.

Islam, M. R., and Gianetto, A. (1993). Mathematical modeling and scaling up of microbial enhanced oil recovery. *J. Can. Pet. Technol.* **32,** 30–36.

Ivanov, M. V., and Belyaev, S. S. (1983). Microbial activity in water-flooded oil fields and its possible regulation. *In* "Proceedings of the 1982 International Conference on Microbial Enhancement of Oil Recovery" (E. C. Donaldson and J. B. Clark, Eds.), pp. 48–57. U. S. Dept. of Energy, Bartellesville, Okla. (CONF-8205140).

Ivanov, M. V., Belyaev, S. S., Borzenkov, I. A., Glumov, I. F., and Ibatullin, R. R. (1993). Additional oil production during field trials in Russia. *Dev. Pet. Sci.* **39,** 373–381.

Jack, T. R. (1988). Microbially enhanced oil recovery. *Biorecovery* **1,** 59–73.

Jack, T. R., and DiBlasio, E. (1985). Selective plugging for heavy oil recovery. *In* "Microbes and Oil Recovery" (J. E. Zajic and E. C. Donaldson, Eds.), pp. 205–212. Bioresources, El Paso, Texas.

Jack, T. R., and Steheier, G. L. (1988). Selective plugging in watered out oil reservoirs. *In* "Proceedings of the Symposium on Applications of Microorganisms to Petroleum Technology" (T. E. Burchfield and R. S. Bryant, Eds.), pp. VII 1–VII 13. National Technical Information Service, Springfield, VA.

Jack, T. R., Thompson, B. G., and DiBlasio, E. (1983). The potential for use of microbes in the production of heavy oil. *In* "Proceedings of the 1982 International Conference on Microbial Enhancement of Oil Recovery" (E. C. Donaldson and J. B. Clark, Eds.), pp. 88–93. U. S. Dept. of Energy, Bartellesville, Okla. (CONF-8205140).

Jang, L.-K., Chang, P. W., Findley, J. E., and Yen, T. F. (1983). Selection of bacteria with favorable transport properties through porous rock for the application of microbial-enhanced oil recovery. *App. Environ. Microbiol.* **46,** 1066–1072.

Janssen, P. H. (2006). Identifying the dominant soil bacterial taxa in libraries of 16S rRNA and 16S rRNA genes. *Appl. Environ. Microbiol.* **72,** 1719–1728.

Jaranyi, I. (1968). Beszamolo a nagylengyel terzegeben elvegzett koolaj mikrobiologiai Kiserletkrol. *M. All. Foldtani Intezet Evi Jelentese A.* 423–426.

Javaheri, M., Jenneman, G. E., McInerney, M. J., and Knapp, R. M. (1985). Anaerobic production of a biosurfactant by *Bacillus licheniformis* JF-2. *Appl. Environ. Microbiol.* **50,** 698–700.

Jeanthon, C., Nercessian, O., Corre, E., and Graboeski-Lux, A. (2005). Hyperthermophilic and methanogenic archaea in oil fields. *In* "Petroleum Microbiology" (B. Ollivier and M. Magot, Eds.), pp. 55–70. ASM, Washington, DC.

Jenneman, G. E., and Clark, B. (1994a). Injection of phosphorus nutrient sources under acid conditions for subterranean microbial processes. U.S. Patent office. Patent No. 5341875.
Jenneman, G. E., and Clark, J. B. (1994b). Utilization of phosphite salts as nutrients for subterranean microbial processes. U.S. Patent office. Patent No. 5327967.
Jenneman, G. E., Gevertz, D., Davey, M. E., Clark, J. B., Wood, W. A., Stevens, J. C., and Tankersley, C. (1995). Development and application of microbial selective plugging processes. *In* "Proceedings of Fifth International Conference on Microbial Enhanced Oil Recovery and Related Problems for Solving Environmental Problems (CONF-9509173)" (R. S. Bryant and K. L. Sublette, Eds.), pp. 7–26. National Technical Information Service, Springfield, VA.
Jenneman, G. E., Knapp, R. M., McInerney, M. J., Menzie, D. E., and Revus, D. E. (1984). Experimental studies of *in-situ* microbial enhanced oil recovery. *Soc. Pet. Eng. J.* **24,** 33–37.
Jenneman, G. E., Lappan, R. E., and Webb, R. H. (2000). Bacterial profile modification with bulk dextran gels produced by *in-situ* growth and metabolism of *Leuconostoc* species. *Soc. Pet. Eng. J.* **5,** 466–473.
Jenneman, G. E., McInerney, M. J., and Knapp, R. M. (1985). Microbial penetration through nutrient-saturated Berea sandstone. *Appl. Environ. Microbiol.* **50,** 383–391.
Jenneman, G. E., McInerney, M. J., and Knapp, R. M. (1986). Effect of nitrate on biogenic sulfide production. *Appl. Environ. Microbiol.* **51,** 1205–1211.
Jenneman, G. E., McInerney, M. J., Knapp, R. M., Clark, J. B., Feero, J. M., Revus, D. E., and Menzie, D. E. (1983). A halotolerant, biosurfactant-producing *Bacillus* species potentially useful for enhanced oil recovery. *Dev. Ind. Microbiol.* **24,** 485–492.
Jenneman, G. E., Moffitt, P. D., Bala, G. A., and Webb, R. H. (1999). Sulfide removal in reservoir brine by indigenous bacteria. *Soc. Pet. Eng. Prod. Facil.* **14,** 219–225.
Jenneman, G. E., Moffitt, P. D., and Young, G. R. (1996). Application of a microbial selective-plugging process at the North Burbank unit: Prepilot tests. *SPE Prod. Facil.* **11,** 11–17.
Jinfeng, L., Lijun, M., Bozhong, M., Rulin, L., Fangtian, N., and Jiaxi, Z. (2005). The field pilot of microbial enhanced oil recovery in a high temperature petroleum reservoir. *J. Pet. Sci. Eng.* **48,** 265–271.
Johnson, S. J., Salehi, M., Eisert, K. E., Liang, J.-T., Bala, G. A., and Fox, S. L. (2007). "Biosurfactants Produced from Agricultural Process Waste Streams to Improve Recovery in Fractured Carbonate Reservoirs." SPE 106078. *Proceedings of SPE International Symposium on Oilfield Chemistry*. Society of Petroleum Engineers, Richardson, Texas.
Jones, D. M., Head, I. M., Gray, N. D., Adams, J. J., Rowan, A. K., Aitken, C. M., Bennett, B., Huang, H., Brown, A., Bowler, B. F., Oldenburg, T., Erdmann, M., *et al.* (2008). Crude-oil biodegradation via methanogenesis in subsurface petroleum reservoirs. *Nature* **451,** 176–180.
Joshi, S., Bharucha, C., and Desai, A. J. (2008a). Production of biosurfactant and antifungal compound by fermented food isolate *Bacillus subtilis* 20B. *Biores. Technol.* **99,** 4603–4608.
Joshi, S., Bharucha, C., Jha, S., Yadav, S., Nerurkar, A., and Desai, A. J. (2008b). Biosurfactant production using molasses and whey under thermophilic conditions. *Biores. Technol.* **99,** 195–199.
Kantzas, A., Burger, D., Pow, M., and Cheung, V. (1995). A study of improving reservoir conformance using polymer gels in producer wells. *In Situ* **19,** 41–87.
Karaskiewicz, J. (1975). Studies on increasing petroleum oil recovery from carpathian deposits using bacteria. *Nafta* (*Petroleum*) **21,** *144–149.*
Kashefi, K., and Lovley, D. R. (2003). Extending the upper temperature limit for life. *Science* **301,** 934.
Keller, M., and Zengler, K. (2004). Tapping into microbial diversity. *Nat. Rev. Microbiol.* **2,** 141–150.
Khachatoorian, R., Petrisor, I. G., Kwan, C.-C., and Yen, T. F. (2003). Biopolymer plugging effect: Laboratory-pressurized pumping flow studies. *J. Pet. Sci. Eng.* **38,** 13–21.

Kieft, T. L., Phelps, T. J., and Frederickson, J. K. (2007). Drilling, coring, and sampling subsurface environments. *In* "Manual of Environmental Microbiology" (C. J. Hurst, R. L. Crawford, J. L. Garland, D. A. Lipson, A. L. Mills and L. D. Stetzenbach, Eds.), pp. 799–817. ASM, Washington, DC.

Kielemoes, J., De Boever, P., and Verstraete, W. (2000). Influence of denitrification on the corrosion of iron and stainless steel powder. *Environ. Sci. Technol.* **34,** 663–671.

Kjellerup, B. V., Veeh, R. H., Sumithraratne, P., Thomsen, T. R., Buckingham-Meyer, K., Frolund, B., and Sturman, P. (2005). Monitoring of microbial souring in chemically treated, produced-water biofilm systems using molecular techniques. *J. Ind. Microbiol. Biotechnol.* **32,** 163–170.

Knapp, R. M., Chisholm, J. L., and McInerney, M. J. (1990). "Microbially Enhanced Oil Recovery." *Proceedings of the SPE/UH emerging technologies conference*, pp. 229–234. Institute for Improved Oil Recovery, University of Houston, Houston, Texas.

Knapp, R. M., McInerney, M. J., Coates, J. D., Chisholm, J. L., Menzie, D. E., and Bhupathiraju, V. K. (1992). "Design and Implementation of a Microbially Enhanced Oil Recovery Field Pilot, Payne County, Oklahoma." SPE 24818. *Proceedings of SPE Annual Technical Conference.* Society of Petroleum Engineers, Richardson, Texas.

Knapp, R. M., McInerney, M. J., Nagle, D. P., Folmsbee, M. M., and Maudgalya, S. (2002). Development of a microbially enhanced oil recovery process for the Delaware Childers Field, Nowata County. *Okla. Geol. Survey Circ.* **108,** 193–200.

Knapp, R. M., Silfanus, N. J., McInerney, M. J., Menzie, D. E., and Chisholm, J. L. (1991). Mechanisms of microbial enhanced oil recovery in high salinity core environments. *Am. Inst. Chem. Eng. Symp. Ser.* **87,** 134–140.

Kotlar, H. K., Wetzel, A., Thorne-Holst, M., Zotchev, S., and Ellingensen, T. (2007). "Wax Control by Biocatalytic Degradation in High-Paraffinic Crude Oils." SPE 106420. *Proceedings of International Symposium on Oilfield Chemistry.* Society of Petroleum Engineers, Richardson, Texas.

Kowalewski, E., Rueslatten, I., Boassen, T., Sunde, E., Stensen, J. A., Lillebo, B.-L.P., Bødtker, G., and Torsvik, T. (2005). "Analyzing Microbial Improved Oil Recovery Processes from Core Floods." SPE 10924. *Proceedings of International Petroleum Technology Conference.* Society of Petroleum Engineers, Richardson, Texas.

Kowalewski, E., Rueslatten, I., Steen, K. H., Bødtker, G., and Torsæter, O. (2006). Microbial improved oil recovery-bacterial induced wettability and interfacial tension effects on oil production. *J. Pet. Sci. Eng.* **52,** 275–286.

Krumholz, L., McKinley, J. P., Ulrich, G. A., and Suflita, J. M. (1997). Confined subsurface microbial communities in Cretaceous rock. *Nature* **386,** 64–66.

Kulik, E. S., Somov, Y. P., and Rozanova, E. P. (1985). Oxidation of hexadecane in a porous system with the formation of fatty acids. *Microbiologiya* **54,** 381–386.

Kuyukina, M. S., Ivshina, I. B., Makarov, S. O., Litvinenko, L. V., Cunningham, C. J., and Philp, J. C. (2005). Effect of biosurfactants on crude oil desorption and mobilization in a soil system. *Environ. Int.* **31,** 155–161.

Kuznetsov, S. I. (1962). "Geological Activity of Microorganisms (English Translation)." Consultants Bureau, New York, NY.

Kuznetsov, S. I., Ivanov, M. V., and Lyalikova, N. N. (1963). "Introduction to Geological Microbiology (English Translations)." McGraw-Hill, New York, NY.

Lappan, R. E., and Fogler, H. S. (1992). Effect of bacterial polysaccharide production on formation damage. *Soc. Pet. Eng. Prod. Eng* **7,** 167–171.

Lappan, R. E., and Fogler, H. S. (1994). *Leuconostoc mesenteroides* growth kinetics with application to bacterial profile modification. *Biotechnol. Bioeng.* **43,** 865–873.

Lappan, R. E., and Fogler, H. S. (1996). Reduction of porous media permeability from *in situ Leuconostoc mesenteriodes* growth and dextran production. *Biotechnol. Bioeng.* **50,** 6–15.

Lappin-Scott, H. M., Cusack, F., and Costerton, J. W. (1988). Nutrient resuscitation and growth of starved cells in sandstone cores: A novel approach to enhanced oil recovery. *Appl. Environ. Microbiol.* **54,** 1373–1382.

Larsen, J., Rod, M. H., and Zwolle, S. (2004). "Prevention of Reservoir Sourcing in the Halfdan Field by Nitrite Injection." Corrosion 04761. *Proceedings of Corrosion 2004.* NACE, Houston, Texas.

Lazar, I. (1987). Research on the microbiology of MEOR in Romania. *In* "Proceedings of the First International MEOR Workshop" (J. W. King and D. A. Stevens, Eds.), pp. 124–153. National Technical Information Service, Springfield, VA.

Lazar, I. (1991). MEOR field trials carried out over the world during the last 35 years. *Dev. Pet. Sci.* **31,** 485–530.

Lazar, I. (1992). "MEOR, the Suitable Bacterial Inoculum According to the Kind of Technology Used: Results from Romania's Last 20 Years Experience." SPE/DOE 24207. *Proceedings of SPE/DOE Eighth Symposium on Enhanced Oil Recovery.* Society of Petroleum Engineers, Richardson, Texas.

Lazar, I. (1993). The microbiology of MEOR: Practical experience in Europe. *Biohydrometall. Technol. Proc. Int. Biohydrometall. Symp.* **2,** 329–338.

Lazar, I. (1998). Microbial systems for enhancement of oil recovery used in Romanian oil fields. *Miner. Process. Extr. Metall. Rev.* **19,** 379–393.

Lazar, I., and Constantinescu, P. (1985). Field trial results of microbial enhanced oil recovery. *In* "Microbes and Oil Recovery" (J. E. Zajic and E. C. Donaldson, Eds.), pp. 122–143. Bioresources, El Paso, TX.

Lazar, I., Dobrota, S., and Stefanescu, M. (1988). Some considerations concerning nutrient support injected into reservoirs subjected to microbiological treatment. *In* "Proceedings of the Symposium on Applications of Microorganisms to Petroleum Technology" (T. E. Burchfield and R. S. Bryant, Eds.), pp. XIV 1–XIV 6. National Technical Information Service, Springfield, VA.

Lazar, I., Dobrota, S., Stefanescu, M., Sandulescu, L., Constantinescu, P., Morosanu, C., Botea, N., and Iliescu, O. (1991). Preliminary results of some recent MEOR field trials in Romania. *Dev. Pet. Sci.* **33,** 365–386.

Lazar, I., Dobrota, S., Stefanescu, M. C., Sandulescu, L., Paduraru, R., and Stefanescu, M. (1993). MEOR, recent field trials in Romania: Reservoir selection, type of inoculum, protocol for well treatment and line monitoring. *Dev. Pet. Sci.* **39,** 265–288.

Lazar, I., Voicu, A., Nicolescu, C., Mucenica, D., Dobrota, S., Petrisor, I. G., Stefanescu, M., and Sandulescu, L. (1999). The use of naturally occurring selectively isolated bacteria for inhibiting paraffin deposition. *J. Pet. Sci. Eng.* **22,** 161–169.

Leu, J. Y., McGovern-Traa, C. P., Porter, A. J. R., and Hamilton, W. A. (1999). The same species of sulphate-reducing *Desulfomicrobium* occur in different oil field environments in the North Sea. *Lett. Appl. Microbiol.* **29,** 246–252.

L'Haridon, S., Miroshnichenko, M. L., Hippe, H., Fardeau, M. L., Bonch-Osmolovskaya, E., Stackebrandt, E., and Jeanthon, C. (2001). *Thermosipho geolei* sp. nov., a thermophilic bacterium isolated from a continental petroleum reservoir in Western Siberia. *Int. J. Syst. Evol. Microbiol.* **51,** 1327–1334.

L'Haridon, S., Miroshnichenko, M. L., Hippe, H., Fardeau, M. L., Bonch-Osmolovskaya, E. A., Stackebrandt, E., and Jeanthon, C. (2002). *Petrotoga olearia* sp. nov. and *Petrotoga sibirica* sp. nov., two thermophilic bacteria isolated from a continental petroleum reservoir in Western Siberia. *Int. J. Syst. Evol. Microbiol.* **52,** 1715–1722.

L'Haridon, S., Reysenbach, A. L., Glenat, P., Prieur, D., and Jeanthon, C. (1995). Hot subterranean biosphere in a continental oil reservoir. *Nature* **377,** 223–224.

Li, Q., Kang, C., Wang, H., Liu, C., and Zhang, C. (2002). Application of microbial enhanced oil recovery technique to the Daqing Oilfield. *Biochem. Eng. J.* **11,** 197–199.

Li, H., Yang, S.-Z., and Mu, B.-Z. (2007a). Phylogenetic diversity of the archaeal community in a continental high-temperature, water-flooded petroleum reservoir. *Curr. Microbiol.* **55,** 382–388.

Li, H., Yang, S.-Z., Mu, B.-Z., Rong, Z.-F., and Zhang, J. (2006). Molecular analysis of the bacterial community in a continental high-temperature and water-flooded petroleum reservoir. *FEMS Microbiol. Lett.* **257,** 92–98.

Li, H., Yang, S.-Z., Mu, B.-Z., Rong, Z.-F., and Zhang, J. (2007b). Molecular phylogenetic diversity of the microbial community associated with a high-temperature petroleum reservoir at an offshore oilfield. *FEMS Microbiol. Ecol.* **60,** 74–84.

Lien, T., and Beeder, J. (1997). *Desulfobacter vibrioformis* sp. nov., a sulfate reducer from a water-oil separation system. *Int. J. Syst. Bacteriol.* **47,** 1124–1128.

Lien, T., Madsen, M., Rainey, F. A., and Birkeland, N. K. (1998a). *Petrotoga mobilis* sp. nov., from a North Sea oil-production well. *Int. J. Syst. Bacteriol.* **48,** 1007–1013.

Lien, T., Madsen, M., Steen, I. H., and Gjerdevik, K. (1998b). *Desulfobulbus rhabdoformis* sp. nov., a sulfate reducer from a water-oil separation system. *Int. J. Syst. Bacteriol.* **48,** 469–474.

Lin, S.-C., Minton, M. A., Sharma, M. M., and Georgiou, G. (1994). Structural and immunological characterization of a biosurfactant produced by *Bacillus licheniformis* JF-2. *Appl. Environ. Microbiol.* **60,** 31–38.

Lin, L. H., Wang, P. L., Rumble, D., Lippmann-Pipke, J., Boice, E., Pratt, M. L., Lollar, B. S., Brodie, E. L., Hazen, T. C., Andersen, G. L., DeSantis, T. Z., Moser, D. P., *et al.* (2007). Long-term sustainability of a high-energy, low-diversity crustal biome. *Science* **314,** 497–482.

Londry, K. L., Suflita, J. M., and Tanner, R. S. (1999). Cresol metabolism by the sulfate-reducing bacterium *Desulfotomaculum* sp. strain Groll. *Can. J. Microbiol.* **45,** 458–463.

Lozupone, C. A., and Knight, R. (2007). Global patterns in bacterial diversity. *Proc. Natl. Acad. Sci. USA* **104,** 11436–11440.

Ludwig, W., Strunk, O., Westram, R., Richter, L., Meier, H., Kumar, Y., Buchner, A., Lai, T., Steppi, S., Jobb, G., Förster, W., Brettske, I., *et al.* (2004). ARB: A software environment for sequence data. *Nucleic Acids Res.* **32,** 1363–1371.

Lundquist, A., Cheney, D., Powell, C. L., O'Niell, P., Norton, G., Veneman, A. M., Evans, D. L., Minda, N. Y., Abraham, S., Allbaugh, J. M., Whitman, C. T., Bolten, J. A., *et al.* (2001). Energy for a new century: Increasing domestic energy supplies. *In* "National Energy Policy. Report of the National Energy Policy Development Group," pp. 69–90. U.S. Government printing office, Washington, DC.

Luton, P. E., Wayne, J. M., Sharp, R. J., and Riley, P. W. (2002). The *mcrA* gene as an alternative to 16S rRNA in the phylogenetic analysis of methanogen populations in landfill. *Microbiology* **148,** 3521–3530.

MacLeod, F. A., Lappin-Scott, H. M., and Costerton, J. W. (1988). Plugging of a model rock system by using starved bacteria. *Appl. Environ. Microbiol.* **54,** 1365–1372.

Maezumi, S., Ono, K., Shutao, H. K., Chao, Z. S., Enomoto, H., and Hong, C.-X. (1998). "The MEOR Field Pilot in Fuyu Oilfield, China: A Status Review." *Proceedings of 19th IEA Workshop and Symposium, Carmel, CA.* International Energy Agency, Paris, France.

Magot, M. (2005). Indigenous microbial communities in oil fields. *In* "Petroleum microbiology" (B. Ollivier and M. Magot, Eds.), pp. 21–34. ASM, Washington, DC.

Magot, M., Basso, O., Tardy-Jacquenod, C., and Caumette, P. (2004). *Desulfovibrio bastinii* sp. nov. and *Desulfovibrio gracilis* sp. nov., moderately halophilic, sulfate-reducing bacteria isolated from deep subsurface oilfield water. *Int. J. Syst. Evol. Microbiol.* **54,** 1693–1697.

Magot, M., Caumette, P., Desperrier, J. M., Matheron, R., Dauga, C., Grimont, F., and Carreau, L. (1992). *Desulfovibrio longus* sp. nov., a sulfate-reducing bacterium isolated from an oil-producing well. *Int. J. Syst. Bacteriol.* **42,** 398–403.

Magot, M., Fardeau, M.-L., Arnauld, O., Lanau, C., Ollivier, B., Thomas, P., and Patel, B. K. C. (1997a). *Spirochaeta smaragdinae* sp. nov., a new mesophilic strictly anaerobic spirochete from an oil field. *FEMS Microbiol. Lett.* **155,** 185–191.

Magot, M., Ollivier, B., and Patel, B. K. C. (2000). Microbiology of petroleum reservoirs. *Antonie van Leeuwenhoek* **77,** 103–116.

Magot, M., Ravot, G., Campaignolle, X., Ollivier, B., Patel, B. K., Fardeau, M. L., Thomas, P., Crolet, J. L., and Garcia, J. L. (1997b). *Dethiosulfovibrio peptidovorans* gen. nov., sp. nov., a new anaerobic, slightly halophilic, thiosulfate-reducing bacterium from corroding offshore oil wells. *Int. J. Syst. Bacteriol.* **47,** 818–824.

Maier, R. M., and Soberon-Chavez, G. (2000). *Pseudomonas aeruginosa* rhamnolipids: Biosynthesis and potential applications. *Appl. Microbiol. Biotechnol.* **54,** 625–633.

Makkar, R. S., and Cameotra, S. S. (1997). Utilization of molasses for biosurfactant production by two *Bacillus* strains at thermophilic conditions. *J. Am. Oil Chem. Soc.* **74,** 887–889.

Makkar, R. S., and Cameotra, S. S. (1998). Production of biosurfactant at mesophilic and thermophilic conditions by a strain of *Bacillus subtilis*. *J. Ind. Microbiol. Biotechnol.* **20,** 48–52.

Malachosky, E., and Herd, M. (1986). Polymers reduce water production. *Pet. Eng. Int.* **58,** 48.

Maneerat, S. (2005). Production of biosurfactants using substrates from renewable-resources. *Songklanakarin J. Sci. Technol.* **27,** 675–683.

Marsh, T. L., Zhang, X., Knapp, R. M., McInerney, M. J., Sharma, P. K., and Jackson, B. E. (1995). Mechanisms of microbial oil recovery by *Clostridium acetobutylicum* and *Bacillus* strain JF-2. *In* "Proceedings of Fifth International Conference on Microbial Enhanced Oil Recovery and Related Problems for Solving Environmental Problems (CONF-9509173)" (R. S. Bryant and K. L. Sublette, Eds.), pp. 593–610. National Technical Information Service, Springfield, VA.

Matz, A. A., Borisov, A. Y., Mamedov, Y. G., and Ibatulin, R. R. (1992). "Commercial (pilot) Test of Microbial Enhanced Oil Recovery Methods." SPE/DOE 24208. *Proceedings of the Eighth Symposium on Enhanced Oil Recovery*. Society of Petroleum Engineers, Richardson, Texas.

Maudgalya, S., Knapp, R. M., and McInerney, M. J. (2007). "Microbial Enhanced Oil Recovery: A Review of the Past, Present and Future." SPE 106978. *Proceedings of Society of Petroleum Engineers Production and Operations Symposium*. Society of Petroleum Engineers, Richardson, Texas.

Maudgalya, S., Knapp, R. M., McInerney, M. J., Nagle, D. P., and Folsmbee, M. J. (2004). "Development of a Bio-Surfactant-Based Enhanced Oil Recovery Procedure." SPE 89473. *Proceedings of the SPE/DOE Improved Oil Recovery Symposium*. Society of Petroleum Engineers, Richardson, Texas.

Maudgalya, S., McInerney, M. J., Knapp, R. M., Nagle, D. P., and Folmsbee, M. M. (2005). "Tertiary Oil Recovery with Microbial Biosurfactant Treatment of Low-Permeability Berea Sandstone Cores." SPE 94213. *Proceedings of Society of Petroleum Engineers Production and Operations Symposium*. Society of Petroleum Engineers, Richardson, Texas.

Maure, A., Dietrich, F., Diaz, V. A., and Arganaraz, H. (1999). "Microbial Enhanced Oil Recovery Pilot Test in Piedras Coloradas Field, Argentina." SPE 53715. *Proceedings of SPE Latin American and Caribbean Petroleum Engineering Conference*. Society of Petroleum Engineers, Richardson, Texas.

Maure, A., Dietrich, F., Gómez, U., Vallesi, J., and Irusta, M. (2001). "Waterflooding Optimization Using Biotechnology: 2-Year Field Test, La Ventana Field, Argentina." SPE 69652. *Proceedings of SPE Latin American and Caribbean Petroleum Engineering Conference*. Society of Petroleum Engineers, Richardson, Texas.

Maure, A., Saldana, A. A., and Juarez, A. R. (2005). "Biotechnology Application to EOR in Talara Off-Shore Oil Fields, Northwest Peru." SPE 94934. *Proceedings of SPE Latin American and Caribbean Petroleum Engineering Conference*. Society of Petroleum Engineers, Richardson, Texas.

McInerney, M. J., Bhupathiraju, V. K., and Sublette, K. L. (1992). Evaluation of microbial method to reduce hydrogen sulfide levels in porous rock biofilm. *J. Ind. Microbiol.* **11,** 53–58.

McInerney, M. J., Duncan, K. E., Youssef, N., Fincher, T., Maudgalya, S. K., Folmsbee, M. J., Knapp, R., Simpson, D. R., Ravi, N., and Nagle, D. P. (2005b). "Development of Microorganisms with Improved Transport and Biosurfactant Activity for Enhanced Oil Recovery." Report to the Department of Energy, DE-FE-02NT15321, Washington, DC.

McInerney, M. J., Javaheri, M., and Nagle, D. P. (1990). Properties of the biosurfactant produced by *Bacillus licheniformis* strain JF-2. *J. Ind. Microbiol.* **5,** 95–102.

McInerney, M. J., Jenneman, G. E., Knapp, R. M., and Menzie, D. E. (1985a). *In situ* microbial plugging process for subterrranean formations. U.S. Patent office. Patent No. 4,558,739.

McInerney, M. J., Jenneman, G. E., Knapp, R. M., and Menzie, D. E. (1985b). Biosurfactant and enhanced oil recovery. US patent Office. Patent No. 4,522,261.

McInerney, M. J., Maudgalya, S., Nagle, D. P., and Knapp, R. M. (2002). Critical assessment of the use of microorganisms for oil recovery. *Recent Res. Dev. Microbiol.* **6,** 269–284.

McInerney, M. J., Nagle, D. P., and Knapp, R. M. (2005a). Microbially enhanced oil recovery: Past, present, and future. *In* "Petroleum Microbiology" (B. Ollivier and M. Magot, Eds.), pp. 215–238. ASM, Washington, DC.

McInerney, M. J., and Sublette, K. L. (1997). Petroleum microbiology: Biofouling, souring, and improved oil recovery. *In* "Manual of Environmental Microbiology" (C. J. Hurst, G. R. Knudsen, M. J. McInerney, L. D. Stetzenbach and M. V. Walter, Eds.), pp. 600–607. ASM, Washington, DC.

McInerney, M. J., Sublette, K. L., and Montgomery, A. D. (1991). Microbial control of the production of sulfide. *Dev. Pet. Sci.* **31,** 441–449.

McInerney, M. J., Voordouw, G., Jenneman, G. E., and Sublette, K. L. (2007). Oil field microbiology. *In* "Manual of Environmental Microbiology" (C. J. Hurst, R. L. Crawford, J. L. Garland, D. A. Lipson, A. L. Mills and L. D. Stetzenbach, Eds.), pp. 898–911. ASM, Washington, DC.

McInerney, M. J., Wofford, N. Q., and Sublette, K. L. (1996). Microbial control of hydrogen sulfide production in a porous medium. *Appl. Biochem. Biotechnol.* **57–58,** 933–944.

Miranda-Tello, E., Fardeau, M. L., Fernandez, L., Ramirez, F., Cayol, J. L., Thomas, P., Garcia, J. L., and Ollivier, B. (2003). *Desulfovibrio capillatus* sp. nov., a novel sulfate-reducing bacterium isolated from an oil field separator located in the Gulf of Mexico. *Anaerobe* **9,** 97–103.

Miranda-Tello, E., Fardeau, M.-L., Joulian, C., Magot, M., Thomas, P., Tholozan, J.-L., and Ollivier, B. (2007). *Petrotoga halophila* sp. nov., a thermophilic, moderately halophilic, fermentative bacterium isolated from an offshore oil well in Congo. *Int. J. Syst. Evol. Microbiol.* **57,** 40–44.

Miranda-Tello, E., Fardeau, M.-L., Thomas, P., Ramirez, F., Casalot, L., Cayol, J.-L., Garcia, J.-L., and Ollivier, B. (2004). *Petrotoga mexicana* sp. nov., a novel thermophilic, anaerobic and xylanolytic bacterium isolated from an oil-producing well in the Gulf of Mexico. *Int. J. Syst. Evol. Microbiol.* **54,** 169–174.

Misra, S., Baruah, S., and Singh, K. (1995). Paraffin problems in crude oil production and transportation: A review. *Soc. Pet. Eng. Prod. Facil.* **10,** 50–54.

Montgomery, A. D., McInerney, M. J., and Sublette, K. L. (1990). Microbial control of the production of hydrogen sulfide by sulfate-reducing bacteria. *Biotechnol. Bioeng.* **35,** 533–539.

Mormile, M. R., Biesen, M. A., Gutierrez, M. C., Ventosa, A., Pavlovich, J. B., Onstott, T. C., and Fredrickson, J. K. (2003). Isolation of *Halobacterium salinarum* retrieved directly from halite brine inclusions. *Environ. Microbiol.* **5,** 1094–1102.

Moses, V., Brown, M. F., Burton, C. C., Gralla, D. S., and Cornelius, C. (1993). Microbial hydraulic acid fracturing. *Dev. Pet. Sci.* **39,** 207–229.

Mukherjee, S., Das, P., and Sen, R. (2006). Towards commercial production of microbial surfactants. *Trends Biotechnol.* **24,** 509–515.

Muller, J. A., Galushko, A. S., Kappler, A., and Schink, B. (2001). Initiation of anaerobic degradation of *p*-cresol by formation of 4-hydroxybenzylsuccinate in *Desulfobacterium cetonicum. J. Bacteriol.* **183,** 752–757.

Muller, J. A., Galushko, A. S., Kappler, A., and Shink, B. (1999). Anaerobic degradation of *m*-cresol by *Desulfobacterium cetonicum* is initiated by formation of 3-hydroxybenzylsuccinate. *Arch. Microbiol.* **172,** 287–294.

Murygina, V. P., Mats, A. A., Arinbassrov, M. U., Salamov, Z. Z., and Cherkasov, A. B. (1995). Oil field experiments of microbial improved oil recovery in Vyngapour, West Siberia, Russia. *In* ''Proceedings of Fifth International Conference on Microbial Enhanced Oil Recovery and Related Problems for Solving Environmental Problems (CONF-9509173)'' (R. S. Bryant and K. L. Sublette, Eds.), pp. 87–94. National Technical Information Service, Springfield, VA.

Musslewhite, C. L., Swift, D. J. P., Gilpen, J., and McInerney, M. J. (2007). Spatial variability of sulfate reduction in a shallow aquifer. *Environ. Microbiol.* **9,** 2810–2819.

Myhr, S., Lillebo, B.-L. P., Sunde, E., Beeder, J., and Torsvik, T. (2002). Inhibition of microbial H_2S production in oil reservoir model column by nitrate injection. *Appl. Microbiol. Biotechnol.* **58,** 400–408.

Myhr, S., and Torsvik, T. (2000). *Denitrovibrio acetiphilus,* a novel genus and species of dissimilatory nitrate-reducing bacterium isolated from an oil reservoir model column. *Int. J. Syst. Evol. Microbiol.* **50,** 1611–1619.

Nagase, K., Zhang, S. T., Asami, H., Yazawa, N., Fujiwara, K., Enomoto, H., Hong, C. X., and Laing, C. X. (2001). "Improvement of Sweep Efficiency by Microbial EOR Process in Fuyu Oilfield, China." SPE 68720. *Proceedings of SPE Asia Pacific Oil and gas Conference and Exhibition.* Society of Petroleum Engineers, Richardson, Texas.

Nagase, K., Zhang, S. T., Asami, H., Yazawa, N., Fujiwara, K., Enomoto, H., Hong, C. X., and Laing, C. X. (2002). "A Successful Test of Microbial EOR Process in Fuyu Oilfeild, China." SPE 75238. *Proceedings of SPE/DOE Improved Oil Recovery Symposium.* Society of Petroleum Engineers, Richardson, Texas.

Nakano, M. M., Dailly, Y. P., Zuber, P., and Clark, D. P. (1997). Characterization of anaerobic fermentative growth of *Bacillus subtilis*: Identification of fermentation end products and genes required for growth. *J. Bacteriol.* **179,** 6749–6755.

Nazina, T. N., Griror'yan, A. A., Feng, Q., Shestakova, N. M., Babich, T. L., Pavlova, N. K., Ivoilov, V. S., Ni, F., Wang, J., She, Y., Xiang, T., Mai, B., *et al.* (2007a). Microbiological and production characteristics of the high-temperature Kongdian petroleum reservoir revealed during field trial of biotechnology for enhancement of oil recovery. *Microbiology* **76,** 297–309.

Nazina, T. N., Griror'yan, A. A., Shestakova, N. M., Babich, T. L., Ivoilov, V. S., Feng, Q., Ni, F., Wang, J., She, Y., Xiang, T., Luo, Z., Belyaev, S. S., *et al.* (2007b). Microbiological investigations of high-temperature horizons of the Kongdian petroleum reservoir in connection with field trial of a biotechnology for enhancement of oil recovery. *Microbiology* **76,** 287–296.

Nazina, T. N., Ivanova, A. E., Borzenkov, I. A., Belyaev, S. S., and Ivanov, M. V. (1995a). Occurrence and geochemical activity of microorganisms in high-temperature, water-flooded oil fields of Kazakhstan and western Siberia. *Geomicrobiol. J.* **13,** 181–192.

Nazina, T. N., Ivanova, A. E., Golubeva, O. V., Ibatullin, R. R., Belyaev, S. S., and Ivanov, M. V. (1995b). Occurrence of sulfate- and iron-reducing bacteria in stratal waters of the Romashkinskoe oil field. *Microbiology (Eng. Tr.)* **64,** 203–208.

Nazina, T. N., Ivanova, A. E., Ivoilov, V. S., Kandaurova, G. F., Ibatullin, R. R., Belyaev, S. S., and Ivanov, M. V. (1999a). Microbiological investigation of the stratal waters of the Romashkinskoe oil field in the course of a trial of a biotechnology for the enhancement of oil recovery. *Microbiology (Eng. Tr.)* **68,** 214–221.

Nazina, T. N., Ivanova, A. E., Ivoilov, V. S., Miller, Y. M., Kandaurova, G. F., Ibatullin, R. R., Belyaev, S. S., and Ivanov, M. V. (1999b). Results of the trial of the microbiological method for the enhancement of oil recovery at the carbonate collector of the Romashkinskoe oil field: Biogeochemical and productional characteristics. *Microbiology (Eng. Tr.)* **68,** 222–226.

Nazina, T. N., Ivanova, A. E., Kanchaveli, L. P., and Rozanova, E. P. (1988). A new spore-forming thermophilic methylotrophic sulfate-reducing bacterium, *Desulfotomaculum Kuznetsovii*. *Microbiology (Eng. Tr.)* **57,** 659–663.

Nazina, T. N., Rozanova, E. P., and Kuznetsov, S. I. (1985). Microbial oil transformation processes accompanied by methane and hydrogen-sulfide formation. *Geomicrobiol. J.* **4,** 103–130.

Nazina, T. N., Tourova, T. P., Poltaraus, A. B., Novikova, E. V., Grigoryan, A. A., Ivanova, A. E., Lysenko, A. M., Petrunyaka, V. V., Osipov, G. A., Belyaev, S. S., and Ivanov, M. V. (2001). Taxonomic study of aerobic thermophilic bacilli: Descriptions of *Geobacillus subterraneus* gen. nov., sp. nov. and *Geobacillus uzenensis* sp. nov. from petroleum reservoirs and transfer of *Bacillus stearothermophilus, Bacillus thermocatenulatus, Bacillus thermoleovorans, Bacillus kaustophilus, Bacillus thermoglucosidasius* and *Bacillus thermodenitrificans* to *Geobacillus* as the new combinations *G. stearothermophilus, G. thermocatenulatus, G. thermoleovorans, G. kaustophilus, G. thermoglucosidasius* and *G. thermodenitrificans*. *Int. J. Syst. Evol. Microbiol.* **51,** 433–446.

Nazina, T. N., Xue, Y.-F., Wang, X.-Y., Belyaev, S. S., and Ivanov, M. V. (2000a). Microorganisms of the high-temperature Liaohe oil field of China and their potential for MEOR. *Resource Environ. Biotechnol* **3,** 149–160.

Nazina, T. N., Xue, Y.-F., Wang, X.-Y., Grigoriyan, A. A., Ivoilov, V. S., Belyaev, S. S., and Ivanov, M. V. (2000b). Diversity and activity of microorganisms in the Daqing oil field of China and their potential for biotechnological applications. *Resource Environ. Biotechnol.* **3,** 161–172.

Nelson, S. J., and Launt, P. D. (1991). Stripper well production increased with MEOR treatment. *Oil Gas J.* **89,** 114–116.

Nelson, L., and Schneider, D. R. (1993). Six years of paraffin control and enhanced oil recovery with the microbial product, Para-BacTM. *Dev. Pet. Sci.* **39,** 355–362.

Nemati, M., Jenneman, G. E., and Voordouw, G. (2001a). Mechanistic study of microbial control of hydrogen sulfide production in oil reservoirs. *Biotechnol. Bioeng.* **74,** 424–434.

Nemati, M., Jenneman, G. E., and Voordouw, G. (2001c). Impact of nitrate-mediated microbial control of souring in oil reservoirs on the extent of corrosion. *Biotechnol. Prog.* **17,** 852–859.

Nemati, M., Mazutinec, T. J., Jenneman, G. E., and Voordouw, G. (2001b). Control of biogenic H_2S production with nitrite and molybdate. *J. Ind. Microbiol. Biotechnol.* **26,** 350–355.

Nesbo, C. L., Dlutek, M., Zhaxybayeva, O., and Doolittle, W. F. (2006). Evidence for existence of "Mesotogas," members of the order *Thermotogales* adapted to low-temperature environments *Appl. Environ. Microbiol.* **72,** 5061–5068.

Ng, T. K., Weimer, P. J., and Gawel, L. J. (1989). Possible nonanthropogenic origin of two methanogenic isolates from oil-producing wells in the San Miguelito field, Ventura county, California. *Geomicrobiol. J.* **7,** 185–192.

Nga, D. P., Ha, D. T. C., Hien, L. T., and Stan-Lotter, H. (1996). *Desulfovibrio vietnamensis* sp. nov., a halophilic sulfate-reducing bacterium from Vietnamese oil fields. *Anaerobe* **2,** 385–392.

Nguyen, T., Youssef, N., McInerney, M. J., and Sabatini, D. (2008). Rhamnolipid biosurfactant mixtures for environmental remediation. *Water Res.* **42,** 1735–1743.

Ni, S. S., and Boone, D. R. (1991). Isolation and characterization of a dimethyl sulfide-degrading methanogen, *Methanolobus siciliae* HI350, from an oil well, characterization of *M. siciliae* T4/MT, and emendation of *M. siciliae*. *Int. J. Syst. Bacteriol.* **41,** 410–416.

Nilsen, R. K., and Torsvik, T. (1996a). *Methanococcus thermolithotrophicus* isolated from North sea oil field reservoir water. *Appl. Environ. Microbiol.* **62,** 728–731.

Nilsen, R. K., Torsvik, T., and Lien, T. (1996b). *Desulfotomaculum thermocisternum* sp. nov., a sulfate reducer isolated from a hot North Sea oil reservoir. *Int. J. Syst. Bacteriol.* **46,** 397–402.

Nourani, M., OPanahi, H., Biria, D., Roosta, R., Haghighi, M., and Mohebbi, M. (2007). "Laboratory Studies of MEOR in the Micromodel as a Fractured System." SPE 110988. *Proceedings of Eastern Regional Meeting*. Society of Petroleum Engineers, Richardson, Texas.

Obrazstova, A. Y., Shipin, O. V., Bezrukova, L. V., and Belyaev, S. S. (1988). Properties of the coccoid methylotrophic methanogen. *Microbiology (Eng. Tr.)* **56,** 523–527.

Obraztsova, A. Y., Tsyban, V. E., Vichus, K. S. L., Bezrukova, L. V., and Belyaev, S. S. (1987). Biological properties of *Methanosarcina* not utilizing carbonic acid and hydrogen. *Microbiology (Eng. Tr.)* **56,** 807–812.

Ohno, K., Maezumi, S., Sarma, H. K., Enomoto, H., Hong, C., Zhou, S. C., and Fujiwara, K. (1999). "Implementation and Performance of a Microbial Enhanced Oil Recovery Field Pilot in Fuyu Oilfield, China." SPE 54328. *Proceedings of the SPE Asia Pacific Oil and Gas Conferecne and Exhibition*. Society of Petroleum Engineers, Richardson, Texas.

Okpokwasili, G. C., and Ibiene, A. A. (2006). Enhancement of recovery of residual oil using a biosurfactant slug. *Afr. J. Biotechnol.* **5,** 453–456.

Ollivier, B., Cayol, J. L., Patel, B. K. C., Magot, M., Fardeau, M. L., and Garcia, J. L. (1997). *Methanoplanus petrolearius* sp. nov., a novel methanogenic bacterium from an oil producing well. *FEMS Microbiol. Lett.* **147,** 51–56.

Ollivier, B., Fardeau, M. L., Cayol, J. L., Magot, M., Patel, K. C., Prensier, G., and Garcia, J. L. (1998). *Methanocalculus halotolerans* gen. nov., sp. nov., isolated from an oil-producing well. *Int. J. Syst. Bacteriol.* **48,** 821–828.

Ollivier, B., and Magot, M. (2005). "Petroleum Microbiology." ASM, Washington, DC.

Ommedal, H., and Torsvik, T. (2007). *Desulfotignum toluenicum* sp. nov., a novel toluene-degrading, sulphate-reducing bacterium isolated from an oil-reservoir model column. *Int. J. Syst. Evol. Microbiol.* **57,** 2865–2869.

Orphan, V. J., Goffredi, S. K., DeLong, E. F., and Boles, J. R. (2003). Geochemical influence on diversity and microbial processes in high temperature oil reservoirs. *Geomicrobiol. J.* **20,** 295–311.

Orphan, V. J., Taylor, L. T., Hafenbradl, D., and Delong, E. F. (2000). Culture-dependent and culture-independent characterization of microbial assemblages associated with high-temperature petroleum reservoirs. *Appl. Environ. Microbiol.* **66,** 700–711.

Pace, N. R. (1997). A molecular view of microbial diversity and the biosphere. *Science* **276,** 734–740.

Partidas, C. J., Trebbau, G., and Smith, T. L. (1998). Microbes aid heavy oil recovery in Venezuela. *Oil Gas J.* **96,** 62–64.

Peihui, H., Fengrong, S., and Mei, S. (2001). "Microbial EOR Laboratory Studies on the Microorganisms Using Petroleum Hydrocarbon as a Sole Carbon Source." SPE 72128. *Proceedings of SPE Asia Pacific Improved Oil Recovery Conference*. Society of Petroleum Engineers, Richardson, Texas.

Pelger, J. W. (1991). Microbial enhanced oil recovery treatments and wellbore stimulation using microorganisms to control paraffin, emulsion, corrosion, and scale formation. *Dev. Pet. Sci.* **31,** 451–466.

Petri, R., Podgorsek, L., and Imhoff, J. F. (2001). Phylogeny and distribution of the soxB gene among thiosulfate-oxidizing bacteria. *FEMS Microbiol. Lett.* **197,** 171–178.

Petzet, G. A., and Williams, R. (1986). Operators trim basic EOR research. *Oil Gas J.* **84,** 41–46.

Peypoux, F., and Michel, G. (1992). Controlled biosynthesis of Val7- and Leu7-surfactins. *Appl. Microbiol. Biotechnol.* **36,** 515–517.

Planckaert, M. (2005). Oil reservoirs and oil production. *In* "Petroleum Microbiology" (B. Ollivier and M. Magot, Eds.), pp. 3–20. ASM, Washington, DC.

Portwood, J. T. (1995a). A commercial microbial enhanced oil recovery process: Statistical evaluation of a multi-project database. *In* "The Fifth International Conference on Microbial Enhanced Oil Recovery and Related Problems for Solving Environmental Problems (CONF-9509173)" (R. S. Bryant and K. L. Sublette, Eds.), pp. 51–76. National Technical Information Service, Springfield, VA.

Portwood, J. T. (1995b). "A Commercial Microbial Enhanced Oil Recovery Technology: Evaluation of 322 Projects." SPE 29518. *Proceedings of the SPE Production Operation Symposium*. Society of Petroleum Engineers, Richardson, Texas.

Portwood, J. T., and Hiebert, F. K. (1992). "Mixed Culture Microbial Enhanced Waterflood: Tertiary MEOR Case Study." SPE 24820. *Proceedings of SPE Annual Technical Conference*. Society of Petroleum Engineers, Richardson, Texas.

Raiders, R. A., Freeman, D. C., Jenneman, G. E., Knapp, R. M., McInerney, M. J., and Menzie, D. E. (1985). "The Use of Microrganisms to Increase the Recovery of Oil from Cores." SPE 14336. *Proceedings of SPE Annual Technical Conference*. Society of Petroleum Engineers, Richardson, TX.

Raiders, R. A., Knapp, R. M., and McInerney, M. J. (1989). Microbial selective plugging and enhanced oil recovery. *J. Ind. Microbiol.* **4,** 215–230.

Raiders, R. A., Maher, T. F., Knapp, R. M., and McInerney, M. J. (1986a). "Selective Plugging and Oil Displacement in Crossflow Core Systems by Microorganisms." SPE 15600. *Proceedings of 61st Annual Meeting and Technical Exhibition*. Society of Petroleum Engineers, Richardson, Texas.

Raiders, R. A., McInerney, M. J., Revus, D. E., Torbati, H. M., Knapp, R. M., and Jenneman, G. E. (1986b). Selectivity and depth of microbial plugging in Berea sandstone cores. *J. Ind. Microbiol.* **1,** 195–203.

Rappe, M. S., and Giovannoni, S. J. (2003). The uncultured microbial majority. *Ann. Rev. Microbiol.* **57,** 369–394.

Rauf, M. A., Ikram, M., and Tabassum, N. (2003). Enhanced oil recovery through microbial treatment. *J. Trace Microprobe Tech.* **21,** 533–541.

Ravot, G., Magot, M., Fardeau, M. L., Patel, B. K. C., Thomas, P., Garcia, J. L., and Ollivier, B. (1999). *Fusibacter paucivorans* gen. nov., sp. nov., an anaerobic, thiosulfate-reducing bacterium from an oil-producing well. *Int. J. Syst. Bacteriol.* **49,** 1141–1147.

Ravot, G., Magot, M., Ollivier, B., Patel, B. K. C., Ageron, E., Grimont, P. A. D., Thomas, P., and Garcia, J.-L. (1997). *Haloanaerobium congolense* sp. nov., an anaerobic, moderately halophilic, thiosulfate- and sulfur-reducing bacterium from an African oil field. *FEMS Microbiol. Lett.* **147,** 81–88.

Reed, R. L., and Healy, R. N. (1977). Some physical aspects of microemulsion flooding: A review. *In* "Improved Oil Recovery by Surfactant and Polymer Flooding" (D. O. Shah and R. S. Scheckter, Eds.), pp. 383–437. Academic Press, New York.

Rees, G. N., Grassia, G. S., Sheehy, A. J., Dwivedi, P. P., and Patel, B. K. C. (1995). *Desulfacinum infernum* gen. nov., sp. nov., a thermophilic sulfate-reducing bacterium from a petroleum reservoir. *Int. J. Syst. Bacteriol.* **45,** 85–89.

Rees, G. N., Patel, B. K., Grassia, G. S., and Sheehy, A. J. (1997). *Anaerobaculum thermoterrenum* gen. nov., sp. nov., a novel, thermophilic bacterium which ferments citrate. *Int. J. Syst. Bacteriol.* **47,** 150–154.

Reinsel, M. A., Sears, J. T., Stewart, P. S., and McInerney, M. J. (1996). Control of microbial souring by nitrate, nitrite, or glutaraldehyde injection in a sandstone column. *J. Ind. Microbiol.* **17,** 128–136.

Rempel, C. L., Evitts, R. W., and Nemati, M. (2006). Dynamics of corrosion rates associated with nitrite or nitrate mediated control of souring under biological conditions simulating an oil reservoir. *J. Ind. Microbiol. Biotechnol.* **33,** 878–886.

Rizk, T. Y., Scott, J. F. D., Eden, R. D., Davis, R. A., McElhiney, J. E., and Di Iorio, C. (1998). "The Effect of Desulphated Seawater Injection on Microbiological Hydrogen Sulphide Generation and Implication for Corrosion Control." Paper # 287. *Corrosion 98*. NACE International, Houston, TX.

Robertson, E. P. (1998). The use of bacteria to reduce water influx in producing oil wells. *Soc. Pet. Eng. Prod. Facil.* **13,** 128–132.

Rosenberg, E., Perry, A., Gibson, D. T., and Gutnick, D. L. (1979a). Emulsifier of *Arthrobacter* RAG-1: Specificity of the hydrocarbon substrate. *Appl. Environ. Microbiol.* **37,** 409–413.

Rosenberg, E., and Ron, E. Z. (1999). High- and low-molecular-mass microbial surfactants. *Appl. Microbiol. Biotechnol.* **52,** 154–162.

Rosenberg, E., Rosenberg, M., and Gutnick, D. L. (1983). Adherence of bacteria to hydrocarbons. *In* "Proceedings of the 1982 International Conference on Microbial Enhancement of Oil Recovery" (E. C. Donaldson and J. B. Clark, Eds.), pp. 20–28. U. S. Dept. of Energy, Bartellesville, Okla. (CONF-8205140).

Rosenberg, E., Zuckerberg, A., Rubinovitz, C., and Gutnick, D. L. (1979b). Emulsifier of *Arthrobacter* RAG-1: Isolation and emulsifying properties. *Appl. Environ. Microbiol.* **37,** 402–408.

Rosnes, J. T., Torsvik, T., and Lien, T. (1991). Spore-forming thermophilic sulfate-reducing bacteria isolated from North sea oil field waters. *Appl. Environ. Microbiol.* **57,** 2302–2307.

Ross, N., Villemur, R., Deschenes, L., and Samson, R. (2001). Clogging of a limestone fracture by stimulating groundwater microbes. *Water Res.* **35,** 2029–2037.

Rouse, B., Hiebert, F., and Lake, L. W. (1992). "Laboratory Testing of a Microbial Enhanced Oil Recovery Process Under Anaerobic Conditions." SPE 24819. *Proceedings of SPE Annual Technical Conference.* Society of Petroleum Engineers, Richardson, Texas.

Rozanova, E. P., Borzenkov, I. A., Tarasov, A. L., Suntsova, L. A., Dong, C. L., Belyaev, S. S., and Ivanov, M. V. (2001a). Microbial processes in a high temperature oil field. *Microbiology* **70,** 102–110.

Rozanova, E. P., and Nazina, T. N. (1980). Hydrocarbon-oxidizing bacteria and their activity in oil pools. *Microbiology* **51,** 342–348.

Rozanova, E. P., Nazina, T. N., and Galushko, A. S. (1988). Isolation of a new genus of sulfate-reducing bacteria and description of a new species of this genus, *Desulfomicrobium apsheronum* gen. nov., sp. nov. *Microbiology (Eng. Tr.)* **57,** 514–520.

Rozanova, E. P., Tourova, T. P., Kolganova, V., Lysenko, A. M., Mityushina, L. L., Yusupov, S. K., and Belyaev, S. S. (2001b). *Desulfacinum subterraneum* sp. nov., a new thermophilic sulfate-reducing bacterium isolated from a high temperature oil field. *Microbiology (Eng. Tr.)* **70,** 466–471.

Sadeghazad, A., and Ghaemi, N. (2003). "Microbial Prevention of Wax Precipitation in Crude Oil by Biodegradation Mechanism." SPE 80529. *Proceedings of SPE Asia Pacific Oil and Gas conference and exhibition.* Society of Petroleum Engineers, Richardson, Texas.

Santamaria, M. M., and George, R. E. (1991). "Controlling Paraffin-Deposition-Related Problems by the Use of Bacteria Treatments." SPE 22851. *Proceedings of the SPE Annual Technical Conference and Exhibition.* Society of Petroleum Engineers, Richardson, Texas.

Sarkar, A. K., Goursaud, J. C., Sharma, M. M., and Georgiou, G. (1989). A critical evaluation of MEOR [microbial enhanced oil recovery] processes. *In Situ* **13,** 207–238.

Sayyouh, M. H. (2002). "Microbial Enhanced Oil Recovery: Research Studies in the Arabic Area During the Last Ten Years." SPE 75218. *Proceedings of SPE/DOE Improved Oil Recovery Symposium.* Society of Petroleum Engineers, Richardson, Texas.

Sayyouh, M. H., Al-Blehed, M. S., and Hemeida, A. M. (1993). Possible applications of MEOR to the arab oil fields. *J. King Saud Univ.* **5,** 291–302.

Schaller, K. D., Fox, S. L., Bruhn, D. F., Noah, K. S., and Bala, G. A. (2004). Characterization of surfactin from *Bacillus subtilis* for application as an agent for enhanced oil recovery. *Appl. Biochem. Biotechnol.* **113–116,** 827–836.

Senyukov, V. M., Yulbarisov, É. M., Taldykina, N. N., and ShiShenina, E. P. (1970). Microbiological method of treating a petroleum deposit containing highly mineralized stratal waters. *Microbiology* **39,** 612–616.

Sharma, P. K., and McInerney, M. J. (1994). Effect of grain size on bacterial penetration, reproduction, and metabolic activity in porous glass beads chambers. *Appl. Environ. Microbiol.* **60,** 1481–1486.

Sharma, P. K., McInerney, M. J., and Knapp, R. M. (1993). *In situ* growth and activity and modes of penetration of *Escherichia coli* in unconsolidated porous materials. *Appl. Environ. Microbiol.* **59,** 3686–3694.

Shaw, J. C., Bramhill, B., Wardlaw, N. C., and Costerton, J. W. (1985). Bacterial fouling in a model core system. *Appl. Environ. Microbiol.* **49,** 693–701.

Sheehy, A. J. (1990). "Field Studies of Microbial EOR." SPE 20254. *Proceedings of the SPE/DOE Seventh Symposium on Enhanced Oil Recovery*. Society of Petroleum Engineers, Richardson, Texas.

Sheehy, A. J. (1991). Development and field evaluation of a new microbial EOR concept. *APEA J.* **31,** 386–390.

Sheehy, A. (1992). Recovery of oil from oil reservoirs. US patent office. Patent No. 5,083,610.

Silver, R. S., Bunting, P. M., Moon, W. G., and Acheson, W. P. (1989). Bacteria and its use in a microbial profile modification process. US patent office. Patent No. 4,799,545.

Simpson, D. R., Knapp, R. M., Youssef, N., Duncan, K. E., McInerney, M. J., and Brackin, C. (2007). "Microbial Stimulation Treatment of High Water-Cut Wells in the Viola Formation, Pontotoc County, OK." *Proceedings of Tertiary Oil Recovery Conference*. Teriary Oil Recovery Project, University of Kansas, Lawrence, Kansas.

Singh, A., Van Hamme, J. D., and Ward, O. P. (2007). Surfactants in microbiology and biotechnology: Part 2. Applications aspects. *Biotechnol. Adv.* **25,** 99–121.

Slobodkin, A. I., Jeanthon, C., L'Haridon, S., Nazina, T., Miroshnichenko, M., and Bonch-Osmolovskaya, E. (1999). Dissimilatory reduction of Fe(III) by thermophilic bacteria and archaea in deep-subsurface petroleum reservoirs of western Siberia. *Curr. Mirobiol.* **39,** 99–102.

Smith, T. L., and Trebbau, G. (1998). MEOR treatments boost heavy oil recovery in Venezuela. *Hart Pet. Eng. Int.* **71,** 45–49.

Soudmand-asli, A., Ayatollahi, S. S., Mohabatkar, H., Zareie, M., and Shariatpanahi, S. F. (2007). The *in situ* microbial enhanced oil recovery in fractured porous media. *J. Pet. Sci. Eng* **58,** 161–172.

Spormann, A. M., and Widdel, F. (2001). Metabolism of alkylbenzenes, alkanes, and other hydrocarbons in anaerobic bacteria. *Biodegradation* **11,** 85–105.

Stachelhaus, T., Schneider, A., and Marahiel, M. A. (1995). Rational design of peptide antibiotics by targeted replacement of bacterial and fungal domains. *Science* **269,** 5571–5574.

Stetter, K. O., Huber, R., Blöchl, E., Kurr, M., Eden, R. D., Fielder, M., Cash, H., and Vance, I. (1993). Hyperthermophilic archaea are thriving in deep North Sea and Alaskan oil reservoirs. *Nature* **365,** 743–745.

Stewart, T. L., and Fogler, H. S. (2001). Biomass plug development and propagation in porous media. *Biotechnol. Bioeng.* **72,** 353–363.

Stewart, T. L., and Fogler, H. S. (2002). Pore-scale investigation of biomass plug development and propagation in porous media. *Biotechnol. Bioeng.* **77,** 577–588.

Strand, S., Standnes, D. C., and Austad, T. (2003). Spontaneous imbibition of aqueous surfactant solutions into neutral to oil-wet carbonate cores: Effects of brine salinity and composition. *Energy Fuels* **17,** 1133–1144.

Strappa, L. A., DeLucia, J. P., Maure, M. A., and Lopez Llopiz, M. L. (2004). "A Novel and Successful MEOR Pilot Project in a Strong Water-Drive Reservoir." SPE 89456. *Proceedings of the SPE/DOE Improved Oil Recovery Symposium*. Society of Petroleum Engineers, Richardson, Texas.

Streeb, L. P., and Brown, F. G. (1992). "MEOR-Altamont/Bluebell Field Project." SPE 24334. *Proceedings of the Rocky Mountain Regional Meeting and Exhibition*, Society of Petroleum Engineers, Richardson, Texas.

Sugihardjo, E. H., and Pratomo, S. W. (1999). "Microbial Core Flooding Experiments Using Indigenous Microbes." SPE 57306. *Proceedings of SPE Asia Pacific Improved Oil Recovery Conference*. Society of Petroleum Engineers, Richardson, Texas.

Sunde, E., Beeder, J., Nilsen, R. K., and Torsvik, T. (1992). "Aerobic Microbial Enhanced Oil Recovery for Offshore Use." SPE/DOE 24204. *Proceedings of the Eighth Symposium on Enhanced Oil Recovery*. Society of Petroleum Engineers, Richardson, Texas.

Sunde, E., Lillebø, B.-L. P., Bødtker, G., Torsvik, T., and Thorstenson, T. (2004). "H_2S Inhibition by Nitrate Injection on the Gulfaks Field." Corrosion 04760. *Proceedings of Corrosion 2004*. NACE, Houston, Texas.

Sunde, E., and Torsvik, T. (2005). Microbial control of hydrogen sulfide production in oil reservoirs. *In* "Petroleum Microbiology" (B. Ollivier and M. Magot, Eds.), pp. 201–213. ASM, Washington, DC.

Tabor, J. J. (1969). Dynamic and static forces required to remove a discontinuous oil phase from porous media containing both oil and water. *Soc. Pet. Eng. J.* **9,** 3–12.

Takahata, Y., Nishijima, M., Hoaki, T., and Maruyama, T. (2000). Distribution and physiological characteristics of hyperthermophiles in the Kubiki oil reservoir in Niigata, Japan. *Appl. Environ. Microbiol.* **66,** 73–79.

Takahata, Y., Nishijima, M., Hoaki, T., and Maruyama, T. (2001). *Thermotoga petrophila* sp. nov. and *Thermotoga naphthophila* sp. nov., two hyperthermophilic bacteria from the Kubiki oil reservoir in Niigata, Japan. *Int. J. Syst. Evol. Microbiol.* **51,** 1901–1909.

Takai, K., Hirayama, H., Nakagawa, T., Suzuki, Y., Nealson, K. H., and Horikoshi, K. (2004). *Thiomicrospira thermophila* sp. nov., a novel microaerobic, thermotolerant, sulfur-oxidizing chemolithomixotroph isolated from a deep-sea hydrothermal fumarole in the TOTO caldera, Mariana Arc, Western Pacific. *Int. J. Syst. Evol. Microbiol.* **54,** 2325–2333.

Tanner, R. S., Udegbunam, E. O., Adkins, J. P., McInerney, M. J., and Knapp, R. M. (1993). The potential for MEOR from carbonate reservoirs: Literature review and recent research. *Dev. Pet. Sci.* **39,** 391–396.

Tardy-Jacquenod, C., Magot, M., Laigret, F., Kaghad, M., Patel, B. K., Guezennec, J., Matheron, R., and Caumette, P. (1996). *Desulfovibrio gabonensis* sp. nov., a new moderately halophilic sulfate- reducing bacterium isolated from an oil pipeline. *Int. J. Syst. Bacteriol.* **46,** 710–715.

Tardy-Jacquenod, C., Magot, M., Patel, B. K. C., Matheron, R., and Caumette, P. (1998). *Desulfotomaculum halophilum* sp. nov., a halophilic sulfate-reducing bacterium isolated from oil production facilities. *Int. J. Syst. Bacteriol.* **48,** 333–338.

Tasaki, M., Kamagata, Y., Nakamura, K., and Mikami, E. (1991). Isolation and characterization of a thermophilic benzoate-degrading, sulfate-reducing bacterium, *Desulfotomaculum thermobenzoicum* sp. nov. *Arch. Microbiol.* **155,** 348–352.

Telang, A. J., Ebert, S., Foght, J. M., Westlake, D. W. S., Jenneman, G. E., Gevertz, D., and Voordouw, G. (1997). Effect of nitrate on the microbial community in an oil field as monitored by reverse sample genome probing. *Appl. Environ. Microbiol.* **63,** 1785–1793.

Thomas, C. P., Bala, G. A., and Duvall, M. L. (1993). Surfactant-based enhanced oil recovery mediated by naturally occurring microorganisms. *Soc. Pet. Eng. Reservoir Eng* **11,** 285–291.

Thorstenson, T., Bødtker, G., Lillebo, B. L. P., Torsvik, T., Sunde, E., and Beeder, J. (2002). "Biocide Replacement by Nitrate in Seawater Injection Systems." Corrosion 02033. *Proceedings of Corrosion 2002*. NACE, Houston, Texas.

Torbati, H. M., Raider, R. A., Donaldson, E. C., Knapp, R. M., McInerney, M. J., and Jenneman, G. E. (1986). Effect of microbial growth on the pore entrance size distribution in sandstone cores. *J. Ind. Microbiol.* **1,** 227–234.

Trebbau, G., Nunezm, G. J., Caira, R. L., Molina, N. Y., Entzeroth, L. C., and Schneider, D. R. (1999). "Microbial Stimulation of Lake Maracaibo Oil Wells." SPE 56503. *Proceedings of the Annual Technical Conference and Exhibition*. Society of Petroleum Engineers, Richardson, Texas.

Trebbau de Acevedo, G., and McInerney, M. J. (1996). Emulsifying activity in thermophilic and extremely thermophilic microorganisms. *J. Ind. Microbiol.* **16,** 1–7.

Tringe, S. G., Mering, C. V., Kobayashi, A., Salamov, A. A., Chen, K., Chang, H. M., Podar, M., Short, J. M., Mathur, E. J., Detter, J. C., Bork, P., Hugenholtz, P., *et al.* (2005). Comparative metagenomics of microbial communities. *Science* **308,** 554–557.

Trushenski, S. P., Dauben, D. L., and Parrish, D. R. (1974). Micellar flooding-fluid propagation, interaction, and mobility. *Soc. Pet. Eng. J.* **14,** 633–645.

Udegbunam, E. O., Adkins, J. P., Knapp, R. M., and McInerney, M. J. (1991). "Assessing the Effects of Microbial Metabolism and Metabolites on Reservoir Pore Structure." SPE 22846. *Proceedings of SPE annual Technical Conference and Exhibition.* Society of Petroleum Engineers, Richardson, Texas..

Udegbunam, E. O., Knapp, R. M., McInerney, M. J., and Tanner, R. S. (1993). Potential of microbial enhanced oil recovery (MEOR) in the petroleum reservoirs of the midcontinent region. *Okla. Geol. Survey Circ.* **95,** 173–181.

Updegraff, D. M. (1956). Recovery of petroleum oil. U. S. Patent Office. Patent No. 2,807,570.

Updegraff, D. M. (1990). Early research on microbial enhanced oil recovery. *Dev. Ind. Microbiol.* **31,** 135–142.

Urum, K., Pekdemir, T., and Gopur, M. (2003). Optimum conditions for washing of crude oil-contaminated soil with biosurfactant solutions. *Trans. Inst. Chem. Eng.* **81,** 203–209.

Vadie, A. A., Stephens, J. O., and Brown, L. R. (1996). "Utilization of Indigenous Microflora in Permeability Profile Modification of Oil Bearing Formations." SPE/DOE 35448. *Proceedings of the 1996 10th Symposium on Improved Oil Recovery.* Society of Petroleum Engineers, Richardson, Texas.

Van Hamme, J. D., Singh, A., and Ward, O. P. (2003). Recent advances in petroleum microbiology. *Microbiol. Mol. Biol. Rev.* **67,** 503–549.

Vance, I., and Thrasher, D. R. (2005). Reservoir souring: Mechanisms and prevention. *In* "Petroleum Microbiology" (B. Ollivier and M. Magot, Eds.), pp. 123–142. ASM, Washington, DC.

Venter, J. C., Remington, K., Heidelberg, J. F., Halpern, A. L., Rusch, D., Eisen, J. A., Wu, D., Paulsen, I., Nelson, K. E., Nelson, W., Fouts, D. E., Levy, S., *et al.* (2004). Environmental genome shotgun sequencing of the Sargasso sea. *Science.* **304,** 66–74.

Vetriani, C., Speck, M. D., Ellor, S. V., Lutz, R. A., and Starovoytov, V. (2004). *Thermovibrio ammonificans* sp. nov., a thermophilic, chemolithotrophic, nitrate-ammonifying bacterium from deep-sea hydrothermal vents. *Int. J. Syst. Evol. Microbiol.* **54,** 175–181.

von Heiningen, J., Jan de Haan, H., and Jensen, J. D. (1958). Process for the recovery of petroleum from rocks. Netherlands Patent office. Patent No. 89580.

Voordouw, G., Armstrong, S. M., Reimer, M. F., Fouts, B., Telang, A. J., Shen, Y., and Gevertz, D. (1996). Characterization of 16S rRNA genes from oil field microbial communities indicates the presence of a variety of sulfate-reducing, fermentative, and sulfide-oxidizing bacteria. *Appl. Environ. Microbiol.* **62,** 1623–1629.

Voordouw, G., Shen, Y., Harrington, C. S., Telang, A. J., Jack, T. R., and Westlake, D. W. S. (1993). Quantitative reverse sample genome probing of microbial communities and its application to oil field production waters. *Appl. Environ. Microbiol.* **59,** 4101–4114.

Voordouw, G., Voordouw, J. K., Jack, T. R., Foght, J., Fedorak, P. M., and Westlake, D. W. S. (1992). Identification of distinct communities of sulfate-reducing bacteria in oil fields by reverse sample genome probing. *Appl. Environ. Microbiol.* **58,** 3542–3552.

Voordouw, G., Voordouw, J. K., Karkhoff-Schweizer, R. R., Fedorak, P. M., and Westlake, D. W. S. (1991). Reverse sample genome probing, a new technique for identification of bacteria in environmental samples by DNA hybridization, and its application to the identification of sulfate-reducing bacteria in oil field samples. *Appl. Environ. Microbiol.* **57,** 3070–3078.

Vossoughi, S. (2000). Profile modification using *in situ* gelation technology–a review. *J. Pet. Sci. Eng.* **26,** 199–209.

Vreeland, R. H., Jones, J., Monson, A., Rosenzweig, W. D., Lowenstein, T. K., Timofeeff, M., Satterfield, C., Cho, B. C., Park, J. S., Wallace, A., and Grant, W. D. (2007). Isolation of live cretaceous (121–112 million years old) halophilic *Archaea* from primary salt crystals. *Geomicrobiol. J.* **24,** 275–282.

Vreeland, R. H., Straight, S., Krammes, J., Dougherty, K., Rosenzweig, W. D., and Kamekura, M. (2002). *Halosimplex carlsbadense* gen. nov., sp. nov., a unique halophilic archaeon, with three 16S rRNA genes, that grows only in defined medium with glycerol and acetate or pyruvate. *Extremophiles* **6,** 445–452.

Wagner, M. (1985). Microbial enhancement of oil recovery from carbonate reservoir with complex formation characteristics. *In* "Micobial Enhancement of Oil Recovery-Recent Advances" (E. C. Donaldson, Ed.), pp. 387–398. Elsevier, Amsterdam.

Wagner, M. (1991). Microbial enhancement of oil recovery from carbonate reservoirs with complex formation characteristics. *Dev. Pet. Sci.* **31,** 387–398.

Wagner, M., Lungerhausen, D., Murtada, H., and Rosenthal, G. (1995). Development and application of a new biotechnology of the molasses *in-situ* method; detailed evaluation for selected wells in the Romashkino carbonate reservoir. *In* "The Fifth International Conference on Microbial Enhanced Oil Recovery and Related Problems for Solving Environmental Problems (CONF-9509173)" (R. S. Bryant and K. L. Sublette, Eds.), pp. 153–174. National Technical Information Service, Springfield, VA.

Wagner, M., Roger, A. J., Flax, J. L., Brusseau, G. A., and Stahl, D. A. (1998). Phylogeny of dissimilatory sulfite reductases supports an early origin of sulfate respiration. *J. Bacteriol.* **180,** 2975–2982.

Wang, Q., Fang, X., Bai, B., Liang, X., Shuler, P. J., Goddard, W. A., and Tang, Y. (2007). Engineering bacteria for production of rhamnolipid as an agent for enhanced oil recovery. *Biotechnol. Bioeng.* **98,** 842–853.

Wang, X.-Y., Xue, Y.-F., Dai, G., Zhao, L., Wang, Z.-S., Wang, J.-L., Sun, T.-H., and Li, X.-J. (1995). Application of bio-huff-"n"-puff technology at Jilin oil field *In* "The Fifth International Conference on Microbial Enhanced Oil Recovery and Related Problems for Solving Environmental Problems (CONF-9509173)" (R. S. Bryant and K. L. Sublette, Eds.), pp. 15–128. National Technical Information Service, Springfield, VA.

Wang, X., Xue, Y., and Xie, S. (1993). Characteristics of enriched cultures and their application to MEOR field tests. *Dev. Pet. Sci.* **39,** 335–348.

Wankui, G., Chengfang, S., Zhenyu, Y., Zhaowei, H., Rui, J., Ying, W., Jiyuan, Z., and Guoghao, S. (2006). "Microbe-Enhanced Oil Recovery Technology Obtains Huge Success in Low-Permeability Reservoirs in Daqing Oilfield." SPE 104281. *Proceedings of SPE Eastern Regional Meeeting.* Society of Petroeum Engineers, Richardson, TX.

Weihong, Q. D., Liangjun, D., Zhongkui, Z., Jie, Y., Huamin, L., and Zongshi, L. (2003). Interfacial behavior of pure surfactants for enhanced oil recovery–Part 1: A study on the adsorption and distribution of cetylbenzene sulfonate. *Tenside, Surfactants, Detergents* **40,** 87–89**.

Wentzel, A., Ellingsen, T. E., Kotlar, H. K., Zotchev, S., and Thorne-Holst, M. (2007). Bacterial metabolism of long-chain n-alkanes. *Appl. Microbiol. Biotechnol.* **76,** 1209–1221.

Wilson, J. J., Chee, W., O'Grady, C., and Bishop, M. D. (1993). "Field Study of Downhole Microbial Paraffin Control." *Proceedings of Annual Southwestern Petroleum Short Course.* Lubbock, Texas.

Wilson, K. H., Wilson, W. J., Radosevich, J. L., DeSantis, T. Z., Viswanathan, V. S., Kuczmarski, T. A., and Andersen, G. L. (2002). High-density microarray of small-subunit ribosomal DNA probes. *Appl. Environ. Microbiol.* **68,** 2535–2541.

Wolf, B. F., and Fogler, H. S. (2001). Alteration of the growth rate and lag time of *Leuconostoc mesenteroides* NRRL-B523. *Biotechnol. Bioeng.* **72,** 603–610.

Yakimov, M. M., Amro, M. M., Bock, M., Boseker, K., Fredrickson, H. L., Kessel, D. G., and Timmis, K. N. (1997). The potential of *Bacillus licheniformis* strains for *in situ* enhanced oil recovery. *J. Pet. Sci. Eng.* **18,** 147–160.

Yakimov, M. M., Fredrickson, H. L., and Timmis, K. N. (1996). Effect of heterogeneity of hydrophobic moieties on surface activity of lichenysin A, a lipopeptide biosurfactant from *Bacillus licheniformis* BAS50. *Biotechnol. Appl. Biochem.* **23,** 13–18.

Yang, Y.-G., Niibori, Y., Inoue, C., and Chida, T. (2005). A fundamental study of microbial attachment and transport in porous media for the design of MEOR. *J. Jpn. Assoc. Pet. Technol.* **70,** 459–468.

Yarbrough, H. F., and Coty, V. F. (1983). Microbially enhanced oil recovery from the upper cretaceous nacatoch formation, Union County, Arkansas. *In* "Proceedings of the 1982 International Conference on Microbial Enhancement of Oil Recovery" (E. C. Donaldson and J. B. Clark, Eds.), pp. 149–153. U. S. Dept. of Energy, Bartellesville, Okla (CONF-8205140).

Youssef, N., Duncan, K., and McInerney, M. (2005). Importance of the 3-hydroxy fatty acid composition of lipopeptides for biosurfactant activity. *Appl. Environ. Microbiol.* **71,** 7690–7695.

Youssef, N., Duncan, K., Nagle, D., Savage, K., Knapp, R., and McInerney, M. (2004). Comparison of methods to detect biosurfactant production by diverse microorganisms. *J. Microbiol. Methods* **56,** 339–347.

Youssef, N., Nguyen, T., Sabatini, D. A., and McInerney, M. J. (2007a). Basis for formulating biosurfactant mixtures to achieve ultra low interfacial tension values against hydrocarbons. *J. Ind. Microbiol. Biotechnol.* **34,** 497–507.

Youssef, N., Simpson, D. R., Duncan, K. E., McInerney, M. J., Folmsbee, M., Fincher, T., and Knapp, R. M. (2007b). *In situ* biosurfactant production by *Bacillus* strains injected into a limestone petroleum reservoir. *Appl. Environ. Microbiol.* **73,** 1239–1247.

Yu, M. S., and Schneider, D. R. (1998). Microbial enhanced oil recovery tested in China. *Hart Pet. Eng. Int.* **71,** 31–33.

Yulbarisov, E. M. (1976). Evaluation of the effectiveness of the biological method for enhancing oil recovery of a reservoir. *Neftyanoe Khozyaistvo* **11,** 27–30.

Yulbarisov, E. M. (1981). On the enhancement of oil recovery of flooded oil strata. *Neftyanoe Khozyaistvo* **3,** 36–40.

Yulbarisov, E. M. (1990). Microbiological method for EOR. *Revue de l'Institut Francais du Petrole* **45,** 115–121.

Yulbarisov, E. M., and Zhdanova, N. V. (1984). On the microbial enhancement of oil recovery of flooded oil strata. *Neftyanoe Khozyaistvo* **3,** 28–32.

Yusuf, A., and Kadarwati, S. (1999). "Field Test of the Indigenous Microbes for Oil Recovery, Ledok field, Central Java." SPE 57309. *Proceedings of SPE Asia Pacific Improved Oil Recovery Conference.* Society of Petroleum Engineers, Richardson, TX.

Zaijic, E. J. (1987). Scale up of microbes for single well injection. *In* "Proceedings of the First International MEOR Workshop" (J. W. King and D. A. Stevens, Eds.), pp. 241–246. National Technical Information Service, Springfield, VA.

Zekri, A. Y., Almehaideb, R., and Chaalal, O. (1999). "Project of Increasing Oil Recovery from UAE Reservoirs Using Bacteria Flooding." SPE 56827. *Proceedings of SPE Annual Technical Conference.* Society of Petroluem Engineers, Richardson, TX.

Zengler, K. (2006). "Genomics and Cultivation: Friend or Foe." *Proceedings of 106th General Meeting of the American Society for Microbiology.* Orlando, FL. ASM, Washington, DC.

Zengler, K., Toledo, G., Rappe, M., Elkins, J., Mathur, E. J., Short, J. M., and Keller, M. (2002). Cultivating the uncultured. *Proc. Natl. Acad. Sci. USA* **99,** 15681–15686.
Zhang, C. Y., and Zhang, J. C. (1993). A pilot field test of EOR by *in-situ* microorganism fermentation in the Daqing Oilfield. *Dev. Pet. Sci.* **39,** 231–244.
ZoBell, C. E. (1947a). Bacterial release of oil from oil-bearing materials, part I. *World Oil* **126,** 36–47.
ZoBell, C. E. (1947b). Bacterial release of oil from oil-bearing materials, part II. *World Oil* **127,** 35–41.
ZoBell, C. E. (1947c). Bacterial release of oil from sedimentary materials. *Oil Gas J.* **46,** 62–65.

INDEX

T

U

V

Y

CONTENTS OF PREVIOUS VOLUMES

Volume 43

Volume 44

Volume 45

Volume 46

Volume 47

Volume 48

Volume 49

Volume 50

Volume 51

Volume 52

Volume 53

Volume 54

Volume 55

Volume 56

Volume 57

Volume 58

Volume 59

Volume 60

Volume 61

Volume 62

Volume 63

Volume 64

Volume 65